GUIDE TO RELIABILITY ENGINEERING

GUIDE TO RELIABILITY ENGINEERING

Data, Analysis, Applications, Implementation, and Management

C. (Raj) Sundararajan, Ph.D.

Consultant
Houston, Texas

VNR VAN NOSTRAND REINHOLD
New York

Library of Congress Catalog Card Number: 91-12227
ISBN: 0-442-20722-0

Manufactured in the United States of America

Published by Van Nostrand Reinhold
115 Fifth Avenue
New York, New York 10003

Chapman and Hall
2-6 Boundary Row
London, SE1 8HN

Thomas Nelson Australia
102 Dodds Street
South Melbourne 3205
Victoria, Australia

Nelson Canada
1120 Birchmount Road
Scarborough, Ontario M1K 5G4, Canada

16   15   14   13   12   11   10   9   8   7   6   5   4   3   2   1

Library of Congress Cataloging-in-Publication Data

Sundararajan, C.
    Guide to reliability engineering: data, analysis, applications, implementation, and management/C. (Raj) Sundararajan.
        p.   cm.
    Includes bibliographical references and index.
    ISBN 0-442-20722-0
    1. Reliability (Engineering)   I. Title.
TA169.S96   1991   91-12227
620′.00452—dc20   CIP

To

(in alphabetical order)

A. C. Gangadharan
Livingston, New Jersey

V. G. Pandarinathan
Coimbatore, Tamil Nadu

D. V. Reddy
Boca Raton, Florida

who influenced my professional career significantly

Preface

Reliability engineering has its origin in the aerospace industry. Simple statistical methods were used during the Second World War (1940s) to estimate the success probability of rockets. The next decade saw the application of more-advanced methods to estimate failure probabilities and life expectancies of mechanical, electrical, and electronic components used in the defense and aerospace industries. Reliability engineering became a distinct and well-developed branch of engineering in the 1960s, with the emergence of fault trees, reliability block diagrams, and other system reliability analysis tools. The 1970s saw reliability engineering principles being applied to the probabilistic risk assessment of nuclear power plants. Reliability engineering, which was once the exclusive preserve of the *aerospace* and *defense* industries, came to the attention of the *nuclear power* industry, and from it to the *fossil-fuel power*, *chemical*, *petroleum*, and *manufacturing* industries.

With the spread of reliability engineering applications to a wider range of industries, a need arose for engineers knowledgeable in both reliability engineering and the specifics of their respective industries (power, chemical, petroleum, manufacturing, etc.). During the past 10 years, the author of this book has taught one-week short courses on reliability engineering to practicing engineers and engineering managers. Engineers from the power, chemical, petroleum, manufacturing, and aerospace industries, with education and training in *mechanical*, *electrical*, *electronic*, *chemical*, and *civil* engineering, have attended these courses. This book is based on the course notes and the feedback from those who attended the courses.

This book is suitable as a text for short courses on reliability engineering. Practicing engineers and engineering managers will also find this book suitable as a self-study text; the book is self-contained. This book may also be used as a textbook for graduate-level and senior undergraduate-level courses on reliability engineering.

A number of example problems are solved in the book. These are purposely kept simple so that readers can repeat the calculations manually. The purpose of these examples is to bring out the implications of the various assumptions and how they affect the final results. How changes in data affect the final results is also brought forth through the examples.

A number of useful equations are provided throughout the book. Users should assure themselves that results obtained using these equations are reasonable for the purposes they are being used.

Some useful commercial and public-domain computer programs are listed and described. Data bases for component failure probability data and human error probability data are also identified and described.

No prior knowledge of reliability engineering, probability theory, or statistics is needed to understand and use this book.

Contents

Preface   vii

1   Introduction   1

2   Fundamentals of Probability and Statistics   5

    2.1   Introduction   5
    2.2   Set Theory   5
    2.3   Probability Theory   12
        Exercise Problems   44
        Reference   47

3   Fundamentals of Reliability Engineering   48

    3.1   Introduction   48
    3.2   General Terminology   48
    3.3   Nonrepairable Components   50
    3.4   Repairable Components   71
        Exercise Problems   85
        References   88

4   Fundamentals of Systems Engineering   89

    4.1   Introduction   89
    4.2   Definitions   89
    4.3   Basic System Configurations   89
    4.4   Types of Parallel Systems   91

5   Failure Data   94

    5.1   Introduction   94
    5.2   Hardware Failure Data   95
    5.3   Software Error Data   127
    5.4   Human Error Data   128
    5.5   Comparison of Hardware and Human Failure Probabilities   129

5.6   Documentation   130
        Exercise Problems   131
        References   134

6   Preliminary Hazard Analysis   136

6.1   Introduction   136
6.2   Analysis Procedure   136
6.3   Benefits and Limitations   142
6.4   Documentation   143
6.5   Example Problem   143
        Exercise Problem   145
        Reference   145

7   Failure Modes and Effects Analysis   146

7.1   Introduction   146
7.2   Analysis Procedure   147
7.3   Benefits and Limitations   151
7.4   Documentation   151
7.5   Example Problem   152
        Exercise Problem   152
        Reference   152

8   Fault Tree Construction   153

8.1   Introduction   153
8.2   An Example Fault Tree   154
8.3   System Description   156
8.4   Symbols   158
8.5   Gates   158
8.6   Events   169
8.7   Transfer Symbols   183
8.8   Common Cause Failures   187
8.9   Propagating Failures   188
8.10   Dependence between Terminal Events   190
8.11   Maintenance Outage   196
8.12   Fault Tree Reduction (Fault Tree Simplification)   199
8.13   Computerized Fault Tree Construction   204
8.14   Complementary Trees   204
8.15   Limitations   205

8.16	Documentation	206
8.17	Example Problem	207
	Exercise Problems	210
	References	210

9 Qualitative Fault Tree Analysis — 212

9.1	Introduction	212
9.2	Cut Sets and Minimal Cut Sets	212
9.3	Path Sets and Minimal Path Sets	216
9.4	Qualitative Ranking of Minimal Cut Sets	219
9.5	Qualitative Ranking of Terminal Events	220
9.6	Methods of Qualitative Analysis	221
9.7	Selective Determination of Minimal Cut Sets	229
9.8	Computer Programs	230
9.9	Documentation	230
9.10	Example Problem	231
	Exercise Problems	232
	References	232

10 Quantitative Fault Tree Analysis — 233

10.1	Introduction	233
10.2	Coherent Structure	234
10.3	Nonrepairable Systems	236
10.4	Repairable Systems	261
10.5	Periodically Maintained Systems	277
10.6	Other Methods	278
10.7	Limitations of Quantitative Fault Tree Analysis	279
10.8	Uncertainty Analysis	280
10.9	Computer Programs	281
10.10	Documentation	281
	Exercise Problems	282
	References	285

11 Common Cause Analysis — 286

11.1	Introduction	286
11.2	Definitions	286
11.3	Preliminary Common Cause Analysis	290
11.4	Qualitative Common Cause Analysis	291

11.5	Qualitative Common Link Analysis	296
11.6	Quantitative Common Cause Analysis	296
11.7	Trees with Large Minimal Cut Sets	315
11.8	Corrective Actions Against Common Cause Failures	321
11.9	Computer Programs	325
11.10	Documentation	326
	Exercise Problems	326
	References	330

12 Importance Analysis   **331**

12.1	Introduction	331
12.2	Qualitative Ranking	331
12.3	Quantitative Measures of Importance	332
12.4	Vesely–Fussell Measure of Importance for Minimal Cut Sets	333
12.5	Vesely–Fussell Measure of Importance for Terminal Events	334
12.6	Vesely–Fussell Measure of Importance for Components	341
12.7	Computer Programs	343
12.8	Documentation	343
	Exercise Problems	343
	References	344

13 Applications   **345**

13.1	Introduction	345
13.2	Reliability Engineering Results	345
13.3	Reliability Assessment	346
13.4	Reliability Improvement	346
13.5	Hardware Improvements	347
13.6	Design to Reliability	363
13.7	Reliability Optimization	365
13.8	Optimal Sensor Locations	366
13.9	Postfailure Inspection Strategies	369
13.10	Management Decisions	380
	Exercise Problems	385
	References	387

14 Implementation and Management   **388**

14.1	Introduction	388
14.2	Charter of the Reliability Program	389

14.3   Reliability Program Plan   390
14.4   Scope of the Program   391
14.5   Organizational Structure   391
14.6   Data Sources   392
14.7   Analysis Procedures   393
14.8   Computer Software   393
14.9   Quality Assurance   393
14.10 Documentation   394
14.11 Engineering and Management Reviews   396
14.12 Components Acquisition   396
14.13 Failure Reporting   397
14.14 Concluding Remarks   398

Appendix I   Computer Programs for Reliability Analysis   399

Appendix II Component Reliability Data Sources   404

Author Index   407

Subject Index   409

GUIDE TO RELIABILITY
ENGINEERING

Chapter 1
Introduction

The scope of reliability engineering is the systematic reduction, elimination, and/or control of potential hardware, software, and human failures throughout the life of a plant (or product) so that the reliability of the plant (or product) is an optimum within the cost and functional constraints of the plant (or product).

Reliability affects both the safety and productivity of the plant (or product). A reliable plant (or product) means fewer accidents and thus less risk to the public. Also, a reliable plant (or product) means fewer repairs and less downtime, thus decreasing operating costs and increasing productivity.

Reliability engineering techniques are used primarily to estimate accident probabilities and plant (or product) reliabilities. As a corollary, reliability engineering techniques are also used to design, build, and operate plants (or products) with specified levels of reliability. But applications of reliability engineering are by no means limited to these. Present and potential applications of reliability engineering are numerous: It can be used to design plants (or products) to meet specified levels of reliability at minimum possible cost (cost optimization); it can also be used to develop optimum preventive maintenance schedules and optimal postfailure inspection strategies. Reliability engineering techniques can be used to arrive at rational management decisions, such as whether to repair a component upon its failure or to wait until the next scheduled maintenance, how many spare parts to keep on-site, how many maintenance personnel to keep on-site, etc. Some applications are discussed in Chapter 13. Possible applications of reliability engineering are many, and an engineering manager or an engineer with some training in reliability engineering could identify opportunities in his or her project where reliability engineering could be fruitfully employed to achieve the desired project objectives at minimum cost and maximum safety and productivity.

Because of the wide and increasing use of reliability engineering not only in the defense, aerospace, and nuclear-power industries but also in the fossil-fuel-power, chemical, petroleum, and manufacturing industries, there is a need to provide reliability engineering training to sufficient numbers of mechanical, electrical, electronic, chemical, and civil engineers working in those industries so that full advantage can be taken of reliability engineering techniques in the design, construction, operation, and maintenance of industrial facilities. This book is aimed at such practicing engineers. They may use it as a self-study text or as a text for short courses. This book may also serve as a text for regular university courses in reliability engineering. The book is self-contained: No prior knowledge of reliability engineering, probability theory, or statistics is necessary.

Chapter 2 presents the fundamentals of probability and statistics. This information is necessary for understanding the remainder of the book. Chapter 3 presents the fundamentals of reliability engineering. Component reliability parameters such as reliability, availability, failure probability, failure rate, repair rate, expected number of failures, and hazard function are defined and discussed. Chapter 4 presents the fundamentals of systems engineering. The three basic system configurations, namely, the series, parallel, and m-out-of-n configurations are described.

An engineering system is composed of a number of components arranged and interlinked in a specified manner. Some of the components or the system as a whole may be controlled by computers that receive their "instructions" from computer software. Human intervention may also be required for routine system operations and emergency shutdowns. Maintenance personnel are also an integral part of the operation of engineering systems. Accidents and system failures occur because of component failures, computer software errors, and/or human errors. So, we need the probabilities of component failures, software errors, and human errors to estimate accident probabilities and system reliability. Chapter 5 deals with failure data (probabilities of component failures, software errors, and human errors).

There are two types of system reliability analysis: inductive analysis and deductive analysis. Inductive analyses are less rigorous and usually are carried out during the earlier stages of system development and design. These analyses do not require the probabilities of component failures, software errors, or human errors as data and they do not compute system reliability or accident probabilities.

Preliminary hazard analysis (PHA) and failure modes and effects analysis (FMEA) are inductive analyses. Chapter 6 discusses preliminary hazard analysis, which identifies hazardous elements in a system and assesses their safety implications. Such information could be useful in eliminating many safety problems at an early stage of system design. Any remaining safety issues may be studied in more detail at a later stage via failure modes and effects analysis, fault tree analysis, or other appropriate methods. Some reliability projects do end at the preliminary hazard analysis and no further analyses are carried out; this depends on the scope and purpose of the reliability project.

Chapter 7 describes failure modes and effects analysis, which identifies possible component failures and their effects at the systems level. This method helps system designers to identify and correct weak areas of the system at an early stage of the design. Preliminary hazards analysis is not a prerequisite for failure modes and effects analysis. Reliability analysts may stop at the failure modes and effects analysis or may proceed to conduct a more rigorous deductive analysis, depending on the scope and purpose of the reliability project.

Failure modes and effects analysis is not well-suited for studying the effects of multiple component failures on the system; it is best-suited for identifying component failures and assessing how each failure, by itself, could affect the system. The effects of multiple component failures could be important in complex systems. Deductive analyses are necessary to identify the combinations of component failures that could cause accidents or system failures. System reliability and accident probabilities may also be computed using deductive methods.

Deductive methods of system reliability analysis include fault tree analysis, reliability block diagram analysis, GO analysis, and Markov analysis. Fault tree analysis is the most widely used and one of the most versatile methods of deductive analysis. This method can analyze the vast majority of system reliability problems encountered in any industry, so we limit our discussions to fault tree analysis.

Neither a preliminary hazard analysis nor a failure modes and effects analysis is a prerequisite for fault tree analysis, although information from preliminary hazard analyses and failure modes and effects analyses could be of some use to fault tree analysts.

Fault tree analysis consists of fault tree construction, qualitative fault tree analysis, and quantitative fault tree analysis. Fault tree construction is a prerequisite for both qualitative and quantitative analyses. Qualitative analysis is not a prerequisite for quantitative analysis; methods of quantitative analysis that utilize the results of qualitative analysis and methods that do not are available. A quantitative analysis is necessary only if an estimate of accident probabilities or system reliability is required. Fault tree construction, qualitative analysis, and quantitative analysis are the topics of Chapters 8, 9, and 10, respectively.

There are certain events (for example, earthquakes and fires) and mechanisms (for example, corrosion and dust) that can degrade or cause the failure of a number of components at about the same time. Such events and mechanisms are called common causes. Common causes could severely reduce the system reliability if they affect a number of components and negate or reduce the benefits of redundancies built into the system. So, if common-cause-induced failures are possible and their effects could seriously affect system reliability, a qualitative and/or quantitative common cause analysis needs to be performed. The common cause analysis procedures described in this book utilize the results of qualitative fault tree analysis, and so a qualitative fault tree analysis is a prerequisite. Chapter 11 discusses common cause analysis.

Some component failures contribute more to system failure than others. A formal ranking of component failures according to their potential contribution to system failure is possible through importance analysis. Chapter 12 discusses importance analysis. Whether or not to perform an importance analysis depends on the scope and purpose of the reliability project: Importance analysis is unnecessary if the purpose of the reliability project is just to compute accident probabilities or system reliability. Importance analysis will be useful if the purpose of the reliability analysis is to design systems to meet specified reliabilities, to develop reliability-improvement strategies, to develop reliability-based inspection–maintenance strategies, etc.

Chapter 13 discusses some specific applications of reliability engineering, for example, reliability assessment, design to reliability, reliability improvement, reliability-based installation of component-failure sensors, and reliability-based postfailure inspection strategies. In addition to these specific, goal-oriented applications, the very process of conducting a reliability analysis forces the engineer to look for and to identify the strengths and weaknesses of the system, and that in itself is useful in improving system reliability.

The final chapter (Chapter 14) discusses the implementation and management of reliability programs within a corporate structure or a project. Development of reliability program charter and plan, organizational structure of reliability groups, interfaces between the reliability group and other engineering groups, quality assurance and independent reviews of reliability programs, and documentation of reliability analysis, results, and recommendations are described.

The book ends with two appendices: Appendix I provides a list and descriptions of computer programs used in fault tree analysis, common cause analysis, and importance analysis. Appendix II lists and describes some sources for component failure data.

Chapter 2
Fundamentals of Probability and Statistics

2.1 INTRODUCTION

A basic knowledge of probability theory and statistics is essential for the understanding and use of reliability engineering. This chapter provides the fundamentals of probability theory and statistics necessary to understand the materials covered in this book. Readers who have a fair knowledge of probability and statistics may skip this chapter. This chapter may also serve as a reference source while reading the other chapters. Readers may refer back to this chapter whenever they encounter an unfamiliar term or equation in probability and statistics.

Section 2.2 provides the basic concepts of *set theory*, such as definitions, operations, and laws relating to sets. Section 2.3 covers the basic concepts of probability theory, such as definitions, axioms, and theorems relating to probability and statistics. Probability distributions, moments, and other descriptors of discrete and continuous random variables are also discussed.

Most theorems and equations are provided here without proof or derivation. Readers interested in the proofs and derivations may refer to textbooks on probability and statistics, for example, Freund (1962).

2.2 SET THEORY

Some fundamental concepts of set theory are presented in this section. As will be evident later, a knowledge of set theory is necessary for the understanding of some aspects of probability theory.

2.2.1 Definitions

Experiment: Any situation where a variety of outcomes (at least two) are possible and an element of chance is involved is called an experiment. An experiment need not necessarily be a laboratory experiment: The performance of a piece of equipment in an industrial plant is an experiment.

How the equipment will perform is not known a priori; it may operate as expected or deviate from the expected operation, and thus there is an element of chance. A horse race is another example of an experiment.

Sample space and sample point: The collection of all possible outcomes of an experiment is called its sample space. It is also referred to as the *universe of discourse* and the *possibilities space*. The sample space is often denoted by the symbol *S*.

Consider a horse race with eight horses in the race. If our interest is limited to who the winner of the race will be, there are eight possible outcomes. Each outcome is called a sample point in the sample space. Together, all the eight sample points constitute the sample space. A sample point is also referred to as an *element* of the sample space.

Events: An event is a collection of one or more sample points of a sample space. If the event consists of only one sample point, it is called a *simple event*; otherwise it is a *compound event*. An event may include all the sample points of a sample space.

Consider the experiment of starting an old car on a cold morning. The car may start in one or more attempts. For the sake of simplicity, let us assume that the car will start in no more than four attempts. So there are four possible outcomes (four sample points), namely, $Q1$ = car starts at the first attempt, $Q2$ = car starts at the second attempt, $Q3$ = car starts at the third attempt, and $Q4$ = car starts at the fourth attempt. Each one of these sample points may be considered an event. One may also define events such as 'car starts in less than three attempts' or 'car starts in more than one attempt,' etc. The event 'car starts in less than three attempts' consists of two sample points ($Q1$ and $Q2$). Similarly, the event 'car starts in more than one attempt' consists of three sample points ($Q2$, $Q3$, and $Q4$).

Some authors use the term *subsets* (or even *sets*) to refer to events. We may use either terminology in this book.

A set *B* is said to be a subset of another set *C* if each element of *B* is also an element of *C*. It is worth noting that each set is a subset of itself.

A set with no elements is called an *empty set* and is often denoted by $\varnothing$. An empty set is a subset of every set.

Multidimensional sample space: The sample space of the car-starting experiment discussed earlier is a one-dimensional sample space because each sample point is definable by a single parameter, namely, the number of attempts to start the car. Consider another experiment, in which two cars (*B* and *C*) are being started. Assuming, as before, that each car will start in 4 or less attempts, there are 16 possible outcomes: *B* starts at the first attempt and *C* starts at the first attempt, *B* starts at the second attempt and C starts at the first attempt, etc. So there are 16 sample points, and each sample point is defined by two parameters, namely, the number of attempts to start car B and the number of attempts to start car C. Because two parameters are required to define each sample point, this sample space is called a *two-dimensional sample space*. In general, a sample space is called an *n-dimensional sample space* if n parameters are needed to define each sample point.

Discrete sample space: In the experiment discussed so far, the possible outcomes are discrete (for example, the number of attempts to start the car is 1, 2, 3, or 4), so the sample space is called a discrete sample space.

Continuous sample space: Consider the example of testing a metal rod for its yield stress. The yield stress could possibly take any value between zero and infinity (for example, 1247.57, 2600.0, 2769.045, etc.). In this case, the sample space of the experiment is continuous and the number of sample points is infinite.

A subset or event in the preceding continuous sample space could be 'yield stress is greater than 2000' or 'yield stress is between 1250 and 1500.' Note that even in the subset where there is a finite lower bound (1250) and a finite upper bound (1500) the number of sample points is infinite.

If the metal rod is tested for both yield stress and modulus of elasticity, then each outcome is defined by two parameters (yield stress and modulus of elasticity), so the sample space of that experiment is a *two-dimensional, continuous sample space*.

Conditional sample space: In the experiment of starting two cars, discussed earlier, the sample space consists of 16 points. If we are interested in only those instances where car B has started at the very first attempt, then we have a conditional sample space with four sample points: (i) car B starts at the first attempt and car C starts at the first attempt; (ii) car B starts at the first attempt and car C starts at the second attempt; (iii) car B starts at the first attempt and car C starts at the third attempt; and (iv) car B starts at the first attempt and car C starts at the fourth attempt. This conditional sample space is defined by the condition that car B starts at the first attempt. Another conditional sample space may be defined by the condition that car C starts only after at least three attempts; this conditional sample space has eight points. Conditional sample spaces may also be defined in a continuous sample space. For example, in the example of testing the metal rod for both yield stress and modulus of elasticity, a conditional sample space may be defined by the condition that the modulus of elasticity is between 20,000 and 30,000. Here the conditional sample space is also continuous and has infinite number of sample points.

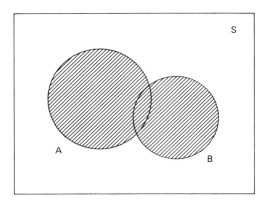

Figure 2.1. Union of two sets A and B.

2.2.2 Operations on Sets

Operations on sets are sometimes called *Boolean operations* and the subset of set operations is known as *Boolean algebra*.

Union of sets: The union of two sets (or subsets or events) A and B is a set C, which consists of all the elements belonging to A or B or to both A and B. Union is denoted by the symbol $\cup$,

$$C = A \cup B$$

If A consists of elements (sample points) $E1$, $E2$, $E7$, and $E9$ and B consists of elements $E2$, $E6$, and $E7$, then C consists of elements $E1$, $E2$, $E6$, $E7$, and $E9$.

A diagrammatic representation of union is shown in Figure 2.1. Such a diagrammatic representation of sample spaces and sets is called a *Venn diagram*.

The rectangular area in Figure 2.1 represents the sample space S. The circular area on the left represents the set A and the circular area on the right represents the set B. The hatched area represents the set C.

The concept of union may be extended to more than two sets, for example

$$X = A \cup B \cup C = Y \cup C = A \cup Z$$

where

$$Y = A \cup B \quad \text{and} \quad Z = B \cup C$$

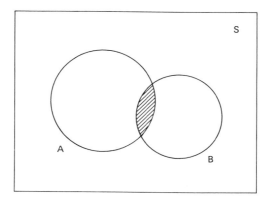

Figure 2.2. Intersection of two sets A and B.

Intersection of sets: The intersection of two sets (subsets or events) A and B is a set C, which consists of all elements belonging to both A and B. Intersection is denoted by the symbol $\cap$,

$$C = A \cap B$$

If A consists of elements $E1$, $E2$, $E7$, and $E9$ and B consists of $E2$, $E6$, and $E7$, then C consists of the elements $E2$ and $E7$.

A Venn diagram showing an intersection of A and B is given in Figure 2.2. The hatched area represents the set C.

The concept of intersection may be extended to more than two sets. For example,

$$X = A \cap B \cap C = Y \cap C = A \cap Z$$

where

$$Y = A \cap B \quad \text{and} \quad Z = B \cap C$$

Mutually exclusive sets: Two sets A and B are said to be mutually exclusive, or *disjoint*, if they have no elements in common. If A and B are mutually exclusive, then $A \cap B = \varnothing$, where $\varnothing$ is an empty set. Figure 2.3 shows a Venn diagram of two mutually exclusive sets. Notice that the two circular areas representing the mutually exclusive sets A and B do not overlap or intersect.

Collectively exhaustive sets: A group of n sets $A_1, A_2, \ldots, A_n$, are said to be collectively exhaustive if, together, these n sets contain all the sample points of the sample space; here $n > 1$.

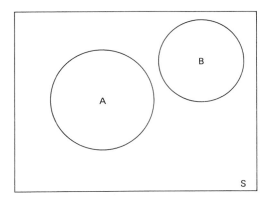

Figure 2.3. Two mutually exclusive sets A and B.

Equality of sets: Two sets B and C are said to be equal $(B = C)$ if and only if each element of B is also an element of C and each element of C is also an element of B.

Complement of a set: A set $\bar{B}$ is said to be the complement of set B, with respect to the sample space S, if and only if B and $\bar{B}$ have no elements in common and B and $\bar{B}$, together, contain all the elements of S. Consider a sample space S that consists of eight sample points $E1, E2, \ldots, E8$. Let a set B consist of $E1$, $E2$, and $E5$. The complement of B with respect to the sample space S will consist of $E3$, $E4$, $E6$, $E7$, and $E8$.

The concept of the complement of a set is shown through a Venn diagram in Figure 2.4. The area inside the circle represents the set B and the hatched area represents its complement $\bar{B}$.

The based on the definitions of union and intersection, we have

$$B \cup \bar{B} = S \quad \text{and} \quad B \cap \bar{B} = \emptyset$$

where $\emptyset$ is an empty set. Also, note that B and $\bar{B}$ are mutually exclusive.

Identities: The following identities are useful:

$$B \cup B = B$$
$$B \cap B = B$$

These two equations are the direct result of the definitions of union and intersection. If B consists of n elements $E_1, E_2, \ldots, E_n$, then the union of B with itself will also consist of the same n elements. This is exactly what

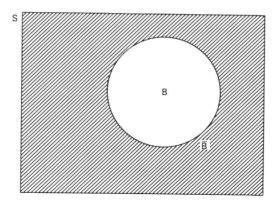

Figure 2.4. Set B and its complement $\bar{B}$.

the first of the two equations states. Similarly, the intersection of B with itself will consist of the same n elements only. This is what the second of the two equations states.

2.2.3 Basic Laws of Sets

Law of closure: For every pair of sets A and B, there exist unique sets $A \cup B$ and $A \cap B$.

Identity law: There exist unique sets S and $\varnothing$ such that, for every set A, we have $A \cap S = A$ and $A \cup \varnothing = A$. The set S is known as the sample space and the set $\varnothing$ is known as the empty set (see Section 2.2.1).

Complementative law: For every set A, there exists a unique complementary set $\bar{A}$, which satisfies the relationships $A \cap \bar{A} = \varnothing$ and $A \cup \bar{A} = S$.

Commutative laws:

$$A \cup B = B \cup A$$
$$A \cap B = B \cap A$$

Associative laws:

$$(A \cup B) \cup C = A \cup (B \cup C)$$
$$(A \cap B) \cap C = A \cap (B \cap C)$$

Distributive laws:

$$A \cap (B \cup C) = (A \cap B) \cup (A \cap C)$$
$$A \cup (B \cap C) = (A \cup B) \cap (A \cup C)$$

Laws of absorption:

$$A \cup (A \cap B) = A$$
$$A \cap (A \cup B) = A$$

2.3 PROBABILITY THEORY

2.3.1 Definitions

Each sample point in the sample space of an experiment has a *probability* associated with it. There are two interpretations of probability, namely, an objective interpretation and a subjective interpretation.

Objective interpretation: Consider an experiment whose sample space consists of M sample points $Q_1, Q_2, \ldots, Q_M$. Let the experiment be repeated n times (each repetition is called a *trial*). Let the number of times Q_i occurs be n_i, where $i = 1, 2, \ldots, M$. We have

$$\sum_{i=1}^{M} n_i = n$$

The ratio (n_i/n) is called the *relative frequency* of the ith sample point. The probability of a sample point, with respect to the sample space, is its relative frequency as the number of trials (n) tends to infinity, that is,

$$P[Q_i] = p_i = \lim_{n \to \infty} \left(\frac{n_i}{n} \right) \qquad (2.1)$$

where $p_i = P[Q_i]$ is the probability of the ith sample point.

The summation of the probabilities of all the sample points in a sample space is unity, that is,

$$\sum_{i=1}^{M} p_i = 1$$

It should be remembered that the probability of a sample point is defined with respect to a specific sample space.

In practice, one cannot repeat the experiment an infinite number of times, so an *approximate value* of the probability may be obtained by repeating the experiment a large number of times. How many repetitions is sufficient to get a fairly accurate probability value? This depends on the value of the probability itself. In general, the lower the probability, the higher is the required number of repetitions. There is no definite rule for the required number of repetitions. As a guide, one may conduct two groups of n trials each, compute the probability value from each group of trials and accept the average of the two values if the two probabilities are within the required accuracy; if not, conduct more trials in each of the two groups and make the comparison again.[1]

Objective interpretation is sometimes referred to as the *frequentist interpretation*.

Subjective interpretation: Consider a situation in which repeating the experiment is not possible, for example, a horse race in which eight horses are running. It is practically impossible to conduct the same horse race a large number of times. So the probability of horse A winning the race or horse B winning the race cannot be assessed through objective interpretation. In such a case, a probability is assigned to each sample point by one or more experts (who are knowledgeable about the experiment) on the basis of their experience with similar situations (for example, their knowledge about the past history of each horse in other races). Such a *subjective probability* depends on the experts' judgement and is, of course, less rigorous than an objective probability.

Subjective probabilities have engineering applications also. For example, what is the probability that a plant operator will respond correctly during an accident in an industrial plant? It is not possible to subject the same operator to the same accident again and again (or, alternatively, subject a large number of operators with the same qualifications and experience to the same accident) and compute the probability through an objective interpretation. The more practical approach is for one or more experts who are knowledgeable about the situation to estimate the probability subjectively. The experts may base their estimates, in part, on any available data on operator responses in similar, or somewhat similar, accidents.

---

[1]There are some rigorous methods for estimating the accuracy of computed probabilities via confidence intervals. [Freund (1962)]. As will be discussed in Section 5.1, component failure probabilities used in system reliability analyses are usually known only within an order-of-magnitude accuracy, so the rigorous methods of accuracy estimation are unnecessary in most reliability engineering applications. (The concept of confidence intervals is described in Section 5.2.5.2.)

Derived probability: Although this is not considered as a distinct interpretation of probability, the probability of a complex experiment is often derived on the basis of known probabilities of similar experiments. For example, if one is interested in the expected probability of failure of a newly built industrial plant, it is of course impossible to predict an objective probability from observed frequencies of that plant failure. One way is to develop a mathematical model that expresses the plant failure as a function of the failures of the various components (equipment, instruments, structures, piping, etc.) of the plant. From such a mathematical model, either an exact or an approximate mathematical expression for the plant failure probability as a function of component failure probabilities may be developed. Then the plant failure probability may be computed by substituting component failure probabilities into the mathematical model. The attractiveness of this approach lies in the fact that it is easier to determine component failure probabilities than the plant failure probability by actual testing. For example, it is possible to test a number of components (say, valves) and compute the failure probability at least approximately. On the contrary, it is impossible to build a number of identical plants and test them all.

Probability of an event: The probability of a sample point is discussed in the preceding paragraphs. The probability of an event (subset or set) is the sum of the probabilities of all the sample points contained in that event. The probability of an event A is denoted by $P[A]$ and is given by

$$P[A] = \sum_{i=1}^{n} P[Q_i]$$

where Q_i, for $i = 1, 2, \ldots, n$, are the sample points contained in event A. Because the probabilities of the sample points are defined in relation to a sample space, the probability of the event is also defined in relation to that same sample space.

Probability of a variable: In many experiments the outcomes (sample points) are numerical values. For example, in the car-starting experiment, discussed earlier, the outcomes are 1, 2, 3, and 4; in the metal-rod-testing experiment, the outcomes are any positive number. The outcomes in such cases are called *random variables* because their specific values cannot be predicted with certainty before the experiment is conducted. A probability may be associated with each possible numerical value.

The random variable may be either *discrete* (that is, it can take only discrete numerical values) or *continuous*. In the car-starting experiment, the outcomes are discrete variables; in the metal-rod-testing experiment, the outcomes are continuous variables. In either case, a probability is associated with each possible numerical value. If the variable is discrete, the probabilities associated with the possible values are given by a *probability mass function* (PMF). If the variable is continuous, the probabilities associated with the possible values are given by a *probability density function* (PDF). These functions are discussed in more detail in Sections 2.3.4.1 and 2.3.4.2.

Random function: A function $G(x)$ is said to be a random function if the values of $G(x)$ at specified values of x (say x_1, x_2, x_3, etc.) are random variables. Note that the variable x is deterministic.

Stochastic process: A random function in time is called a stochastic process or a *random process*. Random functions of other types of variables such as spatial coordinates, for example, are also sometimes referred to as stochastic processes or random processes.

Conditional probability: As noted earlier, a probability is assigned to a sample point or event in relation to a sample space. For the sake of brevity and convenience, the sample space is seldom explicitly mentioned when referring to a probability. For example, we say "the probability of event B" rather than "the probability of event B in relation to the sample space S."

However, when the probability is assigned in relation to a conditional sample space, the condition *should be* mentioned.

For example, going back to the example of starting two cars (see Section 2.2.1), if we are interested in the probability that car B starts at the first attempt given that car C has started only after at least three attempts, the probability is now assigned in relation to the conditional sample space that car C started only after at least three attempts. The corresponding probability is specified as "the probability that car B starts at the first attempt given that car C has started only after at least three attempts." This probability is denoted by $P[G|H]$, where event G is 'car B starts at the first attempt' and event H is 'car C starts only after at least three attempts.' The notation $P[G|H]$ is read as "the conditional probability of event G given that event H has occurred" or "the conditional probability of G given H." Computation of conditional probabilities is illustrated in Example 2.1.

Independence of events: Two events are said to be independent of each other if they are totally unrelated. Independence between two events may be determined on the basis of the underlying physical situation.

Independence between events may also be defined in the following way. An event A is said to be independent of another event B if the probability of A is unaffected by whether B has occurred (or will occur) and the probability of B is unaffected by whether A has occurred (or will occur). Mathematical expressions that may be used to check the independence of events are given in Section 2.3.3.

Example 2.1

An experiment was conducted to determine the number of attempts required to start two old cars B and C in cold weather. The experiment was repeated 100 times. Results are tabulated in Table 2.1. For example, the entry "6" in the row starting with "$B = 2$" under the column headed by "$C = 3$" indicates that car B started at the second attempt and car C started at the third attempt 6 times out of the 100.

Compute the following probabilities: (i) the probability that car B starts at the first attempt; (ii) the probability that car C starts at more than two attempts; (iii) the conditional probability that car B starts at the second attempt given that car C has started at the second attempt; (iv) the conditional probability that car C starts in less than three attempts given that car B has started at more than two attempts.

Solution

(i) The probability that car B starts at the first attempt

$$= \frac{\text{the number of times car } B \text{ started at the first attempt}}{\text{the total number of trials}}$$

$$= \frac{8 + 6 + 5 + 2}{100} = \frac{21}{100} = 0.21$$

Note that this probability is independent of the behavior of car C: It is independent of the number of attempts required to start car C.

Table 2.1.   Data for Example 2.1

	$C = 1$	$C = 2$	$C = 3$	$C = 4$
$B = 1$	8	6	5	2
$B = 2$	8	7	6	3
$B = 3$	9	8	6	4
$B = 4$	10	7	7	4

(ii) The probability that car C starts at more than two attempts

$$= \frac{\left\{ \begin{array}{l} \text{the number of times in which car } B \text{ started} \\ \text{at the third attempt or the fourth attempt} \end{array} \right\}}{\text{the total number of trials}}$$

$$= \frac{(5 + 6 + 6 + 7) + (2 + 3 + 4 + 4)}{100}$$

$$= \frac{47}{100} = 0.47$$

(iii) The conditional probability that car B starts at the second attempt given that car C has started at the second attempt

$$= \frac{\left\{ \begin{array}{l} \text{the number of times in which car } B \text{ started} \\ \text{at the second attempt and car } C \text{ started} \\ \text{at the second attempt simultaneously} \end{array} \right\}}{\left\{ \begin{array}{l} \text{the number of times car } C \text{ started} \\ \text{at the second attempt} \end{array} \right\}}$$

$$= \frac{7}{(6 + 7 + 8 + 7)} = \frac{7}{28} = 0.25$$

(iv) The conditional probability that car C starts at less than three attempts given that car B has started at more than two attempts

$$= \frac{\left\{ \begin{array}{l} \text{the number of times in which car } C \text{ started} \\ \text{at the first or second attempt and car } B \text{ started} \\ \text{at the third or fourth attempt simultaneously} \end{array} \right\}}{\left\{ \begin{array}{l} \text{the number of times in which car } B \text{ started} \\ \text{at the third or fourth attempt} \end{array} \right\}}$$

$$= \frac{(9 + 8) + (10 + 7)}{(9 + 8 + 6 + 4) + (10 + 7 + 7 + 4)}$$

$$= \frac{34}{55} = 0.618$$

2.3.2 Axioms of Probability

Whether a probability is based on objective interpretation or subjective interpretation, it should satisfy the following three axioms.

Axiom 1

The probability of an event lies between 0 and 1. That is,

$$0 \leq P[B] \leq 1 \qquad (2.2)$$

where B is an event.

Axiom 2

The probability of an event that is certain to occur is 1. Because the sample space is a certain event (per definition, one of its sample points is certain to occur), the probability of a sample space is 1, that is,

$$P[S] = 1 \qquad (2.3)$$

where S is a sample space.

Axiom 3

If $A_1, A_2, \ldots$ are a finite or infinite number of mutually exclusive events of the sample space S, then the probability of the union of these events is given by

$$P[A_1 \cup A_2 \cup \cdots] = P[A_1] + P[A_2] + \cdots \qquad (2.4)$$

2.3.3 Theorems of Probability

Theorem 1

The probability of an empty set is zero. That is,

$$P[\varnothing] = 0 \qquad (2.5)$$

where $\varnothing$ is an empty set.

Theorem 2

The probability of the complement of an event can be expressed in terms of the probability of the event as follows:

$$P[\bar{A}] = 1 - P[A] \qquad (2.6)$$

where event $\bar{A}$ is the complement of event A.

Theorem 3

The probability of the intersection of two mutually exclusive events is zero. If A and B are two mutually exclusive events, then

$$P[A \cap B] = 0 \qquad (2.7)$$

Theorem 4

The probability of the union of two events is given by

$$P[A \cup B] = P[A] + P[B] - P[A \cap B] \qquad (2.8)$$

If A and B are mutually exclusive events, the preceding equation reduces to

$$P[A \cup B] = P[A] + P[B] \qquad (2.9)$$

because of Equation (2.7).

The probability of the union of three events is given by

$$P[A \cup B \cup C] = P[A] + P[B] + P[C] - P[A \cap B] - P[B \cap C]$$
$$- P[C \cap A] + P[A \cap B \cap C] \qquad (2.10)$$

The preceding equation may be obtained by rewriting $P[A \cup B \cup C]$ as $P[(A \cup B) \cup C]$ and applying Equation (2.8) twice. Alternatively Equation (2.10) may be obtained by drawing a Venn diagram and accounting for the areas contained by the union of A, B, and C.

Equation (2.10) may be generalized for the union of n events as follows:

$$P[A_1 \cup A_2 \cup \cdots \cup A_n]$$
$$= \{P[A_1] + P[A_2] + \cdots + P[A_n]\}$$
$$- \{P[A_1 \cap A_2] + P[A_2 \cap A_3] + \cdots$$
$$\text{(all two-event combinations of the } n \text{ events)}\}$$
$$+ \{P[A_1 \cap A_2 \cap A_3] + \cdots$$
$$\text{(all three-event combinations of the } n \text{ events)}\}$$
$$- \{\cdots\} + \{\cdots\} - \cdots$$
$$+ (-1)^{n-2}\{\text{all } (n-1) - \text{event combinations of the } n \text{ events}\}$$
$$+ (-1)^{n-1}\{P[A_1 \cap A_2 \cap \cdots \cap A_n]\} \qquad (2.11)$$

When $n = 3$, Equation (2.11) will reduce to Equation (2.10), and when $n = 2$, it will reduce to Equation (2.8).

If the n events are mutually exclusive to each other, Equation (2.11) reduces to

$$P[A_1 \cup A_2 \cup \cdots \cup A] = P[A_1] + P[A_2] + \cdots + P[A_n] \quad (2.12)$$

where A_1, A_2, ..., A_n are mutually exclusive events.

Theorem 5

The conditional probability of event A given that event B has occurred is

$$P[A|B] = \frac{P[A \cap B]}{P[B]} \quad (2.13)$$

provided that $P[B] \neq 0$. The conditional probability is undefined if $P[B] \neq 0$. The conditional probability is zero if A and B are mutually exclusive.

Using Equation (2.13), we get

$$P[A \cap B] = P[A|B]P[B] = P[B|A]P[A] \quad (2.14)$$

The preceding equation may be extended to three events as follows:

$$P[A \cap B \cap C] = P[A|B \cap C]P[B|C]P[C] \quad (2.15)$$

A general form of Equation (2.15) is

$$P[A_1 \cap A_2 \cap \cdots \cap A_n]$$
$$= P[A_1|A_2 \cap A_3 \cap \cdots \cap A_n]P[A_2|A_3 \cap A_4 \cap \cdots \cap A_n] \cdots P[A_n] \quad (2.16)$$

Theorem 6

Let A_1, A_2, ..., A_n be n sets that are mutually exclusive and collectively exhaustive. The probability of another event B (set B) belonging to the same sample space of A_1 to A_n is given by

$$P[B] = \sum_{i=1}^{n} P[B \cap A_i] = \sum_{i=1}^{n} P[B|A_i]P[A_i] \quad (2.17)$$

This theorem is known as the *total probability theorem*. A special case of Equation (2.17) is the following:

$$P[B] = P[B|A]P[A] + P[B|\overline{A}]P[\overline{A}] \qquad (2.18)$$

where $\overline{A}$ is the complement of A. Equation (2.18) is true because A and $\overline{A}$ are mutually exclusive and are collectively exhaustive.

Theorem 7

If two events A and B are statistically independent, then

$$P[A|B] = P[A] \quad \text{and} \quad P[B|A] = P[B] \qquad (2.19)$$

Using Equation (2.13) in Equation (2.19), we get

$$P[A \cap B] = P[A]P[B] \qquad (2.20)$$

where A and B are independent events.

Equations (2.19) and/or (2.20) are sometimes used as a definition of *statistical independence* between events.

Equation (2.20) may be generalized as,

$$P[A_1 \cap A_2 \cap \cdots \cap A_n] = P[A_1]P[A_2] \cdots P[A_n] \qquad (2.21)$$

where $A_1, A_2, \ldots, A_n$ are independent of each other.

Theorem 8

If events A and B are statistically independent, then the following hold:

  (i) A and $\overline{B}$ are statistically independent;
 (ii) $\overline{A}$ and B are statistically independent;
(iii) $\overline{A}$ and $\overline{B}$ are statistically independent;

where $\overline{A}$ and $\overline{B}$ are the complements of A and B, respectively.

Example 2.2

Consider the car-starting experiment discussed in Example 2.1. (i) Check the independence between the event 'car B starts at the second attempt' and the event 'car C starts at the second attempt.' (ii) Compute the conditional probability that car B starts at the second attempt given that car C has started at the second attempt. (iii) Compute the conditional probability that car C starts in less than three attempts given that car B has started at more than two attempts.

Solution

Let us use the following notations to denote the different events of interest.

$$\text{Event } B2 = \text{car } B \text{ starts at the second attempt}$$

$$\text{Event } C2 = \text{car } C \text{ starts at the second attempt}$$

$$\text{Event } C12 = \text{car } C \text{ starts at less than three attempts}$$

$$= \text{car } C \text{ starts at the first or second attempt}$$

$$\text{Event } B34 = \text{car } B \text{ starts at more than two attempts}$$

$$= \text{car } B \text{ starts at the third or fourth attempt}$$

(i) Using the data provided in Table 2.1, we compute the following probabilities:

$$P[B2] = \frac{8 + 7 + 6 + 3}{100} = \frac{24}{100} = 0.24$$

$$P[C2] = \frac{6 + 7 + 8 + 7}{100} = \frac{28}{100} = 0.28$$

$$P[B2 \cap C2] = \frac{7}{100} = 0.07$$

$$P[B2]P[C2] = (0.24 \times 0.28) = 0.0672$$

We see that Equation (2.20) is not satisfied by $B2$ and $C2$. So $B2$ and $C2$ are not statistically independent.

(ii) Using Equation (2.13),

$$P[B2|C2] = \frac{P[B2 \cap C2]}{P[C2]} = \frac{0.07}{0.28} = 0.25$$

This result is identical to that obtained in Example 2.1, using a different method.

(iii) Using the data provided in Table 2.1, we get

$$P[B34] = \frac{(9 + 8 + 6 + 4) + (10 + 7 + 7 + 4)}{100}$$

$$= \frac{55}{100} = 0.55$$

Using Equation (2.13), we get

$$P[C12|B34] = \frac{P[C12 \cap B34]}{P[B34]} = \frac{0.34}{0.55} = 0.618$$

This result is identical to that obtained in Example 2.1, using a different method.

2.3.4 Probability Distributions

2.3.4.1 Distributions for discrete variables.   It is a common practice to denote a random variable by capital letters (X, Y, etc.) and the specific values they can take by corresponding lowercase letters (x, y, etc.). For example, in the experiment of testing a metal rod for yield strength, the yield strength (the random variable) is denoted by X and the values it can take (for example, 1485.7, 2467.45, etc.) are denoted by x. Note that x represents a possible outcome of the experiment.

The probability associated with each possible value x of the random variable X is denoted by $P_X[X = x]$. Sometimes it is denoted simply by $P[X = x]$ or $P_X[x]$. Probabilities associated with all the possible values of X are given by a probability distribution. In the case of discrete random variables, this distribution is called the *probability mass function* (PMF) and is denoted by $p_X[x_i]$, where $i = 1, 2, \ldots, n$ and the x_i are the possible values X can take:

$$p_X(x_i) = P_X[X = x_i] \quad \text{for } i = 1, 2, \ldots, n \quad (2.22)$$

For example, let the random variable X denote the number of pumps in a plant that need to be replaced during a 20-year period. Let the total number of pumps be 5. Here, possible values of X are 0, 1, 2, 3, 4, and 5.

A typical probability mass function is of the form

$$p_X(x) = \begin{cases} 0.10 & \text{for } x = 0 \\ 0.15 & \text{for } x = 1 \\ 0.24 & \text{for } x = 2 \\ 0.40 & \text{for } x = 3 \\ 0.10 & \text{for } x = 4 \\ 0.01 & \text{for } x = 5 \end{cases}$$

The preceding equation may also be written in the form

$$p_X(0) = 0.10$$
$$p_X(1) = 0.15$$
$$p_X(2) = 0.24$$
$$p_X(3) = 0.40$$
$$p_X(4) = 0.10$$
$$p_X(5) = 0.01$$

A plot of this probability mass function is given in Figure 2.5a.

We may also specify the probability distribution of a random variable by its *cumulative distribution function* (CDF). The cumulative distribution function, $F_X(x)$, provides the probability that the random variable X is equal to or less than a value x, that is,

$$F_X(x) = P[X \le x] \tag{2.23}$$

The probability mass function and the cumulative distribution function are related by the following equation:

$$F_X(x) = \sum p_X(x_i) \quad \text{summed over all } x_i \le x \tag{2.24}$$

For the numerical example of pump replacement discussed earlier, the cumulative distribution function is as follows:

$$F_X(0) = 0.10$$
$$F_X(1) = 0.10 + 0.15 = 0.25$$
$$F_X(2) = 0.10 + 0.15 + 0.24 = 0.49$$
$$F_X(3) = 0.10 + 0.15 + 0.24 + 0.40 = 0.89$$
$$F_X(4) = 0.10 + 0.15 + 0.24 + 0.40 + 0.10 = 0.99$$
$$F_X(5) = 0.10 + 0.15 + 0.24 + 0.40 + 0.10 + 0.01 = 1.00$$

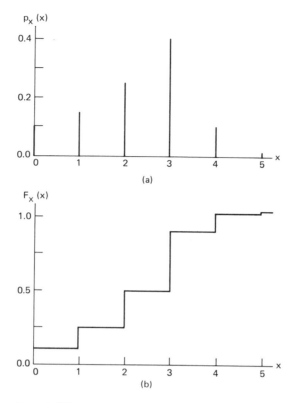

Figure 2.5. Sample probability mass function and cumulative distribution function of a discrete random variable: (a) probability mass function; (b) cumulative distribution function.

A plot of this cumulative distribution function is given in Figure 2.5b.

The probability mass function and the cumulative distribution function are related by the following equation:

$$p_X(x_i) = F_X(x_i) - F_X(x_i - \varepsilon) \qquad (2.25)$$

where ε is a small positive number; ε is so small that $x_{i-1} < x_i - \varepsilon$.

In some cases, the probability mass function and/or the cumulative distribution function may be expressed in a functional form instead of giving a set of values as before. For example,

$$p_X(x) = q(1 - q)^{x-1} \quad \text{for } x = 1, 2, \ldots$$

The corresponding cumulative distribution function is obtained by using Equation (2.24):

$$F_X(x) = \sum_{j=1}^{x} q(1-q)^{j-1}$$

The probability that a random variable X will take values between a and b is given by

$$P_X[a < x \le b] = \sum p_X(x_i) \tag{2.26}$$

The summation in Equation (2.26) is over all values of x_i in the interval $a < x_i \le b$.

Using Equation (2.24), Equation (2.26) may be written as

$$P_X[a < x \le b] = F_X(b) - F_X(a) \tag{2.27}$$

Example 2.3

Consider the car-starting experiment of Example 2.1. Let the random variable Y represent the number of attempts required to start car C. Determine the probability mass function of Y.

Solution

The number of times car C started at the first attempt
    $= 8 + 8 + 9 + 10 = 35$
The number of times car C started at the second attempt
    $= 6 + 7 + 8 + 7 = 28$
The number of times car C started at the third attempt
    $= 5 + 6 + 6 + 7 = 24$
The number of times car C started at the fourth attempt
    $= 2 + 3 + 4 + 4 = 13$
Total number of trials $= 100$
Probability that car C starts at the first attempt $= \frac{35}{100} = 0.35$
Probability that car C starts at the second attempt $= \frac{28}{100} = 0.28$
Probability that car C starts at the third attempt $= \frac{24}{100} = 0.24$
Probability that car C starts at the fourth attempt $= \frac{13}{100} = 0.13$

So the probability mass function of the random variable Y is

$$p_Y(1) = 0.35$$

$$p_Y(2) = 0.28$$

$$p_Y(3) = 0.24$$

$$p_Y(4) = 0.13$$

2.3.4.2 Distributions for continuous variables.   The *cumulative distribution function* (CDF) of a continuous random variable X, denoted by $F_X(x)$, provides the probability that the random variable X is equal to or less than x. That is,

$$F_X(x) = P[X \leq x] \tag{2.28}$$

The cumulative distribution function may be of any functional form, for example,

$$F_X(x) = \begin{cases} 1 - \exp(-cx) & \text{for } x \geq 0 \\ 0 & \text{for } x < 0 \end{cases}$$

where c is a known parameter.

The probability distribution of a continuous random variable may also be specified by its *probability density function* (PDF), which is denoted by $f_X(x)$:

$$f_X(x) = \frac{dF_X(x)}{dx} \tag{2.29}$$

if the first derivative of $F_X(x)$ exists.

The cumulative distribution function may be expressed in terms of the probability density function as follows:

$$F_X(x) = \int_{-\infty}^{x} f_X(u) \, du \tag{2.30}$$

where u is a dummy variable within the integral. Note that $F_X(-\infty) = 0$ because $F_X(-\infty) = P[X \leq -\infty] = 0$.

It is worth noting that a probability density $f_X(x)$ is not a probability, but $f_X(x)\,dx$ is a probability:

$$f_X(x)\,dx = P[x < X \le (x + dx)] \qquad (2.31)$$

that is, $f_X(x)\,dx$ gives the probability that the random variable X lies between x and $x + dx$.

Using Equation (2.29) on the example cumulative distribution function given earlier, we have

$$f_X(x) = \begin{cases} c\exp(-cx) & \text{for } x \ge 0 \\ 0 & \text{for } x < 0 \end{cases}$$

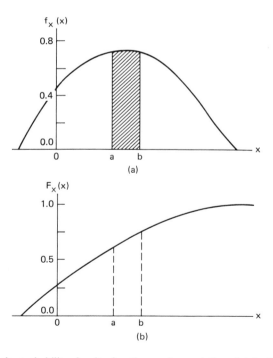

Figure 2.6. Sample probability density function and cumulative distribution function of a continuous random variable: (a) probability density function; (b) cumulative distribution function.

Sample plots of a probability density function and a cumulative density function are given in Figure 2.6.

A closed form or analytical integration of a probability density function [Equation (2.30)] may not always be possible. Numerical integration may be used when necessary.

The probability that the random variable X will have values between a and b is given by

$$P[a < X \le b] = \int_a^b f_X(x)\,dx \qquad (2.32)$$

In other words, the area of the probability density function between a and b is equal to the probability that X will have values between a and b (see hatched area in Figure 2.6). Equation (2.32) may also be written as

$$P[a < X \le b] = F_X(b) - F_X(a) \qquad (2.33)$$

because $\int_a^b f_X(x)\,dx = \int_{-\infty}^b f_X(x)\,dx - \int_{-\infty}^a f_X(x)\,dx$.

2.3.4.3 Basic properties of probability distributions. Cumulative distribution functions of discrete or continuous random variables should satisfy the following conditions:

(i) $F_X(-\infty) = 0$ and $F_X(+\infty) = 1$.
(ii) $0 \le F_X(x) \le 1$.
(iii) $F_X(x)$ is nondecreasing with x.

Because of these properties of cumulative distribution functions, the following conditions are satisfied by probability density functions:

(i) $f_X(x) \ge 0$.
(ii) The total area under a probability density function is 1, that is, $\int_{-\infty}^{\infty} f_X(x)\,dx = 1$.

Similarly, the following conditions are satisfied by probability mass functions:

(i) $p_X(x_i) \ge 0$.
(ii) The sum of all the probability mass function values is 1, that is $\sum p_X(x_i) = 1$, when summed over all x_i.

2.3.4.4 Expectation[2]. The *expectation* (or *expected value*) of a random variable X is given by

$$E[X] = \int_{-\infty}^{\infty} xf_X(x)\, dx \qquad (2.34)$$

and it is equal to the *mean value* (commonly known as the *average value*) of the random variable.

2.3.4.5 Moments and descriptors. A random variable is defined completely by its probability density function or cumulative distribution function. It is not always possible to determine or calculate either of these functions. In such situations, some "generalized average measures" of the distributions may be used to describe the random variable. Also, sometimes it may be convenient to deal with these average measures even if the probability density function or cumulative distribution function is available. These generalized average measures are the *moments* of the random variable. Specific names such as *mean*, *variance*, and *skewness* are given to some of these moments.

The nth moment of a random variable X is given by

$$\mu_n'(X) = E[X^n] = \int_{-\infty}^{\infty} x^n f_X(x)\, dx \qquad (2.35)$$

Because, in this equation, the moment of the probability density function is taken with reference to the origin of the variable (lever arm is measured from the origin $x = 0$), these moments are sometimes referred to as the *nth moment with reference to the origin* or the *nth moment about the origin*; they are usually referred to as simply the *nth moment*. Comparing Equation (2.35) with Equation (2.34), we see that the first moment is identical to the expectation (mean value).

Moments are sometimes taken with reference to the mean value point of the variable. Such moments are called the central moments. The *nth central moment* of a random variable X is given by

$$\mu_n(X) = E\left[(X - \bar{x})^n\right] = \int_{-\infty}^{\infty} (x - \bar{x})^n f_X(x)\, dx \qquad (2.36)$$

where $\bar{x}$ is the mean value of X.

---

[2]The topics in the rest of Section 2.3.4 are discussed with respect to continuous random variables. In most instances, unless otherwise indicated, the discussions are equally applicable to discrete random variables also, provided the integrations are replaced by summations.

If all the moments (either the moments about the origin or the central moments for $n = 1, 2, \ldots, \infty$) of a random variable are known, then the random variable is completely defined. This is equivalent to specifying the probability density function or the cumulative distribution function of the random variable.

Some of the lower-order moments are most commonly used, and special names, such as mean and mean square, are given to them.

Mean, median, and mode: The first moment of a random variable is called the expectation, expected value, or *mean*, of the random variable. This descriptor gives an indication of the central tendency of the random variable.

Two other descriptors, median and mode, are also used to provide an indication of the central tendency of the random variable. The area of the probability density function to the left of the *median* and to the right of the median are equal, and each is equal to 0.5, that is,

$$\int_{-\infty}^{\hat{x}} f_X(x)\, dx = \int_{\hat{x}}^{\infty} f_X(x)\, dx = 0.5 \qquad (2.37)$$

where $\hat{x}$ is the median of X.

The probability that the random variable X takes values below the median is equal to the probability that the random variable X takes values above the median, and each of these two probabilities is equal to 0.5, that is,

$$F_X(\hat{x}) = 0.5 \qquad (2.38)$$

The mean value coincides with the median value if the probability density function is symmetric.

The *mode* of a random variable is defined as the value of the random variable that has the largest probability density. (If it is a discrete random variable, then the value of the random variable that has the largest probability is its mode).

Mean square value, variance, standard deviation, and coefficient of variation: Descriptors associated with the second moment provide a measure of the dispersion of the probability density function. These descriptors include the mean square value, variance, standard deviation, and the coefficient of variation. The higher the value of these descriptors, the broader is the dispersion of the probability density function about the mean.

The second moment about the origin is called the *mean square value* of the random variable and is denoted by $\mu_2'(X)$:

$$\mu_2'(X) = E[X^2] \tag{2.39}$$

The second central moment is called the *variance* and is denoted by $\text{Var}(X)$ or $\mu_2(X)$:

$$\text{Var}(X) = \mu_2(X) = E\left[(X - \bar{x})^2\right] \tag{2.40}$$

The mean square value and the variance of a random variable are related by the following equation:

$$\mu_2(X) = \mu_2'(X) - \bar{x}^2 = E[X^2] - (E[X])^2 \tag{2.41}$$

The square root of variance is called the *standard deviation* and is denoted by σ_X:

$$\sigma_X = [\text{Var}(X)]^{1/2} \tag{2.42}$$

The *coefficient of variation* (COV) is a nondimensional form of the standard deviation. It is denoted by V_X and is defined as

$$V_X = \frac{\sigma_X}{\bar{x}} \tag{2.43}$$

Coefficient of skewness: The third central moment is a descriptor of the skewness or asymmetry of the probability density function. A nondimensional measure of the skewness is given by the *coefficient of skewness*, which is denoted by θ_X:

$$\theta_X = \frac{\mu_3(X)}{\sigma_X^3} \tag{2.44}$$

The coefficient of skewness is zero if the probability density function is symmetric about the mean value. However, a zero value for the coefficient of skewness does not necessarily mean a symmetric probability density function. Positive values of this coefficient usually indicate that the probability density function is skewed to the right (that is, the probability density function has a longer tail on the right side of the mean value). A negative coefficient of skewness usually indicates a longer tail on the left side of the mean value.

2.3.5 Joint Probability Distributions

2.3.5.1 Joint probability density functions and joint cumulative distribution functions. Probability distributions of a single random variable X are considered in the preceding discussions; such distributions are known as *univariate probability distributions*. If we are concerned with the joint behavior of a number of random variables, say n variables, $X_1, X_2, \ldots, X_n$, then their *joint probability distribution* (also known as the *multivariate probability distribution*) is of interest. Either the *n-dimensional joint probability density function* or the *n-dimensional joint cumulative distribution function* is used. (If the n random variables are discrete, the *n-dimensional probability mass function* or the *n-dimensional cumulative distribution function* is used.) The n random variables may or may not be statistically independent.

We will limit our discussion to just two random variables ($n = 2$); the concepts discussed here can be extended to n variables.

Discrete random variables: The joint probability mass function of two discrete random variables X and Y, denoted by $p_{X,Y}(x, y)$, is defined as the probability that $X = x_i$ and $Y = y_j$ simultaneously, that is,

$$P_{X,Y}(x_i, y_j) = P\big[(X = x_i) \cap (Y = y_j)\big] \qquad (2.45)$$

The corresponding cumulative distribution function, denoted by $F_{X,Y}(x, y)$, is defined as

$$F_{X,Y}(x, y) = P\big[(X \le x) \cap (Y \le y)\big] \qquad (2.46)$$

The joint cumulative distribution function may also be computed from the joint probability mass function as follows.

$$F_{X,Y}(x, y) = \sum_i \sum_j p_{X,Y}(x_i, y_j) \quad \text{summed over all } x_i \le x \text{ and } y_j \le y$$

$$(2.47)$$

Continuous random variables: The joint probability density function of two continuous random variable X and Y is denoted by $f_{X,Y}(x, y)$. The probability $f_{X,Y}(x, y)\, dx\, dy$ is defined as the probability that X lies between x and $x + dx$ and Y lies between y and $y + dy$ simultaneously. The corresponding joint cumulative distribution function, denoted by

$F_{X,Y}(x, y)$, is defined as

$$F_{X,Y}(x, y) = P[(X \le x) \cap (Y \le y)] \qquad (2.48)$$

The joint probability density function and the joint cumulative distribution function are related to each other by the following two formulae:

$$f_{X,Y}(x, y) = \frac{\partial^2 F_{X,Y}(x, y)}{\partial x \, \partial y} \qquad (2.49)$$

$$F_{X,Y}(x, y) = \int_{-\infty}^{x} \int_{-\infty}^{y} f_{X,Y}(u, v) \, dv \, du \qquad (2.50)$$

where u and v are dummy variables within the integral.

2.3.5.2 Moments, covariance, and correlation. The joint probability density functions and the joint cumulative distribution functions may not be available in many practical problems; only the lower-order moments are known on the basis of limited amount of available data. These moments and related descriptors provide useful information about the random variables.

We shall discuss some lower-order moments of jointly distributed, continuous random variables by considering just two random variables. The discussion can be extended to n variables. Also, the following discussions may be adapted to discrete random variables by replacing integrations by summations.

Moments: The $(m + n)$th *moment* or the *moment of order* $(m + n)$ of two random variables X and Y is defined as

$$E[X^m Y^n] = \int_{-\infty}^{\infty} \int_{-\infty}^{\infty} x^m y^n f_{X,Y}(x, y) \, dx \, dy \qquad (2.51)$$

When $m = 1$ and $n = 0$, we get

$$E[X] = \int_{-\infty}^{\infty} \int_{-\infty}^{\infty} x \{ f_{X,Y}(x, y) \, dx \, dy \}$$

$$= \int_{-\infty}^{\infty} x \left\{ \int_{-\infty}^{\infty} f_{X,Y}(x, y) \, dy \right\} dx$$

$$= \int_{-\infty}^{\infty} x f_X(x) \, dx = \bar{x}$$

because $\int_{-\infty}^{\infty} f_{X,Y}(x, y) \, dy = f_X(x)$.

The preceding equation shows that the expected value of X is not influenced by Y. It can be shown that similar results are true for higher-order moments of X also. That is, $E[X^m]$ is not influenced by Y. Similarly, $E[Y^n]$ is not influenced by X.

The *central moments of order* $(m + n)$ are defined as follows:

$$E\left[(X - \bar{x})^m(Y - \bar{y})^n\right] = \int_{-\infty}^{\infty}\int_{-\infty}^{\infty}(x - \bar{x})^m(y - \bar{y})^n f_{X,Y}(x, y)\,dx\,dy$$

$$(2.52)$$

As the moments about the origin, it can be shown that the central moments of X are not influenced by Y and the central moments of Y are not influenced by X.

Covariance: When $m = 1$ and $n = 1$ in Equation (2.52), the resulting central moment is called the *covariance* of X and Y. It is denoted by Cov(X, Y) or $\sigma_{X,Y}$. It can be shown that

$$\text{Cov}[X, Y] = E[XY] - E[X]E[Y] \qquad (2.53)$$

Correlation coefficient: The *correlation coefficient* of X and Y is denoted by $\rho_{X,Y}$ and is defined as

$$\rho_{X,Y} = \frac{\text{Cov}[X, Y]}{\sigma_X \sigma_Y} \qquad (2.54)$$

It can be shown that $-1 \leq \rho_{X,Y} \leq 1$.

Two random variables X and Y are said to be *positively correlated* if their correlation coefficient is positive. Two random variables X and Y are said to be *negatively correlated* if their correlation coefficient is negative. Two random variables are said to be *uncorrelated* if the correlation coefficient is zero.

If two random variables are *statistically independent*, their correlation coefficient is zero. The reverse is not necessarily true; that is, if the correlation coefficient of two random variables is zero, it does not necessarily meand that the two variables are statistically independent.

Example 2.4

Consider the car-starting experiment discussed in Example 2.1. Let the discrete random variable X denote the number of attempts required to start car B, and let the discrete random variable Y denote the number of attempts required to start car C. Compute the covariance of X and Y.

Solution

Using Equation (2.45) and the data provided in Table 2.1, we get

$$p_{X,Y}(1,1) = \tfrac{8}{100} = 0.08$$

$$p_{X,Y}(1,2) = \tfrac{6}{100} = 0.06$$

$$p_{X,Y}(1,3) = \tfrac{5}{100} = 0.05$$

$$p_{X,Y}(1,4) = \tfrac{2}{100} = 0.02$$

$$p_{X,Y}(2,1) = \tfrac{8}{100} = 0.08$$

$$p_{X,Y}(2,2) = \tfrac{7}{100} = 0.07$$

$$p_{X,Y}(2,3) = \tfrac{6}{100} = 0.06$$

$$p_{X,Y}(2,4) = \tfrac{3}{100} = 0.03$$

$$p_{X,Y}(3,1) = \tfrac{9}{100} = 0.09$$

$$p_{X,Y}(3,2) = \tfrac{8}{100} = 0.08$$

$$p_{X,Y}(3,3) = \tfrac{6}{100} = 0.06$$

$$p_{X,Y}(3,4) = \tfrac{4}{100} = 0.04$$

$$p_{X,Y}(4,1) = \tfrac{10}{100} = 0.10$$

$$p_{X,Y}(4,2) = \tfrac{7}{100} = 0.07$$

$$p_{X,Y}(4,3) = \tfrac{7}{100} = 0.07$$

$$p_{X,Y}(4,4) = \tfrac{4}{100} = 0.04$$

Now we calculate $E[XY]$ using Equation (2.51). Because X and Y are discrete variables, the integration in that equation is replaced by summation:

$$E[XY] = \sum_i \sum_j x_i y_i p_{X,Y}(x_i, y_j)$$

where $x_i = y_i = i$, for $i = 1, 2, 3,$ and 4.

Substituting for x_i, y_j, and $p_{X,Y}(x_i, y_j)$ and carrying out the summation, we get

$$E[XY] = 5.68$$

We may also calculate $E[X]$ and $E[Y]$ using Equation (2.51).

$$E[X] = E[XY^0] = \sum_i \sum_j x_i y_j^0 p_{X,Y}(x_i, y_j)$$

$$= \sum_i \sum_j x_i p_{X,Y}(x_i, y_j) = 2.62$$

$E[Y]$ is also calculated similarly.

$$E[Y] = 2.15$$

The covariance is computed using Equation (2.53).

$$\text{Cov}[X,Y] = E[XY] - E[X]E[Y] = \{5.68 - (2.62 \times 2.15)\}$$
$$= 0.047$$

2.3.5.3 Statistically independent variables.   If the continuous random variables X and Y are statistically independent, then their joint probability density function is given by the product of the probability density functions of X and Y. That is,

$$f_{X,Y}(x, y) = f_X(x)f_Y(y) \tag{2.55}$$

Also,

$$F_{X,Y}(x, y) = F_X(x)F_Y(y) \tag{2.56}$$

The corresponding equations for statistically independent discrete variables X and Y are as follows.

$$p_{X,Y}(x_i, y_j) = p_X(x_i)p_Y(y_j) \tag{2.57}$$

$$F_{X,Y}(x, y) = F_X(x)F_Y(y) \tag{2.58}$$

In many practical problems, if (based on the physical phenomenon behind X and Y or for other reasons) it can be reasonably assumed that X and Y are statistically independent, then it is better to use Equations (2.55)–(2.58) rather than Equations (2.45)–(2.50). Use of Equations (2.55)–(2.58) will greatly simplify probability computations because the estimation of joint distributions $f_{X,Y}(x, y)$ or $p_{X,Y}(x_i, y_j)$ or $F_{X,Y}(x, y)$ require considerably more data than those required for the estimation of individual distributions $f_X(x)$ and $f_Y(y)$, $p_X(x_i)$ and $p_Y(y_j)$, or $F_X(x)$ and $F_Y(y)$.

2.3.6 Some Specific Probability Distributions

Over the years, statisticians have proposed a number of probability distributions for possible use in different situations. Discrete distributions are used for discrete random variables and continuous distributions are used for continuous random variables. This section discusses a few probability distributions used in reliability engineering.

2.3.6.1 Discrete distributions.   Two popular discrete distributions, namely, the binomial distribution and the Poisson distribution, are discussed here. Other discrete distributions, not discussed here, include the hypergeometric and the multinomial distributions.

Binomial distribution: Consider the following component-testing scenario. We test 1000 components and find that 948 of them successfully perform their intended functions and the remaining 52 components fail to perform their functions correctly. So the failure probability $q = 0.052$. Now, if we have a batch of n components (for example, 40 components), our best estimate is that qn components ($40 \times 0.052 = 2.08 \approx 2$ components) will fail. This is the best estimate but we are dealing with statistics and there is always the possibility that no component, 1 component, 2 components, 3 components, or even all 40 components may fail. The question is, "What is the probability that m components out of the n components will fail (where $0 \le m \le n$)?" This probability is given by the *binomial distribution*, which is defined by the following probability mass function[3]:

$$p_M(m) = \binom{n}{m} q^m (1 - q)^{n-m} \quad \text{for } m = 0, 1, 2, \ldots, n \quad (2.59)$$

where

$$\binom{n}{m} = \frac{n!}{m!(n - m)!}$$

and the random variable M represents the number of successes in n trials. M is a discrete random variable and m is the possible values it can take ($m = 0, 1, 2, \ldots, n$). If we know n and q, then the probability of m

---

[3]Although we discuss here component "success" and "failure," it could be any two mutually exclusive and collectively exhaustive outcomes of an experiment, for example, "head" or "tail" in flipping a coin.

successes is given by $p_M(m)$; n and q are referred to as the *parameters* of the binomial distribution. The binomial distribution is a *two-parameter distribution* because two parameters (n and q) are required to define it.

The corresponding cumulative distribution function is given by

$$F_M(m) = \sum_{j=0}^{m} \left[\binom{n}{j} q^j (1-q)^{n-j} \right] \quad \text{for } m = 0, 1, 2, \ldots, n \quad (2.60)$$

The expectation and variance of M can be derived by methods discussed in Section 2.3.4.5.

$$E[M] = nq \tag{2.61}$$

$$\text{Var}[M] = nq(1-q) \tag{2.62}$$

It is important to understand when binomial distributions can be used. The following assumptions are made in the derivation of the binomial distribution:

1. The probability q is the same in each trial of the experiment. For example, in the case of testing n components, we assume that each component has the same failure probability.
2. The trials are independent, that is, the outcome of one trial does not affect the outcome of another trial. In the example of testing n components, whether a component passes or fails in one test does not affect whether another component passes or fails in another test. Although each component has the same failure probability, the outcome of the test of one component does not affect the outcome of the test of another component.

Example 2.5

We are about to toss a coin four times. The outcome of each toss (trial) is either 'head' or 'tail,' and we know that the probability of 'head' in any toss is 0.5. Compute the probability mass function for the number of times 'head' will occur during the four tosses.

Solution

Let us use the following notation:

M = number of times 'head' will occur during the four tosses
m = possible values of M = 0, 1, 2, 3, and 4
n = number of tosses = 4
q = the probability of 'head' in any toss = 0.5

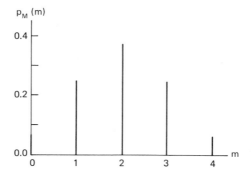

Figure 2.7. Probability mass function (Example 2.5).

Substitution of m, n, and q into Equation (2.59) yields

$$p_M(0) = 0.0625$$

$$p_M(1) = 0.25$$

$$p_M(2) = 0.375$$

$$p_M(3) = 0.25$$

$$p_M(4) = 0.0625$$

The preceding probability mass function is plotted in Figure 2.7.

Example 2.6

There are three similar components in a power boat. Based on past experience with such components, their failure probability during the life of the boat is estimated to be 0.1. Compute the probability mass function of the number of component failures during the life of the boat.

Solution

Let us use the following notation:

M = the number of component failures during the life of the boat
m = the possible values of M = 0, 1, 2, and 3
n = the number of components = 3
q = component failure probability = 0.1

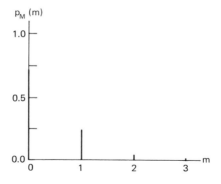

Figure 2.8. Probability mass function (Example 2.6).

Substituting m, n, and q into Equation (2.59), we get

$$p_M(0) = 0.729$$
$$p_M(1) = 0.243$$
$$p_M(2) = 0.027$$
$$p_M(3) = 0.001$$

The preceding probability mass function is plotted in Figure 2.8.

Poisson distribution: The Poisson distribution is used, instead of the binomial distribution, when the number of trials is very large ($n \to \infty$), the probability q is very small ($q \to 0$), and the product nq is finite. In addition, the two assumptions stated previously for binomial distributions should be satisfied.

The probability mass function of Poisson distributions is given by

$$p_M(m) = \frac{\mu^m \exp(-\mu)}{m!} \quad \text{for } m = 0, 1, 2, \ldots \qquad (2.63)$$

where $\mu = nq$.

The expectation and variance of M are given by

$$E[X] = \mu \qquad (2.64)$$
$$\mathrm{Var}[X] = \mu \qquad (2.65)$$

Because a Poisson distribution can be defined by a single parameter (μ), it is a *single-parameter distribution*.

If two random variables X and Y have Poisson distributions with parameters μ_1 and μ_2, respectively, the sum of the two variables ($Z = X + Y$) is also a random variable obeying a Poisson distribution with parameter $\mu = \mu_1 + \mu_2$.

The following recurrence relationship for the probability mass function of a Poisson distribution may be useful in some problems:

$$p_M(m + 1) = \frac{\mu}{(m + 1)} p_M(m) \tag{2.66}$$

Poisson process: The stochastic process $Y(t)$ is a *Poisson process* with parameter λ if $Y(t_1)$ has a Poisson distribution with parameter $\mu = \lambda t_1$. Here t is the independent deterministic variable (such as time or spatial coordinate), t_1 is any value of t, and $Y(t_1)$ is the value of $Y(t)$ at $t = t_1$. The parameter λ is called the *average rate* of the Poisson process.

Poisson processes are used to represent the number of occurrences of an event during a period of time, given that the average rate of occurrence is constant. The number of occurrences is a discrete random variable and is denoted by N. The probability mass function of N over a duration of time t is given by

$$p_N(n) = \frac{(\lambda t)^n \exp(-\lambda t)}{n!} \tag{2.67}$$

where $n = 0, 1, 2, \dots$. Equation (2.67) is obtained by simply replacing μ by λt in the probability mass function of a Poisson distribution [Equation (2.63)].

A physical process may be modelled as a Poisson process if the following conditions are satisfied:

1. The probability of occurrence of the event during a short interval from t to $t + c$ is approximately equal to $c\lambda$, for any value of t.
2. The probability of two occurrences of the event during that same short interval is of a similar order of magnitude than $c\lambda$.
3. The number of occurrences during any interval from t to $t + b$ is independent of the number of occurrences during any other interval from t' to $t' + b$, provided the two intervals are not overlapping. If t is time, then this condition may be stated as follows: Future occurrences of the event are independent of past and present occurrences of the event. The Poisson process is called a process without memory, or a *memoryless process*, because of this property.

Although many physical processes satisfy the preceding conditions only approximately, they are modelled as Poisson processes because of the mathematical ease of using such models.

Poisson processes are widely used to model the occurrence of component failures, earthquakes, hurricanes, floods, etc.

2.3.6.2 Continuous distributions.   Exponential distributions is widely used in reliability engineering and it is discussed here. A number of other distributions, such as the Gaussian distribution, lognormal distribution, Weibull distribution, extreme value distribution, etc., are also available. Interested readers may refer to books on probability and statistics, for example, Freund (1962).

Exponential distribution: Consider an event (for example, component failure or earthquake occurrence) that is modelled as a Poisson process in time. Let the average rate of occurrence be λ. Because it is a stochastic process, we do not know when the first occurrence of the event will happen. Let us represent the time of first occurrence by a continuous random variable T and the number of occurrences of the event between times 0 and t by a discrete random variable N, where $t \geq 0$. If time t is less than T, then the number of occurrences of the event between times 0 and t is zero. That is, $t < T$ means $N = 0$. So the probability that $t < T$ is equal to the probability that the number of occurrences during the interval from time 0 to time t is zero, that is,

$$P[t < T] = p_N(0) \qquad (2.68)$$

We also have,

$$P[t < T] = 1 - P[t \geq T] = 1 - P[T \leq t] \qquad (2.69)$$

Using Equations (2.28), (2.67), and (2.69), we have

$$F_T(t) = 1 - p_N(0) = 1 - \frac{(\lambda t)^0 \exp(-\lambda t)}{0!} = 1 - \exp(\lambda t) \quad (2.70)$$

So the cumulative distribution function of T is

$$F_T(t) = 1 - \exp(-\lambda t) \quad \text{for } t \geq 0 \qquad (2.71)$$

Using Equation (2.29), we get the probability density function for T as follows:

$$f_T(t) = \lambda \exp[-\lambda t] \quad \text{for } t \geq 0 \tag{2.72}$$

The preceding cumulative distribution function and probability density function are known as the *exponential distribution*. This is a single-parameter distribution because a single parameter (λ) fully defines it.

The exponential distribution provides the probability distribution for the *time of first occurrence* of an event, provided the event satisfies the three conditions enumerated for Poisson processes (see Section 2.3.6.1). The mean and standard deviation of the time of first occurrence are given by

$$E[T] = \frac{1}{\lambda} \tag{2.73}$$

$$\sigma_T = \frac{1}{\lambda} \tag{2.74}$$

EXERCISE PROBLEMS

2.1. An independent testing laboratory tested two brands of diesel motors (Brand X and Brand Y). Fifty motors of each brand were tested at different environmental conditions (humidity, temperature, etc.). Tests were always conducted in pairs: One Brand X motor and one Brand Y motor were tested at the same time under the same environmental conditions. Each such test is called a trial and there were fifty such trials. The number of attempts required to start each motor was recorded and the summary is provided in Table 2.2. The entry "8" in the row starting with "$X = 2$" under the column headed by "$Y = 1$" indicates that the Brand X motor started at the second attempt and Brand Y motor started at the first attempt in 8 of the 50 trials.

Compute the following probabilities: (i) The probability that the Brand X motor starts at the first attempt. (ii) The probability that the Brand Y motor starts at the first attempt. (iii) The probability that the Brand X motor starts at the second attempt. (iv) The probability that the Brand Y motor starts at the second attempt. (v) The conditional probability that the Brand X motor starts at the first attempt given that the Brand Y motor has started at the

Table 2.2.   Data for Problem 2.1

	Y = 1	Y = 2	Y = 3
X = 1	7	5	5
X = 2	8	7	5
X = 3	5	5	3

first attempt. (vi) The conditional probability that the Brand Y motor starts at the first attempt given that the Brand X motor has started at the first attempt. (vii) The conditional probability that the Brand X motor starts at the first attempt given that the Brand Y motor has started at the first or second attempt. (viii) The conditional probability that the Brand Y motor starts at the first attempt given that the Brand X motor has started at the first or second attempt.

2.2. There are four components in a piping system. The four components, namely, the pipe, the pump, valve 1, and valve 2, are denoted by $A1$, $A2$, $A3$, and $A4$, respectively, and their failure probabilities are denoted by $P[A1]$, $P[A2]$, $P[A3]$, and $P[A4]$, respectively. These probabilities are as follows: $P[A1] = 10^{-8}$, $P[A2] = 3 \times 10^{-3}$, $P[A3] = 8 \times 10^{-3}$, and $P[A4] = 8 \times 10^{-3}$. Assuming statistical independence between the four failures, compute the following probabilities: (i) The probability of failure of either $A1$, $A2$, $A3$, or $A4$. (ii) The probability of failure of either $A1$, $A2$, or $A3$. (iii) The probability of failure of either $A2$, $A3$, or $A4$. (iv) The probability of failure of either $A3$ or $A4$. [Use Equations (2.11) and (2.21).]

2.3. Redo Problem 2.2, without the assumption of statistical independence between component failures. The following probabilities are available: $P[A2 \cap A3] = P[A2 \cap A4] = 10^{-6}$, $P[A3 \cap A4] = 2 \times 10^{-4}$, $P[A2 \cap A3 \cap A4] = 5 \times 10^{-7}$. All other probabilities of failure combinations required in Equation (2.11) are much smaller; so they may be set to zero.

2.4. There are three identical valves in a petrochemical plant. The failure probabilities of these valves during a magnitude-5 earthquake are denoted by $P[V1]$, $P[V2]$, and $P[V3]$; $P[V1] = P[V2] = P[V3] = 10^{-3}$. We also have the following conditional probabilities: $P[V1|V2] = P[V2|V3] = P[V3|V1] = 0.87$; $P[V1|V2 \cap V3] = P[V2|V3 \cap V1] = P[V3|V1 \cap V2] = 0.95$. Compute the probability of all three valves failing during a magnitude-5 earthquake. [Use Equation (2.16).]

2.5. Consider the diesel motor tests described in Problem 2.1. Determine and plot the probability mass function (PMF) and the cumulative distribution function (CDF) of the number of attempts required to start Brand X motors and the PMF and CDF of the number of attempts required to start Brand Y motors. Also compute the following probabilities: (i) The probability that the number of attempts required to start Brand X motors is less than or equal to 2. (ii) The probability that the number of attempts required to start Brand Y motors is less than or equal to 2. [Use both Equations (2.26) and (2.27).]

2.6. Consider the diesel motor tests described in Problem 2.1. Compute the mean value of the number of attempts required to start Brand X motors and the mean value of the number of attempts required to start Brand Y motors. [Use Equation (2.34) with the integration replaced by the appropriate summation.]

2.7. The probability mass function of a random variable B is given by $p_B(0) = 0.1$, $p_B(1) = 0.9$, and $p_B(i) = 0.0$, where i takes all integer values except 0 and 1. Determine the cumulative distribution function, the mean, and the standard deviation of B.

2.8. The probability density function of a random variable Y is given by

$$f_Y(y) = \begin{cases} 0.0 & \text{for } y < 0 \\ 0.5 & \text{for } 0 \leq y < 2 \\ 0.0 & \text{for } 2 \leq y \end{cases}$$

(i) Find the cumulative distribution function of Y. (ii) What is the probability that Y will take values between 0.2 and 0.8? [Use both Equations (2.32) and (2.33) and compare the results.] (iii) Show that both the probability density function and the cumulative distribution function satisfy the basic properties of probability distributions discussed in Section 2.3.4.3.

2.9. Consider the random variable Y described in Problem 2.8. Calculate the mean, mean square, variance, standard deviation, coefficient of variation, and coefficient of skewness.

2.10. The cumulative distribution function of a random variable C is

$$F_C(c) = \begin{cases} 0 & \text{for } c < 0 \\ \frac{1}{81}c^4 & \text{for } 0 \leq c \leq 4 \\ 0 & \text{for } c > 4 \end{cases}$$

(i) Determine the probability density function. (ii) Compute the mean and median. (iii) What is the probability that C is less than or equal to 2?

2.11. Consider the diesel motor tests described in Problem 2.1. Let the discrete random variable A denote the number of attempts required to start Brand X motors and let the discrete random variable B denote the number of attempts required to start Brand Y motors. (i) Determine the two-dimensional joint probability mass function and the two-dimensional joint cumulative distribution function of A and B. (ii) Calculate the expectation of A [using Equation (2.51) with the integrals replaced by the appropriate summations]. (iii) Calculate the expectation of B. (iv) Calculate the moment of order $(1 + 1)$ for A and B [using Equation (2.51) with the integrals replaced by the appropriate summations]. (v) Calculate the covariance of A and B.

2.12. Consider the random variables A and B described in Problem 2.11. Calculate the correlation coefficient of A and B. Are A and B positively correlated, negatively correlated, or uncorrelated?

2.13. There are four similar components in a missile. Let us call these components Component Z. The missile will fail to function as intended if three or more of these components fail. The failure probability of each component is

0.05. Using the binomial distribution [Equation (2.59)], compute the probability of the missile failing to function as intended because of three or more "Component Z" failures.

2.14. The average rate of occurrence of magnitude-4 or higher earthquakes at a proposed pharmaceutical plant site is 10^{-2} per year. The expected life of the plant is 40 years. Compute the probability that 0, 1, or 2 earthquakes of magnitude 4 or higher will occur during the life of the plant. Assume that earthquake occurrences may be modelled by a Poisson process [Equation (2.67)].

2.15. The average rate of failure of a piece of heavy construction equipment is 0.5 per year. What is the probability that the equipment will fail during the first eight months of operation? Assume that the time to failure can be modelled by an exponential distribution.

REFERENCE

Freund, J. E. (1962). *Mathematical Statistics*. Prentice-Hall, Inc., Englewood Cliffs, NJ.

Chapter 3
Fundamentals of Reliability Engineering

3.1 INTRODUCTION

Components are the building blocks of systems, so an understanding of component reliability is essential before discussing system reliability. This chapter discusses the basic concepts relating to component reliability: Components are classified as repairable and nonrepairable components. Component reliability parameters (such as reliability, unreliability, availability, unavailability, cumulative failure probability, failure probability density, failure rate, repair rate, expected life, and expected number of failures) are defined. Computation of these parameters from historical failure–repair data is discussed.

3.2 GENERAL TERMINOLOGY

In order for the reliability theory to be understood and used in a consistent and precise manner, a number of terms used in this book are defined here.

In the context of reliability analysis, a *component* is the basic unit of a system. What is considered as a "basic unit" depends on the level of the system reliability analysis. What is a component in one reliability analysis could be treated as a system in another reliability analysis. For example, a pump may be treated as a basic unit (component) in the reliability analysis of the feedwater system of an industrial facility; the same pump may be treated as a system in a detailed reliability analysis of the pump, in which the various parts of the pump are treated as the basic units (components).

A *mission* is the objective, task, or purpose of a system or component. The mission of an aircraft (system) is to fly from one airport to another; in this case, the mission includes start-up, takeoff, flight, landing, and shutoff. The mission of a pressure gauge (component) is to measure pressure.

A *fault* is a noncompliance with specifications. For example, if the size of a bolt connecting a turbine casing to its base is less than or greater than the specified size (allowing for specified tolerances), this is a fault. If the yield strength of the material used in the fabrication of a pipe is outside the specified values, this is a fault.

Failure is the inability of a component to perform its intended function as specified, for example, a pump failing to pump water when required, a valve failing to close when required, or a valve failing to open when required. However, a failure need not necessarily be a total functional failure. A component may function but, if it does not function as specified, it may still be considered a failure. A pressure gauge may be measuring the pressure, but if the measurement is not within the specified accuracy, then it is still considered a failure. (*Note:* Sometimes the term "fault" is used to refer to some types of failures, and the term "failure" is used to refer to faults. The meaning of the terms fault and failure should be interpreted in the proper context.)

A component may be required to perform more than one function. It may fail to perform one function as specified but continue to perform the other functions correctly. So failure is related to specific functions. A valve has two functions: It should shut off under certain conditions and open under some other conditions. A valve may fail to open when required but may still be operating normally with respect to the shutoff function.

The term *failure mode* is used to refer to the possible ways in which a component can fail. A component may have one or more failure modes. Consider the possible failures of a piping system. The possible ways through which the piping system could fail (the failure modes) include pipe rupture, pipe clogging, and pipe leakage. One may delve further into each of these failure modes and enumerate more-detailed failure modes. For example, pipe leakage may be due to (i) cracks in pipes or (ii) loose coupling of pipes. Each of these two means of leakage may be treated as a separate failure mode. Whether an analyst lumps together these two failure modes into a single failure mode (pipe leakage) or consider them separately (as pipe leakage due to cracks in pipes and as pipe leakage due to loose coupling of pipes) depends on the level (detail) of the reliability analysis.

A failure may not be detected at the instant it occurs. For example, a thermometer might fail and provide incorrect temperature readings without anyone detecting the failure. It may be noticed only at the time of periodic testing and maintenance. A hydraulic mechanism activating a valve may fail but remain undetected until it is required to close or open a valve or until it is checked during the periodic testing and maintenance.

There are failures that are detected immediately. For example, the failure of an electric pump, as it is operating, may be detected immediately. In some critical components, fault detection alarms are provided so that any failure will be detected immediately and come to the attention of a human operator.

A component is said to be in a *normal state* if it is not in a failed state.

The term *basic failure* is used in this book to refer to failures that are not further broken down to contributory failures. It is the lowest level of failure considered in an analysis, and what is considered as a basic failure depends on the particular analysis. For example, in a system-level analysis of a feedwater system, the event 'pump fails to start' may be treated as the basic failure and the details of why the pump failed to start and what parts of the pump contributed to the failure may not be considered in the system reliability analysis. On the other hand, in a detailed reliability analysis of the pump, the level of analysis will be more detailed. The underlying failures of individual parts of the pump contributing to the pump's failure may be considered. Thus, 'pump fails to start' is the basic failure in the system-level reliability analysis whereas the failures of individual parts are the basic failures in the pump-level reliability analysis. So, what is considered as a basic failure depends on the specific scope and the extent of the reliability analysis.

The term *time* is used in a broad sense—that is, "time" may be stated in terms of temporal units (hours, days, years, etc.) or in terms of the number of missions, number of cycles of operations, number of demands, etc. For example, the life of a heat exchanger may be 20 operating years. The life of a rotor shift may be 1,000,000 cycles. The life of an ignition switch may be 10,000 demands. The life of a shuttle craft may be 40 missions.

The *time interval* between $t = t_1$ and $t = t_2$ is represented as follows: by (t_1, t_2) if it does not include the time instants t_1 and t_2; by $[t_1, t_2]$ if it includes the time instants t_1 and t_2; by $[t_1, t_2)$ if it includes the time instant t_1 but does not include the time instant t_2; by $(t_1, t_2]$ if it does not include the time instant t_1 but includes the time instant t_2.

3.3 NONREPAIRABLE COMPONENTS

3.3.1 Definitions of Repairable and Nonrepairable Components

A component is said to be a *repairable component* if it is repaired upon detection of its failure. (Replacement is equivalent to repair in the context of reliability analysis. Also, a repaired component is considered *as good as new*.)

If it is not possible to repair the component even after its failure is detected, such a component is called a *nonrepairable component*. For example, if an inaccessible component in an airplane fails during flight, it would not be possible to repair it during flight. The component can, of course, be repaired after the end of the flight, but this is irrelevant from the point of view of airplane operation during that flight. Even if it is possible to repair a component upon detection of its failure but operations–maintenance policies require that it not be repaired until the next periodic maintenance, such a component is still considered a nonrepairable component.

3.3.2 Reliability and Availability

Component reliability at time t is the probability that the component is in its normal state from time 0 to time t, given that the component was new or as good as new at time zero. A component may have more than one function, and different reliabilities are associated with the different functions.

Unreliability is the complement of reliability. If the reliability at time t is $r(t)$, then the unreliability at time t, denoted by $u(t)$, is given by

$$u(t) = 1 - r(t) \qquad (3.1)$$

Availability at time t is the probability that the component is in its normal state at time t, given that it was new or as good as new at time zero.

Unavailability is the component of availability. If the availability at time t is $a(t)$, then the unavailability at time t, denoted by $q(t)$, is given by

$$q(t) = 1 - a(t) \qquad (3.2)$$

If a nonrepairable component fails anytime between times 0 and t, it remains in the failed state at time t also, because it is not repaired. So reliability at time t is *identical* to the availability at time t for nonrepairable components.

Consider N supposedly identical components. All the N components are new or as good as new at time zero. Let $N - n$ components fail anytime between times 0 and t. Reliability of the component at time t is given by

$$r(t) = \frac{n}{N} \qquad (3.3)$$

The *cumulative failure probability* at time t (usually referred to as the *failure probability* at time t, in this context) is equal to the unreliability at time t. The failure probability, denoted by $F(t)$, is given by

$$F(t) = u(t) = 1 - r(t) \qquad (3.4)$$

$$F(t) = u(t) = \frac{N - n}{N} \qquad (3.5)$$

Note that n, r, u, and F are functions of time, where the term "time" is used in a more general sense here: It could be in temporal units (years, days, hours, etc.) or in terms of the number of missions, number of cycles, number of demands, etc. (see Section 3.2).

Reliability may also be computed from a different perspective. Consider a component that is new or as good as new at time zero. Let the component fail at time t'. Two supposedly similar components may not fail at the same time because of the inherent random variations in material properties, manufacturing processes, assembly, etc. The time at which a component fails is thus a random variable. The reliability at time t is given by

$$r(t) = P[t < t'] \tag{3.6}$$

That is, the reliability of a component at time t is equal to the probability that time t is less than the random variable t' at which the component fails.

Similarly, the failure probability or unreliability at time t is given by

$$F(t) = u(t) = P[t' \le t] \tag{3.7}$$

Reliability at time t represents the probability that a component is in its normal state from time 0 to time t, given that is was new or as good as new at time zero [Equation (3.6)]. It also represents the ratio

$$\frac{\left(\begin{array}{l} \text{number of components that are in their} \\ \text{normal state from time 0 to time } t \end{array} \right)}{\left(\begin{array}{l} \text{total number of components that} \\ \text{were new or as good as new at time zero} \end{array} \right)}$$

[Equation (3.3)]. Unreliability at time t is the probability that the component will fail between times 0 and t, given that it was new or as good as new at time zero [Equation (3.7)]. It also represents the ratio

$$\frac{\text{number of components that fail between times 0 and } t}{\left(\begin{array}{l} \text{the total number of components that} \\ \text{were new or as good as new at time 0} \end{array} \right)}$$

[Equation 3.5)].

3.3.3 Failure Probability Density Function

The *failure probability density function* $f(t)$ is the derivative of the cumulative failure probability distribution function $F(t)$ with respect to time t:

$$f(t) = \frac{dF(t)}{dt} = \frac{du(t)}{dt} \tag{3.8}$$

Differentiating both sides of Equation (3.4) and using Equation (3.8), we get

$$f(t) = -\frac{dr(t)}{dt} \tag{3.9}$$

The quantity $f(t)\,dt$ is equal to the probability that the component will fail during the time interval between t and $t + dt$, given that the component is new or as good as new at time zero.

3.3.4 Expected Life

The *expected life* of a component is the expected value of the time at which the component fails, given that it was new or as good as new at time zero (no repair is considered). This is also known as the *mean time to failure* (MTTF) and is given by

$$\text{MTTF} = \int_0^\infty r(t)\,dt \tag{3.10}$$

Alternatively, if we test a number of components to failure or observe the failure of a number of components operating in the field and determine the life (time to failure) of each component, the mean life (MTTF) is computed as the average of those values.

3.3.5 Expected Number of Failures

The *expected number of failures* (ENF) over the time interval between t_1 and t_2, given that the component was new or as good as new at time zero, is denoted by $w(t_1, t_2)$ or ENF (t_1, t_2).

The expected number of failures of a nonrepairable component between times 0 and t is equal to the component unreliability at time t, that is,

$$w(0, t) = \text{ENF}(0, t) = u(t) \tag{3.11}$$

Example 3.1

One hundred thousand components were put into service at time zero. The number of components in their normal state at $t = 10{,}000$, 20,000, 30,000, 40,000, and 50,000 cycles of operation are listed in Table 3.1. Compute the reliability and unreliability at $t = 0$, 10,000, 20,000, 30,000, 40,000, and 50,000 cycles of operation and the mean time to failure.

Table 3.1.   Data for Example 3.1

t (CYCLES)	$n(t)$
0	100,000
10,000	80,000
20,000	55,000
30,000	20,000
40,000	5,000
50,000	0

Solution

The following notations is used:

N = the number of components in normal state at time zero
$n(t)$ = the number of components in normal state at time t
$r(t)$ = reliability at time t
$F(t)$ = unreliability at time t

$N = 100,000$; Table 3.1 gives $n(t)$ (the data). We compute $r(t)$ and $u(t)$ using Equations (3.3) and (3.5); the computed values are tabulated in Table 3.2.

Computed reliabilities and unreliabilities are plotted in Figure 3.1. Reliabilities and unreliabilities are computed at only six time points in this example. If these quantities are computed at closer intervals, smoother curves for reliability and unreliability could have been obtained. If necessary, the unreliability values (cumulative failure probability values) computed at discrete time points can be fitted to some standard cumulative distribution function (exponential distribution, Gaussian distribution, etc.). Methods of fitting standard probability distribution functions to data points may be found in textbooks on probability and statistics, for example, Benjamin and Cornell (1970).

Table 3.2.   Computation of Component Reliability and Unreliability (Example 3.1)

t (CYCLES)	$n(t)$	$r(t)$	$u(t)$
0	100,000	1.00	0.00
10,000	80,000	0.80	0.20
20,000	55,000	0.55	0.45
30,000	20,000	0.20	0.80
40,000	5,000	0.05	0.95
50,000	0	0.00	1.00

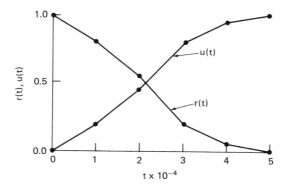

Figure 3.1. Reliability and unreliability plots (Example 3.1).

The mean time to failure (MTTF) is calculated using Equation (3.10). Because the reliability is zero for $t \geq 50{,}000$ cycles, the integration limits of 0 and ∞ may be replaced by the 0 and 50,000. The integration is carried out numerically through trapezoidal integration.

$$\mathrm{MTTF} = 10{,}000\left[\frac{1.0}{2} + (0.80 + 0.55 + 0.20 + 0.05) + \frac{0.0}{2}\right]$$

$$= 21{,}000 \text{ cycles}$$

Example 3.2

Consider a component with linearly decreasing reliability function. The reliability is 1 at time $t = 0$ and is 0 at $t = 10{,}000$ hours. Calculate its expected life.

Solution

The reliability function is 1 at time $t = 0$ and its linearly decreases to 0 at $t = 10{,}000$ hours. This can be expressed as follows:

$$r(t) = \begin{cases} (1.0 - 0.0001t) & \text{for } 0 \leq t \leq 10{,}000 \\ 0 & \text{elsewhere} \end{cases}$$

Using Equation (3.10),

$$\mathrm{MTTF} = \int_0^\infty (1.0 - 0.0001t)\, dt = 5000 \text{ hours}$$

Example 3.3

Consider a component with parabolically decreasing reliability function. The reliability is 1 a time $t = 0$ and is 0 at $t = 10,000$ hours. Calculate its expected life.

Solution

The reliability function is 1 at time $t = 0$, and it parabolically decreases to 0 at $t = 10,000$ hours. This can be explained as follows:

$$r(t) = \begin{cases} (1.0 - 0.00000001t^2) & \text{for } 0 \leq t \leq 10,000 \\ 0 & \text{elsewhere} \end{cases}$$

Using Equation (3.10),

$$\text{MTTF} = \int_0^\infty (1.0 - 0.00000001t^2) \, dt = 6666.67 \text{ hours}$$

Although the reliability values at $t = 0$ and $t = 10,000$ are the same as in Example 3.2, the expected life in this example differs from that in Example 3.2 because the functional form (time dependence) of the reliability function is different.

3.3.6 Failure Rate

The rate at which failure occurs during a specified interval of time is called the *failure rate during that interval*. The failure rate g during the time interval from t_1 to t_2 is given by

$$g(t_1, t_2) = \frac{r(t_1) - r(t_2)}{r(t_1)(t_2 - t_1)} \tag{3.12}$$

Constant hazard rate (defined in Section 3.3.10.1) is also referred to in the literature as the failure rate. Note that two different parameters are referred to by the same name. In the rest of this book, we will mostly use the term "failure rate" to denote a *constant hazard rate* (see Section 3.3.10.1 for more details).

Example 3.4

The reliability of a component after 10,000 cycles of operation is 0.80 and after 12,000 cycles of operation is 0.78. Calculate the failure rate during the interval between 10,000 and 12,000 cycles.

Solution

Using Equation (3.12),

$$g(12{,}000, 10{,}000) = \frac{0.80 - 0.78}{(12{,}000 - 10{,}000) \times 0.80}$$

$$= 0.125 \times 10^{-4} \text{ failures per cycle}$$

3.3.7 Hazard Rate and Hazard Function

The *hazard rate* at time t, denoted by $h(t)$, is the failure rate during the time interval from t to $t + \Delta t$, in the limit when Δt tends to zero. That is,

$$h(t) = \lim_{\Delta t \to 0} \left[\frac{r(t) - r(t + \Delta t)}{\Delta t \, r(t)} \right] = \frac{f(t)}{r(t)} \tag{3.13}$$

The hazard rate is also known as the *instantaneous failure rate* and as the *hazard function*.

The hazard rate of a component at time t is also defined as the number of failures per unit time at time t divided by the number of components in their normal state at time t. This definition may be expressed as

$$h(t) = \lim_{\Delta t \to 0} \left[\frac{n(t) - n(t + \Delta t)}{n(t) \, \Delta t} \right] \tag{3.14}$$

where $n(t)$ is the number of components in their normal state at time t.

Let N be the total number of components, all of which were new or as good as new at time zero (total population $= N$). Equation (3.14) reduces to Equation (3.13) if we divide both the numerator and the denominator by the total number of components N.

A third definition is also used by some analysts: The hazard rate at time t is the rate of change of "the conditional probability of failure at time t given that the component is in its normal state at time t", that is,

$$h(t) = \lim_{\Delta t \to 0} \left\{ \frac{P[(t < t' \le t + t)|t' > t]}{\Delta t} \right\} \tag{3.15}$$

3.3.8 Relationship between Hazard Function and Failure Probability Density Function

The mathematical relationship between hazard function and failure probability density function is given by Equation (3.13). What does it mean?

Let N be the total number of components (total population). All the N components were new or as good as new at time zero. Let the number of components in normal state all the time between times 0 and t be $n(t)$; the number of components that fail between times 0 and t is $N - n(t)$. Let the number of failures per unit time at time t be $\eta(t)$. The failure possibility density function at time t, denoted by $f(t)$, is the number of failures per unit time at time t divided by the total number of components, that is,

$$f(t) = \frac{\eta(t)}{N} \tag{3.16}$$

The hazard function at time t is the number of failures per unit time at time t divided by the number of components in normal state all the time between times 0 and t, that is,

$$h(t) = \frac{\eta(t)}{n(t)} \tag{3.17}$$

The difference between the failure probability density function and the hazard function lies in the denominator (which may be considered the normalizing factor). The failure probability density function uses the total number of components as the normalizing factor whereas the hazard function uses the number of components in normal state all the time between times 0 and t as the normalizing factor. Because the total number of components can never be less than the number of components in normal state all the time between times 0 and t, the hazard function can never be less than the failure probability density function.

Example 3.5

Calculate the failure probability density function and the hazard function for the failure data given in Table 3.1.

Solution

The number of failures at $t = 0$, 10,000, 20,000, 30,000, 40,000, and 50,000 cycles are given as data in Table 3.1. The reliability and unreliability (failure probability) at these time points have already been calculated and are listed in Table 3.2.

Failure probability density is given by the time derivative of the failure probability distribution [Equation (3.8)]. Because, in this example, the failure probability distribution in each time interval is a linear function (see Figure 3.1) and the failure probability density function is equal to the

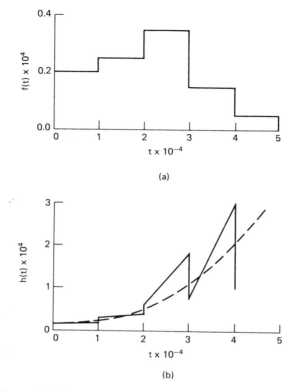

Figure 3.2. Failure probability density function and hazard function (Example 3.5): (a) failure probability density function; (b) hazard function.

slope of the failure probability distribution in each time interval, the failure probability density function is a constant in each time interval: these failure probability density function values are computed and tabulated in Table 3.3; they are plotted in Figure 3.2. Table 3.3, two values each are indicated at 10,000, 20,000, 30,000, and 40,000 cycles; as seen in Figure 3.2, the probability density value jumps from one value to the other because the failure probability density function is a step function in this example. The hazard function is calculated using Equation (3.13). Because the reliability function is zero at t = 50,000 cycles, Equation (3.13) will provide a value of infinity at this time point. A finite value for the hazard function can be obtained only up to a time point for which a nonzero reliability is available. If we had a time point closer to 50,000 cycles (say 49,000 cycles) for which the reliability is not zero, we could have computed a finite value for the hazard function for that point.

Table 3.3.   Failure Probability Density Function
and Hazard Function (Example 3.5)

t (CYCLES)	$f(t)$	$h(t)$
0	0.20×10^{-4}	0.20×10^{-4}
10,000	0.20×10^{-4}	0.25×10^{-4}
	0.25×10^{-4}	0.3125×10^{-4}
20,000	0.25×10^{-4}	0.4545×10^{-4}
	0.35×10^{-4}	0.6364×10^{-4}
30,000	0.35×10^{-4}	1.7500×10^{-4}
	0.15×10^{-4}	0.7500×10^{-4}
40,000	0.15×10^{-4}	3.0000×10^{-4}
	0.05×10^{-4}	1.0000×10^{-4}
50,000	0.05×10^{-4}	

Note: Both $f(t)$ and $h(t)$ are in failures per cycle.

The failure density function and the hazard function would have been smooth if the time points were more closely spaced. If necessary, one could fit a standard probability density function (exponential distribution, Gaussian distribution, etc.) for the failure probability density function. Methods of fitting probability density functions for data points may be found in textbooks on probability and statistics, for example, Benjamin and Cornell (1970). Reliability function may be calculated using Equation (3.3), and the hazard function may then be obtained through Equation (3.13). Alternatively, the hazard function, calculated as in the present example, but at closer time intervals, may be fitted to a standard hazard function (see Sections 3.3.12–3.3.14) and the corresponding reliability functions may then be obtained (see Sections 3.3.9 and 3.3.12–3.3.14).

We have obtained a rather odd shape for the hazard function (Figure 3.2) because we have plotted it on the basis of only a few data points. A smoother curve may be obtained if more data points at closer intervals are used. [A smooth curve is shown in Figure 3.2 (broken lines) to illustrate how a smooth hazard function for this example might look.]

3.3.9 Relationship between Reliability Function, Cumulative Failure Probability Distribution Function, Failure Probability Density Function, and Hazard Function

Substituting Equation (3.9) into Equation (3.13), we get

$$h(t) = -\frac{dr(t)}{dt}\frac{1}{r(t)}$$

This can be written as

$$h(t) = -\frac{d}{dt}(\ln r(t))$$

Multiplying both sides by dt and integrating, we get

$$r(t) = \exp\left[-\int_0^t h(t)\,dt\right] \tag{3.18}$$

Using Equation (3.4), we get

$$F(t) = u(t) = 1 - \exp\left[-\int_0^t h(t)\,dt\right] \tag{3.19}$$

Using Equation (3.9), we get

$$f(t) = h(t)\exp\left[-\int_0^t h(t)\,dt\right] \tag{3.20}$$

Equations (3.18)–(3.20) are important and useful fundamental relationships in reliability engineering.

3.3.10 Constant Hazard Rate and the Exponential Law

3.3.10.1 General formulation and validity. Let the hazard function $h(t)$ be a constant with respect to time. Let the constant hazard rate be equal to λ,

$$h(t) = \lambda$$

In such a case, Equations (3.18)–(3.20) can be simplified to

$$r(t) = \exp(-\lambda t) \tag{3.21}$$
$$F(t) = u(t) = 1 - \exp(-\lambda t) \tag{3.22}$$
$$f(t) = \lambda \exp(-\lambda t) \tag{3.23}$$

Substituting Equation (3.21) into Equation (3.10), we get

$$\text{MTTF} = \frac{1}{\lambda} \tag{3.24}$$

That is, the mean time to failure is equal to the inverse of the hazard rates, provided the hazard rate is a constant.

The constant hazard rate is often referred to as the *constant failure rate* or simply the *failure rate*. (See Section 3.3.6, where a different quantity is referred to as the failure rate; that quantity is defined over a time interval, whereas the quantity referred to here is defined at an instant of time. One should distinguish between the two in order to avoid confusion. In the rest of the book, we use the term "failure rate" to refer to the constant failure rate defined here.)

Referring to Section 2.3.6.2, it is evident that the failure distribution function and the failure density function given by Equations (3.22) and (3.23) are exponential distributions. So the constant hazard rate cases are said to obey the *exponential failure law*. Exponential failure distributions and constant hazard rates are widely used in reliability calculations because system reliability analysis using constant hazard rates is much easier than system reliability analysis using other hazard functions. Moreover, a constant hazard rate assumption is reasonably valid in a broad range of situations.

A constant hazard rate is usually assumed when the component is in its *prime of life* (see Section 3.3.11 for the definition of "prime of life"). Solid-state electronic components are said to have approximately constant hazard rates. Electrical components have fairly constant hazard rates in comparison to most mechanical components. Even mechanical components have approximately constant hazard rates if they are made up of a number of subcomponents (parts) with different failure rates. Herd (1957) and Drenick (1960) have shown that if a component contains many subcomponents, each with a different hazard rate, the overall hazard rate of the component is approximately a constant. Davis (1952) also discusses the applicability to constant hazard rates. Constant hazard rate may also be used if available failure data are so sparse as to allow the determination of more exact hazard functions; however, this assumption should not be used indiscriminately.

It is important to remember that a constant failure rate (constant hazard rate) implies that previous use of the component does not affect the failure rate (failure rate does not change with the age of the component); in other words, it is assumed that there is no damage or degradation accumulation.

Many types of components do deteriorate with time, and this is in direct contradiction with the assumption of constant failure rate. The argument made by proponents of constant failure rate is that when a component is made up of a number of subcomponents with different failure rates, the overall effect is to have a constant failure rate for the component. It has been shown mathematically that component failure rate tends to a constant value as the number of subcomponents increases. It has also been shown mathematically that, for a fixed number of subcomponents, the component failure rate tends to a constant value as the operating time increases, provided the failures are repaired and the component continues to operate.

If a component does not fall under one of the preceding categories, or if there is doubt as to the applicability of an exponential distribution, a check of whether the available failure data fit an exponential distribution may be made. Methods of checking whether a particular probability distribution (here, an exponential distribution) firs data points within specified levels of accuracy are available in many textbooks on probability and statistics, for example, Benjamin and Cornnell (1970). It should be remembered that data points will seldom fit the exponential distribution (or any other standard probability distribution) exactly; the fit will only be approximate.

In spite of the assumptions and approximations involved in a constant-failure-rate model, it is widely used in reliability engineering because it seems to provide *reasonably good results in most applications*. Unless otherwise stated, we will be using constant failure rates in this book.

Computation of the constant failure rate from failure data (operating history) is discussed in Section 5.2.4 (see Examples 5.1 and 5.2).

3.3.10.2 High-reliability cases. The relationships given by Equations (3.21)–(3.23) may be simplified considerably if the reliability of the component at time t is high, say, $r(t) \geq 0.95$. Consider the following Taylor series expansion:

$$\exp(-\lambda t) = 1 - \lambda t + \frac{(\lambda t)^2}{2!} - \frac{(\lambda t)^3}{3!} + \cdots$$

When $\lambda t \leq 0.05$, the second- and higher-order terms may be dropped without losing much accuracy. So,

$$\exp(-\lambda t) = 1 - \lambda t$$

Substituting this into Equations (3.21)–(3.23), we get the following approximate relationships that may be used when $r(t) \geq 0.95$ or $F(t) \leq 0.05$:

$$r(t) = 1 - \lambda t \qquad (3.25)$$

$$F(t) = u(t) = \lambda t \qquad (3.26)$$

$$f(t) = \lambda(1 - \lambda t) \qquad (3.27)$$

Equation (3.26) shows that unreliability increases almost linearly with time until it reaches a value of about 0.05. Similarly, the reliability function decreases almost linearly with time until it reaches a value of about 0.95. Comparing Equations (3.21) and (3.22) with Equations (3.25) and (3.26), component reliability computed using Equation (3.25) is slightly less than that computed using Equation (3.21), and the unreliability computed using Equation (3.26) is slightly higher that that computed using Equation (3.22).

   Note that the mean time to failure (MTTF) should be calculated using Equation (3.24); one should not calculate the expected life by substituting Equation (3.25) into Equation (3.10), because the integration in Equation (3.10) is from $t = 0$ to $t = \infty$ and, invariably, the $1 - \lambda t$ term (reliability) will become less than 0.95 for some values of t in that interval, thus invalidating Equation (3.25).

Example 3.6

The constant hazard rate (failure rate) of a component is 10^{-8} failures per cycle. Compute (i) the reliability, unreliability, and failure density at 10^6 cycles and (ii) the mean time to failure.

Solution

We shall first use Equations (3.21)–(3.24):

$$h(t) = \lambda = 10^{-8} \text{ failures per cycle}$$

$$r(t) = \exp(-10^{-2}) = 0.99004983$$

$$F(t) = u(t) = 1 - \exp(-10^{-2}) = 0.00995017$$

$$f(t) = 10^{-8} \times \exp(-10^{-2}) = 0.99004983 \times 10^{-8} \text{ failures per cycle}$$

$$\text{MTTF} = \frac{1}{10^{-8}} = 10^8 \text{ cycles}$$

We may also use Equations (3.24)–(3.27) because $\lambda t = 0.01 < 0.05$:

$$r(t) = 1 - (10^{-8} \times 10^6) = 0.99$$

$$F(t) = u(t) = 10^{-8} \times 10^6 = 0.01$$

$$f(t) = 10^{-8} \times (1 - 10^{-8} \times 10^6) = 0.99 \times 10^{-8} \text{ failures per cycle}$$

$$\text{MTTF} = \frac{1}{10^{-8}} = 10^8 \text{ cycles}$$

   As can be seen, the differences between the two sets of results are insignificant.

Example 3.7

For Example 3.6, compute the reliability, unreliability, and failure density at $t = 5 \times 10^7$ cycles.

Solution

We shall fist use Equation (3.21)–(3.23):

$$r(t) = \exp(-0.5) = 0.6065$$

$$F(t) = u(t) = 1 - \exp(-0.5) = 0.3935$$

$$f(t) = 10^{-8} \times \exp(-0.5)$$

$$= 0.6065 \times 10^{-8} \text{ failures per cycle}$$

Here, $\lambda t = 0.5 > 0.05$. So Equations (3.25)–(3.27) are not applicable. However, we shall use those equations to examine the differences in results.

$$r(t) = 1 - (10^{-8} \times 5 \times 10^7) = 0.5$$

$$F(t) = u(t) = 10^{-8} \times 5 \times 10^7 = 0.5$$

$$f(t) = 10^{-8} \times \left[1 - (10^{-8} \times 5 \times 10^7)\right]$$

$$= 0.5 \times 10^{-8} \text{ failures per cycle}$$

The differences are significant, and the results obtained using Equations (3.25)–(3.27) shall not be used.

3.3.10.3 Effect of operating and environmental conditions.
The failure rate is a function of operating conditions. For example, the failure rate of a pressure vessel operating at pressure 120 psi and temperature 100°F may be different from the failure rate of an identical pressure vessel operating at 200 psi and 150°F. Similarly, the failure rate of a fan belt may depend on the operating speed of the fan. Environmental conditions such as humidity, corrosive chemicals, dust, etc., may also affect the failure rates of components sensitive to such conditions. So, when failure rates are used in reliability computations, care should be taken that the rate is based on data obtained under similar, if not identical, operating and environmental conditions.

Consider a component operating for t_1 hours under conditions with a corresponding failure rate of λ_1, then t_2 hours under conditions with a corresponding failure rate of λ_2, etc. The reliability of the component at time $t = t_1 + t_2 + \cdots + t_n$ is

$$r(t_1 + t_2 + \cdots + t_n) = \exp(-\lambda_1 t_1)\exp(-\lambda_2 t_2) \cdots \exp(-\lambda_n t_n)$$
$$= \exp[-(\lambda_1 t_1 + \lambda_2 t_2 + \cdots + \lambda_n t_n)] \qquad (3.28)$$

3.3.10.4 Standby and operating modes. Some components are in a standby mode (idle mode) most of the time and are required to operate when a need for their operation arises. For example, a standby redundant pump, which is in parallel to an operating pump, will be required to operate only when the operating pump fails or is undergoing maintenance. Another example is an electric generator that needs to operate only when external electric power fails; the generator stands idle at other times.

Failure rate during standby mode could be different from that during operating mode. If the component is in standby mode for a duration of t_s hours and in operating mode for a duration of t_o hours, the reliability at time $(t_s + t_o)$ is given by [using Equation (3.28)]

$$r(t_s + t_o) = \exp(-\lambda_s t_s)\exp(-\lambda_o t_o) \qquad (3.29)$$

where λ_s and λ_o are the failure rates during standby and operating modes, respectively.

If the standby failure rate is not available, the operating failure rate may be used in its place as an approximation. Some failure data sources do not identify whether the failure rates provided are for standby or operating mode; some data sources may provide failure rates that are a composite of both standby- and operating-mode failures. In such situations, the failure rate provided by the data source may be used for both standby and operating modes as an approximation.

3.3.10.5 Multiple failure modes. As discussed in Section 3.2, a component may have more than one failure mode. Each failure mode has a failure rate associated with it. If the failure modes are statistically independent (that is, the occurrence of one failure mode does not affect the occurrence of other failure modes), the component reliability is given by

$$r(t) = \exp[-(\lambda_1 + \lambda_2 + \cdots + \lambda_n)t] \qquad (3.30)$$

where λ_i is the constant failure rate of the ith failure mode and n is the number of failure modes. Based on Equation (3.30), we conclude that the total failure rate of the component is the sum of the failure rates of the individual failure modes, provided the failure modes are statistically independent:

$$\lambda = \lambda_1 + \lambda_2 + \cdots + \lambda_n \qquad (3.31)$$

3.3.11 Break-In and Wear-Out Phases

A component will have a relatively high hazard rate at the time it is put into service. The hazard rate will start decreasing for a time immediately after it goes into service (see Figure 3.3). The higher hazard rate during this initial period (called the *break-in period* or the *wear-in period*) is primarily due to undetected defects in design, material, manufacture, and installation. The higher hazard during the initial period is sometimes called the *infant mortality*. Once the defective components are weeded out during the break-in period, the surviving components may have a fairly constant hazard rate during a subsequent period (see Figure 3.3); this period is called the *normal useful life* or the *prime of life* of the component. Of course, there are situations where the hazard rate is not constant even during this period. As the component ages, deterioration of its various parts takes place and the hazard rate starts increasing. This is the *wear-out period* (see Figure 3.3). The hazard function shown in Figure 3.3 is referred to as the *bathtub curve* because its shape resembles a bathtub. (Note that the duration of the prime-of-life period is much longer than the durations of the break-in and wear-out periods.) If our interest is restricted to the prime-of-life period and if the hazard rate is approximately constant during this period, the constant hazard rate (failure rate) analysis procedures discussed in Section 3.3.10 may be used.

Quality control testing could eliminate most break-in failures during component use, thus significantly reducing or eliminating the higher hazard rates during the early periods of component use. Proper maintenance could reduce significantly or eliminate the wear-out phase. Thus, with

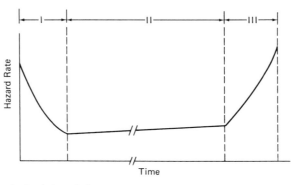

I : Break-in period
II : Prime of life
III : Wear-out period

Figure 3.3. Hazard rate as a function of time (hazard function).

well-executed quality control and maintenance, both the break-in and wear-out phases can be reduced significantly or eliminated. So the component will operate mostly in the prime-of-life phase, which has an approximately constant hazard rate for many types of component.

3.3.12 Piecewise Linear Hazard Function

If the hazard rates in the break-in and wear-out periods, can be approximated by linear functions and the hazard rate during the prime of life is approximately constant, the hazard function may be represented by Figure 3.4. The mathematical representation of such a hazard function is

$$
h(t) = \begin{cases} 1 - bt + \lambda & \text{for } 0 \leq t \leq t_b \\ \lambda & \text{for } t_b \leq t \leq t_b + t_u \\ c(t - t_b - t_u) + \lambda & \text{for } t > t_b + t_u \end{cases} \tag{3.32}
$$

where $t_b = a/b$ and $c = \tan \theta$. The time period from 0 is t_b is the break-in phase, the time period from t_b to $t_b + t_u$ is the prime-of-life phase, and the time period after $t_b + t_u$ is the wear-out phase.

The corresponding reliability function, unreliability function, and failure probability density function are given by [Shooman (1968)]

$$
r(t) = 1 - u(t) = 1 - F(t)
$$

$$
= \begin{cases} \exp\left\{ -\left[(a + \lambda)t - b\left(\dfrac{t^2}{2}\right) \right] \right\} & \text{for } 0 \leq t \leq t_b \\ \exp\{ -[\lambda t + (0.5at_b)] \} & \text{for } t_b \leq t \leq t_b + t_u \\ \exp\left\{ -\left[\left(\dfrac{c}{2}\right)(t - t_b - t_u)^2 + \lambda t + (0.5at_b) \right] \right\} & \text{for } t > t_b + t_u \end{cases}
$$

$$
\tag{3.33}
$$

$$
f(t) = \begin{cases} (a + \lambda - bt)\exp\left\{ -\left[(a + \lambda)t - b\left(\dfrac{t^2}{2}\right) \right] \right\} \\ \qquad\qquad\qquad\qquad\qquad \text{for } 0 \leq t \leq t_b \\ \lambda \exp\{ -[\lambda t + (0.5at_b)] \} \\ \qquad\qquad\qquad\qquad\qquad \text{for } t_b \leq t \leq t_b + t_u \\ [c(t - t_b - t_u) + \lambda]\exp\left\{ -\left[\left(\dfrac{c}{2}\right)(t - t_b - t_u)^2 + (0.5at_b) + \lambda t \right] \right\} \\ \qquad\qquad\qquad\qquad\qquad \text{for } t > t_b + t_u \end{cases}
$$

$$
\tag{3.34}
$$

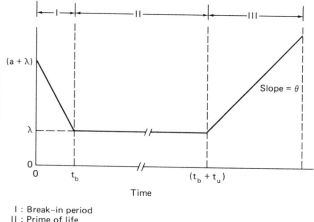

Figure 3.4. Piecewise linear hazard function.

If we assume that the component has a constant hazard rate throughout its life, we have to set $t_b = 0$ because the break-in period is ignored under a constant hazard rate approximation. Also, $t_u = \infty$ because, under a constant hazard rate assumption, we ignore the wear-out period and assume that the component is always in its prime-of-life phase. So, substituting $t_b = 0$ and $t_u = \infty$ in Equations (3.32)–(3.34), we get

$$h(t) = \lambda$$
$$r(t) = \exp(-\lambda t)$$
$$f(t) = \lambda \exp(-\lambda t)$$

These equations are exactly what we had for cases of constant failure rate [see Section 3.3.10.1 and Equations (3.21) and (3.23)].

We discuss different types of hazard function in the following three sections (Sections 3.3.13–3.3.15). However, in the remaining chapters of the book we use almost exclusively constant hazard functions (constant failure rates). So readers who do not have an interest in learning more about different types of hazard function may skip the next three sections.

3.3.13 General Bathtub Curve

Dhillon (1979) proposed the following hazard function for the bathtub curve:

$$h(t) = 0.5kAt^{-0.5} + (1 - k)b \exp(bt) \qquad (3.35)$$

where $A > 0$, $b > 0$, $0 \leq k \leq 1$, and $t \geq 0$; A, b, and k are chosen such that the hazard curve fits the available failure data fairly well; standard curve-fitting techniques may be used.

The corresponding reliability function and unreliability function may be obtained by using Equations (3.18) and (3.19). The reliability function is

$$r(t) = 1 - u(t) = 1 - F(t)$$

$$= \exp\{-kAt^{-0.5} - (1 - k)[\exp(bt) - 1]\} \qquad (3.36)$$

3.3.14 Weibull Distribution

If the failure probability distribution (unreliability function) is defined by a Weibull distribution, the corresponding failure probability density function and hazard function are as follows:

$$F(t) = u(t) = 1 - r(t)$$

$$= 1 - \exp\left\{-\left[\frac{t - a}{b}\right]^{c}\right\} \qquad \text{for } t \geq a$$

$$f(t) = \left(\frac{c}{b}\right)\left[\frac{t - a}{b}\right]^{c-1} \exp\left\{-\left[\frac{t - a}{b}\right]^{c}\right\} \qquad \text{for } t \geq a \qquad (3.37)$$

$$h(t) = \begin{cases} \left(\frac{c}{b}\right)\left[\frac{(t - a)}{b}\right]^{c-1} & \text{for } t \geq a \\ 0 & \text{for } t < a \end{cases}$$

In the preceding equations, a is greater than or equal to zero and b and c are greater than zero.

By choosing the value of c appropriately, constant, decreasing, and increasing hazard functions can be obtained. When $c = 1$ and $a = 0$, the Weibull distribution reduces to an exponential distribution, so the hazard function is a constant. When $c < 1$, the hazard function decreases with time; when $c > 1$, the hazard function increases with time. The hazard functions of many mechanical components increase with time due to wear, corrosion, and other forms of deterioration. In such cases an increasing hazard function may be appropriate (see Figure 3.2b).

The coefficients a, b, and c may be chosen such that the corresponding Weibull probability distribution, $F(t)$, fits the available failure data fairly well. Methods of fitting data points to standard probability distributions (here, the Weibull distribution) are available in books on probability and statistics.

3.3.15 Other Distributions

Other distributions that have been used as failure probability distributions include the normal, lognormal, Gamma, and Gumbel distributions.

3.4 REPAIRABLE COMPONENTS

3.4.1 Life Cycle of a Repairable Component

The complete life cycle of a repairable component consists of the placement of a new component into service, the occurrence of failure at some time, the repair and restoration of the component to service (the repaired component is assumed to be as good as new), the occurrence of a second failure, the repair and restoration of the component to service, and so on. This cycle repeats until the component is retired from service.

3.4.2 Reliability and Availability

Definitions of reliability, availability, unreliability, and unavailability, as given in Section 3.3.2, are applicable to repairable components also.

Because some components that fail anytime between time 0 and time t may be repaired and brought back to their normal state by time t, reliability is not necessarily equal to availability for nonrepairable components. Reliability at time t is either equal to or less than the availability at time t.

Availability at time t, as $t \to \infty$, is called the *steady-state availability* or the *asymptotic availability* and is denoted by $a(\infty)$. The corresponding steady-state unavailability is denoted by $q(\infty)$.

Calculation of repairable component availability and reliability in terms of component failure rate and repair rate is discussed in Section 3.4.15.

Availability and unavailability are defined here for a time instant t and are more specifically referred to as the *point availability* and *point unavailability*, although it is a common practice to call them simply availability and unavailability. Another type of availability called *interval availability*, is defined in Section 3.4.12. Whereas point availability is defined for a specific time instant, interval availability is defined for a specific time interval.

3.4.3 Repair Time and Mean Time to Repair

Repair time is the length of time from the instant of failure to the instant the component is restored to service after repair. This includes the time taken to detect the failure (detection time), the interval from detection to the time repair starts, the time taken to carry out the repair, and the time for preoperational testing of the repaired component. Repair of a component is said to have been completed when it is restored to service. The component is said to be in a failed state until it is restored to service. Repair time, also called the *time to repair* (TTR), is treated as a random variable because the repair time for two supposedly identical components may not necessarily be equal. The expected value of repair time (time to repair) is called the *mean repair time* or the *mean time to repair* (MTTR). If the repair times for the same or similar components in the same or similar plants are recorded over a period of time, the MTTR can be computed as the average of the observed repair times. MTTR is denoted by τ.

If we set the repair time as infinity, availability will coincide with reliability, because, with infinite repair time, failed components cannot be restored to service by a finite time t; in fact, setting repair time to infinity is like setting the component to be nonrepairable.

(*Note:* Some authors do not include detection time as part of the repair time. All discussions in this book include the detection time as part of repair time.)

3.4.4 Detection Time

Detection time is the duration between the instant component failure occurs to the instant failure is detected. Some component failures are detected immediately by plant operating personnel; for example, if an operating pump fails, such a failure may be noticed almost immediately by operating personnel. Some component failures trigger alarms that alert the operators to the failure. Detection time is zero if the component failure is thus detected almost immediately. There are components whose failures are not detected immediately; for example, a standby pump may fail during its standby phase and the failure may not be detected until the next periodic maintenance.

Consider a component whose failure is detected only during periodic maintenance. Let the periodic maintenance interval (the duration between one routine maintenance and the next) be T_I. The component could fail anytime during the periodic maintenance interval. In such cases the detection time is taken as $(T_I/2)$; this value is acceptable in reliability analysis if the maintenance interval is less than 10% of the MTTF (that is, $T_I < 0.1$ MTTF).

3.4.5 Repair Rate

The *repair rate* at time t, denoted by $m(t)$, is the probability density (that is, probability per unit time) that the component is repaired at time t given that the component failed at time zero and had been in a failed state (that

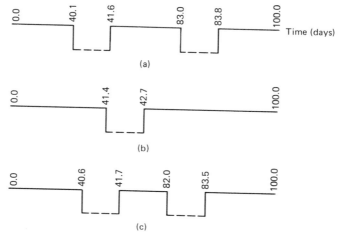

(a)

(b)

(c)

Solid horizontal lines : component in normal state.
Broken horizontal lines : component in failed state (in repair).

Figure 3.5. Failure–repair history of three components: (a) component 1; (b) component 2; (c) component 3.

is, the component is not yet restored to service) to time t. The quantity $m(t)\,dt$ is equal to the probability that the component is repaired during $[t, t + dt)$, given that the component failed at time zero and remained in a failed state to time t. In most cases, we assume that the repair rate is independent of time (constant repair rate); this is discussed in Section 3.4.6.

Example 3.8

Failure–repair histories of three components over a 100-day period are shown in Figure 3.5. All three components were as good as new at time zero. Compute the MTTF, MTTR, and repair rate.

Solution

Examining the failure–repair histories of the three components over a period of 100 days, there are five failures and five repairs during that period. A component is considered as good as new after repair; so the time instant after repair may be set back to $t = 0$, as if it were a new component. Thus Figure 3.5 is transformed to Figure 3.6. In figure 3.6a1, the first failure and repair of the first component is shown. In Figure 3.6a2, the failure–repair history of the first component after the first repair is shown. Essentially, the origin is shifted to $t = 41.6$ days, which is

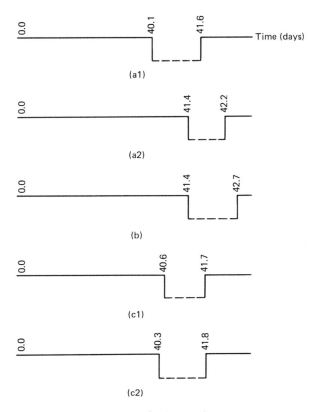

Solid horizontal lines : component in normal state.
Broken horizontal lines : component in failed state (in repair).

Figure 3.6. Transformed failure–repair history of three components: (a1) first failure and repair of component 1; (a2) second failure and repair of component 1; (b) failure and repair of component 2; (c1) first failure and repair of component 3; (c2) second failure and repair of component 3.

the time instant when the first component is restored to service after repair. Failure in Figure 3.6a2 occurs at $t = 41.4$ days (that is, $83.0 - 41.6$). The time the component is restored to service in Figure 3.6a2 is $t = 42.2$ (that is, $83.8 - 41.6$). Figures 3.6b, 3.6c1, and 3.6c2 are drawn similarly.

The MTTF is obtained by averaging the time to failure in each of the five cases (a1, a2, b, c1, and c2) shown in Figure 3.6.

$$\text{MTTF} = \frac{40.1 + 41.4 + 41.4 + 40.6 + 40.3}{5}$$

$$= 40.76 \text{ days}$$

The MTTR is obtained by averaging the repair time in each of the five cases shown in Fig. 3.6:

$$\text{MTTR} = \frac{\left\{ \begin{array}{c} (41.6 - 40.1) + (42.2 - 41.4) + (42.7 - 41.4) \\ + (41.7 - 40.6) + (41.8 - 40.3) \end{array} \right\}}{5}$$

$$= \frac{1.5 + 0.8 + 1.3 + 1.1 + 1.5}{5}$$

$$= 1.24 \text{ days}$$

In order to calculate the repair rate, we consider the duration of repair in each of the five cases. First, we shift the origin to the time when failure occurs. (This satisfies the condition that failure occurs at $t = 0$, which is inherent in the definition of the repair rate.) The time-shifted diagrams are shown in Figure 3.7. For example, the time coordinates in Figure 3.7a1 are obtained by deducting 40.1 (the time of failure) from the time coordinates of Figure 3.6a1.

The repair rate $m(t)$ may be calculated using its definition. Let us divide the time axis into 0.25-day intervals. To compute $m(0.75)$, we consider the time interval $[0.75, 1.00)$ in Figure 3.7. Here, the duration of the interval $dt = (1.00 - 0.75) = 0.25$.

The number of components that failed at $t = 0$ and are in a failed state at $t = 0.75$ is 5.

The number of components repaired during the time interval $[0.75, 1.00)$ is 1.

So the probability that "components that failed at time zero and remained in a failed state to time $t = 0.75$" are repaired during $[0.75, 1.0)$ is $\frac{1}{5}$: $m(0.75) \times 0.25 = \frac{1}{5}$, that is, $m(0.75) = \frac{1}{5}/0.25$, where 0.25 is the duration of the interval $[0.75, 1.00)$. Similar calculations yield

$m(0.00) = 0.0$ $m(0.25) = 0.0$

$m(0.50) = 0.0$ $m(0.75) = 0.8$

$m(1.00) = 1.0$ $m(1.25) = 1.33$

$m(1.50) = 4.0$ $m(1.75) = \text{undefined}$

$m(2.00) = \text{undefined}$

Let us also calculate the repair rate using a different time interval (dt) to see if we get the same results. (*Note:* MTTF and MTTR calculations are independent of dt.)

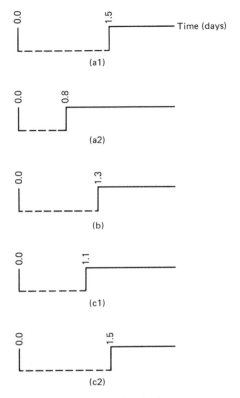

Solid horizontal lines : component in normal state.
Broken horizontal lines : component in failed state (in repair).

Figure 3.7. Time-shifted repair history of three components: (a1) first repair of component 1; (a2) second repair of component 1; (b) repair of component 2; (c1) first repair of component 3; (c2) second repair of component 3.

Following the same procedure as before, but with $dt = 0.5$, we get

$$m(1.0) = \frac{2/4}{0.5} = 1.0$$

Similar calculations yield

$$m(0.0) = 0 \qquad m(0.5) = 0.4 \qquad m(1.0) = 1.0$$
$$m(1.5) = 2.0 \qquad m(2.0) = \text{undefined}$$

Comparing the repair rates using $dt = 0.25$ and $dt = 0.5$, we see that the results are significantly different (as much as 100% difference). It is worth noting that the smaller the dt value, the more accurate are the results. This raises the question, "What is an acceptable dt value?" There is no direct answer. One way of ascertaining the accuracy of the results is to compute the results using two different dt values; if the differences in the results are within limits acceptable to the analyst, the results from the lower dt value may be considered acceptable. If the differences are not within acceptable limits, then the calculations are repeated using a still lower dt value until the differences in results are acceptable. In our problem, we may repeat the repair rate calculations with $dt = 0.125$. If the differences in results with those obtained using $dt = 0.25$ are acceptable, we may use the results from $dt = 0.125$. If not, we shall repeat the calculations with $dt = 0.0625$.

3.4.6 Constant Repair Rate

A constant repair rate may be used if the repair rate is almost independent of time. (This is analogous to the use of constant failure rates in Section 3.3.10.) The constant repair rate is denoted by μ. So $m(t) = \mu$ when $m(t)$ is independent of t.

When we assume a constant repair rate, it can be proven that the constant repair rate is equal to the inverse of the mean time to repair (MTTR), that is

$$\mu = \frac{1}{\text{MTTR}} = \frac{1}{\tau} \tag{3.38}$$

Note the similarity between Equations (3.24) and (3.38).

When the repair rate is constant, the computation of this repair rate is very easy [using Equation (3.38)], compared to the rather tedious procedure we had to use in Example 3.8.

When can we assume a constant repair rate? When the mean time to repair is much smaller than the mean time to failure (which is usually the case), the actual time-dependency of the repair rate is unimportant in reliability–availability calculations. So the assumption of constant repair rate does not introduce any significant errors.

Assuming constant repair rates and constant failure rates for components considerably simplifies the mathematical formulation of system reliability–availability analysis. We assume constant failure rates and constant repair rates in the rest of the book, unless otherwise stated.

3.4.7 Difference between Reliability and Availability

Even when a component is repairable, we may be interested in its *reliability*. For example, if the failure of a component would result in personnel injuries or property damage, we may be interested in computing the probability of such occurrences anytime between times 0 and t (which is equal to the component unreliability at time t), even if the component itself could be repaired and brought back to its normal state before time t.

It is worth comparing the reliability and the availability of a component. Consider that N components are put into service at time zero (new or as good as new at time zero). Let the number of components that have *never* failed from time zero to time t be $n(t)$. Let the number of components in their normal states at time t be $n'(t)$; in this category we include those components that have never failed between times 0 and t and also those components that have failed between times 0 and t but have been repaired and restored to service by time t.

$$
\begin{aligned}
\text{Reliability} \quad r(t) &= \frac{n(t)}{N} \\
\text{Availability} \quad a(t) &= \frac{n'(t)}{N}
\end{aligned}
\tag{3.39}
$$

Based on the definition of $n(t)$ and $n'(t)$, we have

$$
n'(t) \geq n(t)
$$

So,

$$
a(t) \geq r(t) \tag{3.40}
$$

Sample plots of component availability and reliability are shown in Figure 3.8. The reliability will become zero at some point of time, whereas the availability will tend to the asymptotic availability $a(\infty)$ as time increases.

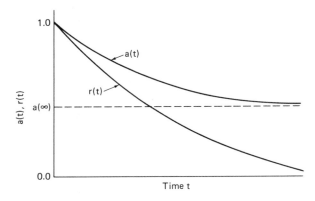

Figure 3.8. Sample availability and reliability plots for a repairable component.

Table 3.4. Component Reliability – Availability Calculations
(Example 3.9)

t	N	$n(t)$	$n'(t)$	$r(t)$	$a(t)$
400	10	10	10	1.0	1.0
530	10	9	9	0.9	0.9
560	10	9	10	0.9	1.0
620	10	8	9	0.8	0.9
650	10	8	10	0.8	1.0

Example 3.9

Ten new components were placed into service at time zero. One component failed at $t = 500$ cycles, and another component failed at $t = 600$ cycles. The first component was repaired and restored to service at $t = 550$ cycles, and the second component was repaired and restored to service at $t = 640$ cycles. Compute the reliability and availability of the components at $t = 400, 530, 560, 620,$ and 650 cycles.

Solution

Results are presented in Table 3.4. Equation (3.39) is used in the computation of $r(t)$ and $a(t)$.

The purpose of this example is to illustrate simply the differences between reliability and availability, so only 10 components are considered. In practical problems, one should consider significantly larger numbers of components in order to get more accurate values for reliability and availability. In such cases $r(t)$ and $a(t)$ will generally be smooth functions of time.

Calculation of reliability and availability as a function of failure rate and repair rate is discussed in Section 3.4.15.

3.4.8 Expected Number of Failures

The *expected number of failures* (ENF) over the time interval between t_1 and t_2, given that the component was as good as new at time zero, is denoted by $w(t_1, t_2)$.

Component unreliability at time t is a lower bound to the expected number of failures between times 0 and t. That is,

$$w(0, t) \geq u(t) \tag{3.41}$$

3.4.9 Mean Time to Failure

The *mean time to failure* (MTTF) is the expected time duration between the instant the component is put into service new or as good as new and the instant of its very next failure. Equation (3.10) may be used to compute the MTTF.

3.4.10 Mean Time between Failures

The *mean time between failures* (MTBF) is the expected time duration between two consecutive failures:

$$MTBF = MTTF + MTTR \qquad (3.42)$$

3.4.11 Mean Time between Repairs

The *mean time between repairs* (MTBR) is the expected time duration between the beginning of two consecutive repairs:

$$MTBR = MTTR + MTTF \qquad (3.43)$$

Comparing Equations (3.42) and (3.43), we have

$$MTBR = MTBF \qquad (3.44)$$

3.4.12 Interval Availability

The availability at a specific time instant t (as we discussed so far) is called the *point availability* at time t (usually it is referred to as simply the availability at time t).

The *interval availability* for a specific time duration from t_1 to t_2 is defined as the ratio of the duration of time the component is in its normal state between t_1 and t_2 to the total duration of the time between t_1 and t_2, where the component is put into service new or as good as new at time zero. Interval availability is denoted by $a(t_1, t_2)$ and is given by

$$a(t_1, t_2) = \frac{t_N}{(t_1 - t_2)}$$

$$= 1 - \frac{t_F}{(t_1 - t_2)} \qquad (3.45)$$

where t_N is the duration of time the component is in its normal state during the time interval between t_1 and t_2, and t_F is the duration of time the component is in the failed state during the time interval between t_1 and t_2.

Interval unavailability is the complement of interval availability:

$$q(t_1, t_2) = 1 - a(t_1, t_2) \qquad (3.46)$$

where $q(t_1, t_2)$ is the interval unavailability between t_1 and t_2.

Interval availability is a function of times t_1 and t_2. In the limiting case when $(t_2 - t_1)$ tends to infinity, the corresponding interval availability is called the *steady-state interval availability*, the *asymptotic interval availability*, or the *limiting interval availability*. It is represented by $a_I(\infty)$. [We use the subscript I to differentiate the steady-state interval availability from the steady-state point availability $a(\infty)$.] The corresponding steady-state interval unavailability is denoted by $a_I(\infty)$ and is given by

$$
\begin{aligned}
q_I(\infty) &= \frac{\text{mean time to repair}}{\text{mean time between failures}} = \frac{\text{MTTR}}{\text{MTBF}} \\
&= \frac{\text{MTTR}}{\text{MTTF} + \text{MTTR}} = \frac{\lambda\tau}{(1 + \lambda\tau)} \qquad (3.47)
\end{aligned}
$$

$$
a_I(\infty) = 1 - q_I(\infty) = \frac{\text{MTTF}}{\text{MTTF} + \text{MTTR}} = \frac{1}{(1 + \lambda\tau)} \qquad (3.48)
$$

Interval availability from t_1 to t_2 may also be defined as the average point availability between t_1 and t_2:

$$
a(t_1, t_2) = \frac{\int_{t_1}^{t_2} a(t)\, dt}{t_2 - t_1} \qquad (3.49)
$$

The *steady-state interval availability* is given by

$$
a_I(\infty) = \lim_{T \to \infty} \left[\frac{1}{T} \int_0^T a(t)\, dt \right] \qquad (3.50)
$$

Interval availabilities are commonly used for standby components that are periodically maintained. Time t is set to zero at the time maintenance is completed and the component is back in standby service (because a component is considered as good as new when it is restored to service after periodic maintenance). We set $t_1 = 0$ and $t_2 = $ the time interval between the time the component is restored to service after the periodic maintenance to the time it is taken out of service for the next periodic maintenance. If the time required for the maintenance (the time during which the component is out of service) is very small compared to the duration between periodic maintenances (T_I), we may set $t_2 = T_I$. (*Note:* Point availabilities may also be used for standby components that are periodically maintained.)

3.4.13 Testing and Maintenance Unavailability

Interval *unavailability due to periodic testing and maintenance* is given by

$$q_{TM}(0, T_I) = \frac{t_{TM}}{T_I} \qquad (3.51)$$

where t_{TM} is the average duration of time the component is out of service for each periodic testing and maintenance and T_I is the interval of time between successive maintenances. [*Note:* This unavailability does not include the unavailability due to component failure. If we are interested in the total unavailability, the unavailability due to failures (as computed in the previous sections) should be added to the testing and maintenance unavailability.]

Some analysts compute the unavailability due to periodic maintenance separately from the unavailability due to periodic testing. The sum of these two unavailabilities is the unavailability due to testing and maintenance. (*Note:* Some authors use the term "maintenance unavailability" to refer to testing and maintenance unavailability.)

3.4.14 Unavailability on Demand

If a component that is on standby mode fails to operate on demand, such a failure is called *failure on demand*. A pump failing to start when required or a valve failing to open or close when required are examples of failure on demand.

The *probability of failure on demand* (or the *unavailability on demand*) may be approached in one of two ways. Consider a standby pump that fails to start when required (on demand). The pump failure could have occurred before the demand, at anytime during the period it was on standby mode (the time between its periodic maintenance and the time of demand). So the probability of failure on demand is the unavailability of the pump at the time of demand; time is set at zero when the pump is put into service new or put into service after a periodic maintenance, and the time of demand is measured from that time point. This logic can be applied to other failure-on-demand scenarios, such as a valve failing to close, a valve failing to open, or a motor failing to start. This is the first approach. (If the time of demand is not known, $T_I/2$ may be used as the time of demand, to obtain an average value for the failure probability; if we want a very conservative result, T_I may be used as the time of demand, where T_I is the duration between periodic maintenances.)

The second approach is to calculate the probability of failure on demand directly as the ratio of the number of failures on demand to the total number of demands, that is,

$$q_D = \frac{n}{N} \tag{3.52}$$

where q_D is the probability of failure on demand (unavailability on demand), n is the number of failures on demand, and N is the number of demands. In order for the calculated probability of failure on demand to be accurate, the number of demands (N) should be sufficiently large.

The probability of failure on demand (unavailability on demand) is a function of time (with the time set at zero when the component is put into service new or as good as new); it usually increases with time. However, it is a common practice to assume a constant unavailability on demand (independent of time) as long as the component is in its prime of life and is periodically maintained. If the failure probability is computed using the first approach, $T_I/2$ may be used as the time of demand and a constant (time-independent) value is obtained for the failure probability. The second approach [Equation (3.52)] does assume a constant failure probability.

3.4.15 Reliability – Availability Computation

Reliability and availability of repairable components can be computed in terms of the failure rate and mean time to repair (MTTR). We provide here, without derivation, the formulae for reliability and availability. Constant failure rate and constant repair rate are assumed.

Component unavailability at time t is given by

$$q(t) = \frac{\lambda\tau}{(1 + \lambda\tau)} \left\{ 1 - \exp\left[-\left(\frac{1 + \lambda\tau}{\tau} \right) t \right] \right\} \tag{3.53}$$

where λ is the failure rate and τ is the mean time to repair.

Component availability is given by

$$a(t) = 1 - q(t)$$

When $t \to \infty$, we get the steady-state unavailability

$$q(\infty) = \frac{\lambda\tau}{(1 + \lambda\tau)} \qquad (3.54)$$

The steady-state availability is given by

$$a(\infty) = 1 - q(\infty)$$

The preceding steady-state unavailability and availability may be used for finite times also if t is fairly large. It has been shown that Equation (3.54) overpredicts $q(t)$ by no more than 14% if $t > 2\tau$, and it overpredicts $q(t)$ by no more than 5% if $t > 3\tau$.

If $\lambda\tau$ is less than 0.1, $q(\infty)$ in Equation (3.54) may be approximated by

$$q(\infty) = \lambda\tau \qquad (3.55)$$

If $\lambda\tau = 0.1$, Equation (3.55) overpredicts $q(\infty)$, as computed by Equation (3.54), by 10%; that is, Equation (3.55) provides a $q(\infty)$ value that is 10% higher than the $q(\infty)$ value provided by Equation (3.54). If $\lambda\tau = 0.05$, the overprediction is 5%; if $\lambda\tau = 0.01$, the overprediction is 1%. Equation (3.55) is widely used in availability analysis.

The reliability of a repairable component is identical to the reliability of the corresponding nonrepairable component, because, in reliability calculations, a failed component is considered failed at all times after the failure; that is, from the perspective of reliability calculations, a component (even a repairable component) is assumed to be unrepaired after failure. The equations given in Section 3.3 for the reliability of nonrepairable components are applicable to repairable components also.

Alternatively, component reliability may be obtained by setting the mean time to repair (τ) to infinity in Equation (3.53). (The rationale behind setting $\tau = \infty$ is discussed in Section 3.4.3.) As $\tau \to \infty$,

$$\left[\frac{\lambda\tau}{(1 + \lambda\tau)} \right] \to 1 \quad \text{and} \quad \left[\frac{(1 + \lambda\tau)}{\tau} \right] \to \lambda$$

Substituting these values into Equation (3.53), we get the component unreliability as

$$u(t) = 1 - \exp(-\lambda t) \qquad (3.56)$$

This is identical to Equation (3.22). As discussed in Section 3.3.10.2, Equation (3.56) may be approximated to

$$u(t) = \lambda t \qquad (3.57)$$

if $u(t)$ is less than 0.05. Component reliability is given by

$$r(t) = 1 - u(t)$$

where $u(t)$ is computed using Equation (3.56) or (3.57), as appropriate.

Note that we shall not set $\tau = \infty$ in Equation (3.54) or (3.55) and expect to obtain the corresponding unreliability. Those equations are a good approximation only for times greater than 3τ. Because we set $\tau = \infty$, those equations are invalid for any finite time.

EXERCISE PROBLEMS

3.1. The performance of 2000 light bulbs as tracked for 1000 hours of use. Table 3.5 summarizes the number of bulbs still operating at time $t = 0, 100, 200, \ldots, 1000$ hours. (i) Compute the reliability and availability of the bulbs at $t = 0, 100, 200, \ldots, 1000$ hours. (ii) Compute the mean time to failure. (Note that light bulbs are nonrepairable.)

3.2. The reliability function of a nonrepairable component is as follows:

$$r(t) = \begin{cases} (1 - 0.2 \times 10^{-6}t^2 - 0.8 \times 10^{-9}t^3) & \text{for } 0 \le t \le 10^3 \text{ hours} \\ 0 & \text{elsewhere} \end{cases}$$

(i) Calculate the unavailability at time $t = 0.7 \times 10^3$ hours. (ii) Calculate the mean time to failure.

3.3. Compute the failure probability density function and the hazard function at time $t = 0, 100, 200, \ldots, 1000$ hours for the light bulbs described in Problem 3.1. Assume that the reliability function is linear between time points $t = 0, 100, 200, \ldots, 1000$ hours.

Table 3.5.   Data for Problem 3.1

TIME t (HOURS)	NUMBER OF BULBS STILL OPERATING AT TIME t
0	2000
100	1914
200	1910
300	1908
400	1902
500	1810
600	1200
700	424
800	112
900	48
1000	12

3.4. The constant failure rate of a nonrepairable pressure sensor at normal temperature conditions is 3×10^{-6} failures per hour and at high temperature conditions is 6×10^{-6} failures per hour. One such sensor (sensor A) is used at normal temperatures for 1000 hours. Another such sensor (sensor B) is used at high temperatures for 1000 hours. (i) Compute the unreliability and failure probability density of sensor A and sensor B at time $t = 1000$ hours. Use Equations (3.21) and (3.23) and Equations (3.25) and (3.27), and compare the results. (ii) After the first 1000 hours of use at normal temperatures, sensor A is used for an additional 1000 hours at high temperatures. Compute the unreliability of sensor A at $t = 2000$ hours. (iii) After the first 1000 hours of use at high temperatures, sensor B is used for an additional 1000 hours at normal temperatures. Compute the unreliability of sensor B at $t = 2000$ hours.

3.5. There are four failure modes associated with an electric charger. The failure modes and their constant failure rates are as follows:

No electrical output: 11×10^{-6} failures per hour
Low electrical output: 80×10^{-6} failures per hour
High electrical output: 6×10^{-6} failures per hour
Erratic electrical output: 23×10^{-6} failures per hour

Assume that the failure modes are statistically independent. (i) Compute the unreliability of the charger at times $t = 300$ and 1200 hours due to each of these failure modes, using Equations (3.21) and (3.25). Compare the results. (ii) Compute the unreliability of the charger at $t = 300$ and 1200 hours due to all the failure modes combined. Use Equation (3.30). (iii) What is the total failure rate of the charger due to all four failure modes combined? Compute the unreliability of the charger at $t = 300$ and 1200 hours due to all failure modes combined, using the total failure rate and Equations (3.21) and (3.25). Compare the results. [*Note:* Equation (3.25) may not be used when the unreliability is not less than 0.05; we request the use of that equation in this and some other problems, even when that condition is not satisfied, to show how the results of Equation (3.25) deviate from the more accurate results of Equation (3.21).]

3.6. The hazard function of a component is piecewise linear; it is given by Equation (3.32) with $a = 10^{-2}$, $b = 10^{-4}$, $c = 2 \times 10^{-5}$, $\lambda = 10^{-3}$, $t_b = 100$, $t_u = 5000$. Units of the hazard function $h(t)$ and time t are failures per hour and hours, respectively. (i) Plot the hazard function for $t = 0$ to 6000 hours. (ii) Compute the reliability function and the failure probability density function at $t = 0$, 10, 100, 1000, and 5000 hours. (iii) Compute the reliability function and the failure probability density function of a component with a constant failure rate of 10^{-3} failures per hour; compute them at $t = 0$, 10, 100, 1000, and 5000 hours.

3.7. Redo Problem 3.6, with $b = 10^{-3}$ and $t_b = 10$; all other values remain the same as in Problem 3.6.

Table 3.6.   Data for Problem 3.12

	COMPONENT 1	COMPONENT 2
put into service	0	0
failure	1,700	2,000
restored to service	1,800	2,080
failure	3,800	4,000
restored to service	3,880	4,070
failure	5,600	6,100
restored to service	5,690	6,200
failure	7,400	8,000
restored to service	7,500	8,090
failure	9,200	9,800
restored to service	9,260	9,910
failure	11,000	11,800
restored to service	11,080	11,890
failure	12,900	13,700

3.8. Redo Problem 3.6, with $a = 10^{-3}$ and $b = 10^{-5}$; all other data remain the same as in Problem 3.6.

3.9. The failure probability distribution of a component is represented by a Weibull distribution and is given by Equation (3.37) with $a = 0$, $b = 1000$, and $c = 0.5$. Units of the hazard function $h(t)$ and time t are failures per hour and hours, respectively. (i) Compute the hazard function, reliability function and the failure probability density function at $t = 0$, 10, 100, 1000, and 5000 hours. (ii) Compute the hazard function, reliability function, and the failure probability density function of a component with a constant failure rate of 10^{-3} failures per hour; compute them at $t = 0$, 10, 100, 1000, and 5000 hours.

3.10. Redo Problem 3.9, with $c = 1.5$; all other values remain the same as in Problem 3.9.

3.11. Redo Problem 3.9, with $c = 1.0$; all other values remain the same as in Problem 3.9.

3.12. The failure–repair history of two similar repairable components is summarized in Table 3.6. Components were as good as new at time $t = 0$. (i) Compute the mean time to failure and the mean time to repair. (ii) Compute the repair rate using a time increment (dt) of 20 hours. (*Note:* The time increment of 20 hours used in this problem may not be small enough to give a good estimate of the repair rate; we use this value here just to illustrate the method of computing repair rates.)

3.13. An electric transformer has a constant failure rate of 3×10^{-7} failures per hour and a constant repair rate of 10^{-2} repairs per hour. (i) Compute the mean time to failure, mean time to repair, mean time between failures, and mean time between repairs. (ii) Compute the unreliability and unavailability at time $t = 100, 200, 300, 500,$ and 1000 hours. (iii) Compute the steady-state unavailability using Equations (3.54) and (3.55). Compare the results.

3.14. A component has constant failure rate and constant repair rate. Its mean time to failure is 1000 hours and its mean time to repair is 200 hours. (i) Compute the constant failure rate and the constant repair rate. (ii) Compute the steady-state unavailability using Equations (3.54) and (3.55). Compare the results. [*Note:* Equation (3.55) may not be used if $\lambda\tau$ is not less than 0.1. We request the use of that equation here just to illustrate the differences in results between Equations (3.54) and (3.55).]

REFERENCES

Benjamin, J. R. and C. A. Cornell (1970). *Probability, Statistics, and Decision for Civil Engineers*. McGraw-Hill Book Company, New York.

Davis, D. J. (1952). An analysis of some failure data. *Journal of the American Statistical Association* 113–150.

Dhillon, B. S. (1979). A hazard rate model. *IEEE Transactions on Reliability* **R-28** 150–154.

Drenick, R. F. (1960). The failure law of complex equipment. *Journal of the Society for Industrial and Applied Mathematics* **8** 680–690.

Herd, G. R. (1957). Estimation of reliability functions. In *Proceedings of the Third National Symposium on Reliability and Quality Control*, pp. 113–122.

Kao, J. H. K. (1958). Computer methods for estimating Weibull parameters in reliability studies. *IRE Transactions on Reliability and Quality Control* **13** 15–22.

Shooman, M. L. (1968). *Probabilistic Reliability: An Engineering Approach*. McGraw-Hill, New York.

Chapter 4
Fundamentals of Systems Engineering

4.1 INTRODUCTION

A *system* is an orderly arrangement of components that interact among themselves and with external components, other systems, and human operators to perform some intended functions. As discussed in Section 3.2, what is considered a system in one reliability analysis could be treated as a component or subsystem in another reliability analysis.

System reliability parameters are defined in Section 4.2. Then a general description of the physical arrangement of the three types of basic system configurations, namely, the series configuration, the parallel configuration and k-out-of-n configuration are provided in Sections 4.3 and 4.4.

4.2 DEFINITIONS

A system is called a *nonrepairable system* if all the components in the system are nonrepairable. A system is called a *repairable system* if at least one of the components is repairable.

System reliability parameters such as reliability, unreliability, availability, unavailability, failure rate, repair rate, mean time to failure, mean time to repair, mean time between failures, expected number of failures, etc. are defined in the same way as these parameters are defined for components (see Chapter 3).

Reliability at time t is equal to the availability at time t for nonrepairable systems.

4.3 BASIC SYSTEM CONFIGURATIONS

There are three basic system configurations. A combination of these three configurations can represent virtually any system.

4.3.1 Series Configuration

The *series configuration* (*series system*) consists to two or more components connected in series. Figure 4.1 shows a series system consisting of n

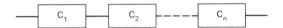

Figure 4.1. Series configuration.

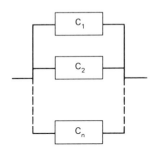

Figure 4.2. Parallel configuration.

components $C_1, C_2, \ldots, C_n$. A series system fails if one or more components fail. (Figure 4.1 is called a schematic diagram, a block diagram, or a reliability block diagram.)

4.3.2 Parallel Configuration

The *parallel configuration* (*parallel system*) consists of two or more components connected in parallel. Figure 4.2 shows a parallel system consisting of n components $C_1, C_2, \ldots, C_n$. A parallel system fails if and only if all the components fail. The system will be capable of performing its intended function even if just one component is in its normal state.

Parallel systems with n components are said to have $(n - 1)$ redundancies. What is *redundancy*? Redundancy is the use of more than the required minimum number of components, in order to increase the system reliability. If a load can be lifted by a single cable (in a crane), providing two cables side by side, with each cable having the capacity to lift the load by itself, is an example of redundancy.

Parallel systems may be classified as active parallel systems and passive parallel systems. This is further discussed in Section 4.4.

4.3.3 *k-out-of-n* Configuration

The *k-out-of-n configuration* (*k-out-of-n system*) is similar to the parallel configuration shown in Figure 4.2, but at least k components out of the total n components (where $0 < k \le n$) should be in their normal state for the system to perform its intended function. The system fails if $(n - k + 1)$ or more components fail.

When $k = 1$, the k-out-of-n system reduces to the parallel system discussed in Section 4.3.2.

4.4 TYPES OF PARALLEL SYSTEMS

In this section, we will restrict our attention to parallel systems with two identical components. However, the concepts discussed here can be generalized to n-component parallel systems.

For convenience, we shall call one of the two components in the parallel system the *primary component* and the other the *duplicate component*. Note that both of the components have identical capabilities to perform the intended component functions.

There are two types of parallel systems, namely, active parallel systems and passive parallel systems.

4.4.1 Active Parallel Systems

A parallel system is called an *active parallel system* if both the primary and duplicate components are actively in operation when the system is in operation. Both the primary and duplicate components are said to be *on-line*. If one of the components (either the primary of the duplicate component) fails, the system will still continue to operate correctly. System failure occurs only if both of the components fail. Redundancy provided by active parallel systems is called *active redundancy* (also known as operating redundancy and on-line redundancy).

There are two types of active parallel systems, namely, shared active parallel systems and pure active parallel systems.

Consider a load being held by (hung from) two identical cables. Each cable is capable of carrying the load by itself. This is an active parallel system because both cables are actively holding the load. The load is shared equally by the cables when both the cables are in their normal states. If one cable breaks, the remaining cable will have to carry the full load. Because the surviving cable is carrying the full load (instead of the half-load it was carrying before) its failure rate may increase. Such systems, where the failure rate of the surviving component increased after the failure of the companion component, are called *shared active parallel systems* or *shared parallel systems*.

If there is no change in the failure rate of the surviving component, such an active parallel system is called a *pure active parallel system* or a *pure parallel system*.

An "exact" reliability analysis of a shared parallel system has to be conducted in two stages: first, with the two components in normal states; then, with only the surviving component in normal state but with a reduced failure rate. If the change in the failure rate of the surviving component is not significant, an approximate reliability analysis may be carried out by assuming the system to be a pure parallel system with both components having the original failure rate (reduction in the failure rate of the surviving component is ignored). A conservative reliability analysis would be to analyze the system as a pure parallel system with both components having the reduced failure rate.

4.4.2 Standby Parallel Systems

Only one component functions during system operation and the other component is idle (but ready to function if the former component fails). The redundancy provided by standby parallel systems is called *standby redundancy* (also known as idle redundancy and off-line redundancy).

As the standby parallel system is put into service, only the primary component will function (operating mode or on-line mode) and the duplicate component will remain idle (standby mode or off-line mode). The duplicate component will start functioning (unless it has failed during its standby mode) if and when the primary component fails. Because the duplicate component has the same functional capabilities as the primary component, the system will continue to perform its intended function correctly. The failed component will be repaired immediately or during the next periodic maintenance and will become the duplicate component in standby mode; the other component, now in operation, will be the primary component in operating mode.

As soon as the primary component fails, the duplicate component has to be switched over from standby mode to operating mode. An automatic switching mechanism that senses the primary component failure and switches the duplicate component to operating mode may be used; or a human operator, on seeing the primary component failure (for example, an alarm sounds in the control room), manually starts the duplicate

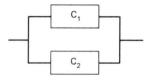

Figure 4.3. Two-component parallel system.

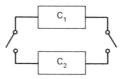

Figure 4.4. Two-component standby parallel system (component C_1 is on-line and component C_2 is off-line).

component. The failure probability of the switching system (whether automatic or manual) should be considered when computing the reliability of standby parallel systems. Switching failures are discussed in Section 4.4.3.

Some authors show a standby parallel system schematically in the same manner as an active parallel system (Figure 4.3), without explicitly showing the switching mechanism; others show the switches also in the schematic diagram (see Figure 4.4).

4.4.3 Switching Failures

Three types of switching failure are possible.

Type 1: *Failure to switch* when required.
Type 2: *Inadvertent or premature switching* (switching occurs when switching is not required).
Type 3: *Contact failure* (switch is unable to maintain the connection).

Unless the probabilities of switching failures are very small compared to component failure probabilities, switching failures should be included in the system reliability analysis. If switching failure probabilities are very small in comparison to component failure probabilities, *failure-free switching* may be assumed, that is, the probabilities of switching failures are assumed to be zero.

Chapter 5
Failure Data

5.1 INTRODUCTION

Data on hardware (component) failures, computer software errors, and human errors are essential for quantitative system reliability analysis. Because the quantitative results are only as good as the failure data used, sufficient care should be taken in the specification of the probabilities of hardware, software, and human failures. Accurate data may not be available for all failures of interest. Usually, failure probabilities are accurate in an order-of-magnitude sense only. We may still use such data as long as we remember that the system reliability results obtained from such data will also be of an approximate nature. Even such approximate results are *very* useful in system design and operation; it is better to have an approximate knowledge of the system reliability than to have no idea at all. Also, in many applications, reliability results are used to compare different designs, different operations–maintenance procedures, etc. Because absolute values of reliabilities are not used in such applications, but only a comparison of the reliabilities of the different alternatives, even approximate results are of great help. Reliability results are also used for ranking different components according to their importance to system reliability so that components may be prioritized in allocating funds for reliability improvements. Approximate reliability results are sufficient in such applications also.

Of the three types of failure (hardware, software, and human), more data is available on hardware failures. This topic is discussed in some detail in Section 5.2. Software reliability assessment is of more recent origin and is of a more specialized nature. A brief discussion of software reliability is provided in Section 5.3. Considerable work has been done in recent years on human failures. This topic is discussed briefly in Section 5.4. Hardware failure probabilities and human error probabilities are compared in Section 5.5. Documentation of failure data is the subject of Section 5.6.

5.2 HARDWARE FAILURE DATA

5.2.1 Overview

Hardware (component) reliability information is needed in quantitative system reliability analysis. Component reliability information of interest includes failure modes, failure rates, failure-on-demand probabilities, the time to carry out the repair procedure, and test–maintenance durations.

The quantity referred to as the time to carry out the repair procedure is different from the repair time defined in Section 3.4.3: Repair time includes, in addition to the time taken to carry out the repair procedure, the time required for detecting the failure (detection time), and the time interval between detection and the start of the repair procedure (see Section 3.4.3). We consider the time to carry out the repair procedure instead of the repair time because the detection time and the time interval between detection and start of repair could be very much dependent on the system configuration, accessibility of the component, on-site repair facilities, etc.

Failure rate discussions in this chapter assume a constant failure rate because it is widely used in system reliability analysis.

Component failure rates and component failure probabilities may be estimated through reliability analysis, reliability testing, or by a statistical analysis of historical failure data. The time to carry out the repair procedure and test–maintenance durations are determined from operational experience. Prediction of component failure rates (or failure probabilities) through reliability analysis is a "bottoms-up" approach. The probability of component failure is computed as a function of the probability distributions of component parameters such as material properties, component dimensions, and operating conditions. Another method of failure probability prediction is through testing a number of components to failure. The third approach is through a statistical analysis of historical failure data. We will restrict our attention to the third approach because it is more widely used by engineers involved in system reliability analysis. The first two approaches are usually used by component manufacturers, who may transmit their results to system designers. Readers interested in predicting component reliabilities through a bottoms-up analysis are referred to Haugen (1980) and Sundararajan (1986). Information on component reliability testing may be found in Mann, Schafer, and Singhpurwalla (1974).

Statistical analysis of historical failure data is the most direct and straightforward approach of estimating component reliabilities. An advantage of this approach is that reliability parameters are obtained from the actual performance of components under actual operating conditions.

Drawbacks of this approach are that (i) sufficient historical (operational) data may not be available, (ii) available data may not pertain to a specific pedigree[1] of component but to components of different manufacture, design, size, and ratings, and (iii) the available data may pertain to different applications under different operating conditions. So component reliability parameters derived from operating histories may not be very accurate. As a general rule, failure rates and probabilities derived from operational experience from a wide spectrum of applications and involving components of different pedigrees are *accurate in an order-of-magnitude sense only*.

If sufficient operational (historical) data pertaining to the component of desired pedigree is available, it may provide more accurate estimates of component reliabilities. However, in many situations, the available pedigree data is very limited and the estimation of component reliabilities from such a limited amount of data may be accurate in an order-of-magnitude sense only.

Failure data may be obtained from a single plant or from a number of plants from a wide spectrum of industries. The former is called *plant-specific data* and the latter, *generic data*. Collection and use of both types of data are discussed in the following subsections.

5.2.2 Types of Component Failure

In the context of collecting failure data, component failures may be classified as *primary failures* and *secondary failures*.

5.2.2.1 Primary failures.
Primary failures are the result of a deficiency in the component, and component failure occurs when the operating and environmental conditions are within design limits. Rupture of a turbine blade when it is operating within the design speed is an example of a primary failure.

5.2.2.2 Secondary failures.
Secondary failures are the result of secondary causes that produce operating and/or environmental conditions that are outside the design limits of the component. There may be no deficiency in the failed component itself and it might not have failed at that time if the operating and/or environmental conditions had been within design limits.

---

[1]The pedigree of a component refers to its specific characteristics (size, manufacture, design, etc.). Components manufactured by the same manufacturer under the same specifications are called components of the same pedigree.

Operating and/or environmental conditions that cause secondary failures include abnormal temperatures, pressures, loads, speeds, vibrations, electric currents, moisture, dust, and chemical corrosion.

Occurrence of secondary causes may not always result in component failure; the component may fail in only a certain percentage of the times a secondary cause occurs. For example, an increase in temperature well above the design limit may cause the failure of a component only 60% of the time, that is, the conditional probability of component failure given that there is an abnormal increase in temperature is 0.6.

Secondary failures may be divided into four categories: (i) common cause failures; (ii) propagating failures; (iii) human error failures; (iv) miscellaneous secondary failures.

(i) *Common Cause Failures:* A secondary failure may be classified as a common cause failure if the secondary cause induces failure in more than one component. For example, an earthquake could produce severe loads on a number of components and thus induce their failures. Common causes include earthquakes, floods, hurricanes, explosions, and fires. Malfunction of other systems or components could also induce failures in a number of components, for example, failure of an air-conditioning system could increase the temperature and thus be the cause of failure of a number of electronic components. Some additional discussion on common cause failures may be found in Section 5.2.9. A detailed discussion of common cause failures within the context of system reliability analysis is provided in Chapter 11.

(ii) *Propagating Failures:* The failure of one component may lead to the failure of one or more additional components. Such failures are called propagating failures. Here, the component failure that induces the other component failures is the secondary cause. If a component failure (the secondary cause) leads to failures of more than one component, such a secondary cause may be treated as a common cause failure.

(iii) *Human Error Failures:* If component failures result from human errors in operation, maintenance, or testing, such human errors could be treated as secondary causes. Component failures due to human errors in design, manufacture, and installation are considered to be primary failures and shall not be included here. If a number of components fail due to the same human error, such failures may be treated as common cause failures.

(iv) *Miscellaneous Secondary Failures:* Secondary failures that do not fall under any of the preceding three categories are grouped here. For example, if only one component in the system is susceptible to earthquake-induced failures, such a failure may not be treated as a common cause failure but may be included under this fourth category.

5.2.3 Statistical Dependence between Failures

If a number of components fail due to a common cause, naturally there is a statistical dependence between those component failures. Propagating failures are also statistically dependent.

There could be statistical dependencies between other types of failure also. Statistical dependence between primary failures of a number of components may arise because of similar design, similar manufacture, or similar installation of those components.

5.2.4 Plant-Specific Data Bases

Failure data from a single plant are called plant-specific data and a data base containing such data is called a *plant-specific data base*. If the plant has just started operation, an ongoing reliability data base should be established that will be updated continually as new failure data become available. If the plant already has been operating for many years, we would (i) establish an ongoing reliability data base from now on and, if possible, (ii) derive the past failure data from past maintenance records, test reports, operator logs, etc.

A good reliability data base should contain the following information for each observed failure:

1. Identification of the component
2. Date of failure
3. Failure mode
4. Failure cause (*)
5. Time to carry out the repair procedure
6. Detection time (*)
7. Time to get spare parts (*)
8. Corrective actions (*)
9. Effect of component failure at the systems level
10. Statistical dependence, if any, with other failures (*)

(The items marked by asterisks (*) are not necessary for reliability computations but may be useful in future design improvements and maintenance–test procedures.)

The preceding information is maintained continually either as a written record or in a computerized data base. At regular intervals (say, every year) the information is summarized in a form similar to Table 5.1.

Table 5.1. Failure Data Summary (plant-specific data base)

COMPONENT	NUMBER OF COMPONENTS (N)	FAILURE MODEL	1989	1990	1991		CUMULATIVE VALUE	FAILURE RATE (FAILURES PER HOUR)	PROBABILITY OF FAILURE ON DEMAND (FAILURES PER DEMAND)	AVERAGE T_R (HOURS)	MTTR (HOURS)
		total number of failures									
		total operating hours									
		total number of demands									
		total T_R (hours)									
		total repair time (hours)									
		total number of failures									
		total operating hours									
		total number of demands									
		total T_R (hours)									
		total repair time (hours)									

T_R = time to carry out the repair procedure (as defined in Section 5.2.1).

The following notes will be useful in preparing Table 5.1.

1. The word "total" indicates that the value entered for the number of failures, operating hours, etc., is the sum total for all N components. For example, if $N = 3$ and during 1989 the first component had two failures, the second component had three failures, and the third component had two failures, we enter 7 as the total number of failures during 1989.
2. If the failure mode relates to a failure on demand, we enter the total number of demands; otherwise it is left blank.
3. Many reliability engineers tend not to record T_R values in the data base; they record only the repair time.
4. The column headed "cumulative value" is the sum total of the number of failures, operating hours, etc., over the years. For example, if the total number of failures for 1989, 1990, 1991, and 1992 are 27, 24, 28, and 31, respectively, the corresponding cumulative value is 110.
5. If the failure mode relates to a failure on demand, the probability of failure on demand is computed and entered in the appropriate column and the failure rate is left blank. Otherwise, the failure rate is computed and entered in the appropriate column and the probability of failure on demand is left blank.
6. The failure rate is obtained by dividing the cumulative number of failures by the cumulative operating hours.
7. The probability of failure on demand is obtained by dividing the cumulative number of failures by the cumulative number of demands.
8. The average of T_R is obtained by dividing the cumulative T_R by the cumulative number of failures.
9. MTTR is obtained by dividing the cumulative repair time by the cumulative number of failures.

It is advantageous to list each and every component separately. That is, if there are six 8-inch valves (valve 1-1, valve 1-2, valve 2-1, valve 3-1, valve 3-2, and valve 3-3), list each one separately and track its failure history separately, rather than grouping all six valves as 8-inch valves. Separate tracking will identify any abnormal behavior of any one valve. For example, if one of the six valves needed repaired eight times during a 2-year period and the other five valves required repair only once during that same period, then the first valve has some inherent defect and it is best replaced or thoroughly checked. If the performance of all six valves is tracked together, this fact may not emerge from the reliability record.

Tracking individual valves may not be advisable if there are a large number of valves in the plant (say 50 8-inch valves). The decision to track the performance individually or as a group is best made on a case-by-case basis.

Failure and repair data may be tracked for each different failure mode or for all failure modes of a component combined. Again, it is advantageous to track by each failure mode but this may not always be practical.

As stated in Section 3.3.10.4, failure rates during standby mode could be different from those during operating mode. Failure rates during standby and operating modes may be computed separately if failures are tracked separately for standby and operating modes.

Example 5.1

Failure data were collected for five similar components in a power plant. The first component was tracked for 2800 hours in standby mode and 400 hours in operation; the second component was tracked for 300 hours in standby mode and 2500 hours in operation; the third component was tracked for 2700 hours in standby mode and 700 hours in operation; the fourth component was tracked for 2500 hours in standby mode and 800 hours in operation; the fifth component was tracked for 0 hours in standby mode and 3100 hours in operation. A total of 7 failures were observed in standby mode and 12 failures in operating mode. Calculate the component failure rates.

Solution

$$\text{Total standby hours} = 2800 + 300 + 2700 + 2500 + 0$$
$$= 8300 \text{ hours}$$
$$\text{number of failures during standby mode} = 7$$
$$\text{failure rate in standby mode} = \frac{7}{8300}$$
$$= 0.00084 \text{ failures per hour}$$
$$\text{total operating hours} = 400 + 2500 + 700 + 800 + 3100$$
$$= 7500 \text{ hours}$$
$$\text{number of failures during operating mode} = 12$$
$$\text{failure rate in operating mode} = \frac{12}{7500}$$
$$= 0.0016 \text{ failures per hour}$$
$$\text{overall failure rate (standby and operating modes)}$$
$$= \frac{7 + 12}{8300 + 7500}$$
$$= 0.0012 \text{ failures per hour}$$

average failure rate (standby and operating modes)

$$= \frac{0.00084 + 0.0016}{2}$$

$$= 0.0012 \text{ failures per hour}$$

Note that the overall failure rate and the average failure rate are not necessarily identical. If we are interested in using a single failure rate (without considering standby and operating modes separately), the former is a better value to use. If data necessary to compute the former are unavailable, the latter may be used. (*Note:* Some authors use the term "average failure rate" to refer to what we have called here the overall failure rate.)

As discussed in Section 5.2.2, component failures may be divided into primary and secondary failures, and the secondary failures may be divided into common cause failures, propagating failures, human error failures and miscellaneous secondary failures. Failure rates and failure-on-demand probabilities may be calculated separately for each type of failure if sufficient failure data are available. This may not always be possible because of a lack of sufficient data for each type of failure.

Example 5.2

Failure data of four similar components in a chemical plant are tracked for 2500 hours each. A total of 40 failures are observed. Of these 40 failures, 25 are primary failures and 15 are secondary failures. Of the 15 secondary failures, 6 are common cause failures, 1 is a propagating failure, and 8 are human error failures. Compute the failure rates.

Solution

Total hours $= 4 \times 2500 = 10,000$ hours

total number of failures $= 40$

overall failure rate (all types of failures) $= \frac{40}{10,000}$

$$= 0.004 \text{ failures per hour}$$

number of primary failures $= 25$

failure rate for primary failures $= \frac{25}{10,000}$

$$= 0.0025 \text{ failures per hour}$$

number of secondary failures $= 15$

failure rate for secondary failures $= \frac{15}{10,000}$

$$= 0.0015 \text{ failures per hour}$$

number of common cause failures $= 6$

failure rate for common cause failures $= \frac{6}{10,000}$

$$= 0.0006 \text{ failures per hour}$$

number of propagating failures $= 1$

failure rate for propagating failures $= \frac{1}{10,000}$

$$= 0.0001 \text{ failures per hour}$$

number of human error failures $= 8$

failure rate for human error failures $= \frac{8}{10,000}$

$$= 0.0008 \text{ failures per hour}$$

It should be remembered that, given a fixed duration of operating hours, the accuracy of the computed failure rate increases with the number of failures during that period. In this example, the failure rate for primary failures is much more accurate than the failure rate for propagating failures.

Statistical dependence between component failures is also of importance. Assuming statistical independence between component failures when there is in fact a significant dependence could result in significant errors in the computed system reliability.

If N failures of component A and n failures of component B are observed during a period of T hours and if n' of the component-B failures occurred simultaneously with n' of the component-A failures, then we may say that there is statistical dependence between component-A and component-B failures. If n' is very small compared to n and N, the simultaneous occurrences may be attributed to mere coincidence and statistical independence may be assumed. Otherwise, the conditional failure probabilities of the components should be calculated and included in the system reliability analysis.

Example 5.3

One hundred failures of component A and 80 failures of component B were observed during a period of 10,000 hours. Among the 80 component-B failures, 30 occurred simultaneously with component-A failures. Compute the conditional probability of component-B failure given that component A has failed. Also calculate the conditional probability of component-A failure given that component B has failed.

Solution

Conditional probability of component-B failure given that component A has failed

$$= P[B|A] = \tfrac{30}{100} = 0.3$$

conditional probability of component-A failure given that component B has failed

$$= P[A|B] = \tfrac{30}{80} = 0.375$$

A summary log or a computerized data base for periodic test–maintenance outages may also be prepared in the same way as the failure record. A sample periodic test–maintenance summary is shown in Table 5.2.

It should be noted that sometimes the system itself is shut down for periodic maintenance and all the components in the system are tested and maintained at the same time. System unavailability due to such systemwide testing and maintenance is computed as the ratio

$$\frac{\text{duration of system shutdown for periodic maintenance}}{\text{operating hours between periodic maintenances}}$$

Table 5.2 is not related to such system shutdown and maintenance.

There is another type of component maintenance in which the system is not shut down: Components are tested and maintained when the system is operating (this is possible only if the component is a redundant component). Table 5.2 deals with such component maintenance in which the system is operating and only the component is unavailable due to the maintenance. The component unavailability due to such test–maintenance is computed from Table 5.2. This test–maintenance unavailability is used in quantitative fault tree analysis (see Section 8.11).

The following notes will be useful in preparing Table 5.2.

1. The word "total" indicates that the value entered for the number of periodic test–maintenances, operating hours, etc., is the sum total for all N components. For example, if $N = 3$ and during 1989 the first component was maintained twice, the second component was maintained twice, and the third component was maintained twice, we enter 6 as the total number of periodic test–maintenances during 1989.

2. The column headed "cumulative value" is the sum total of the number of periodic test–maintenances, operating hours, etc., over the years. If the total number of periodic test–maintenances during 1989, 1990, 1991, and 1992 are 58, 58, 58, and 58, respectively, the cumulative value is 232.

Table 5.2. Periodic Test – Maintenance Summary (plant-specific data base)

COMPONENT	NUMBER OF COMPONENTS (N)	total number of periodic test–maintenances	total operating hours	total component outage hours due to periodic test–maintenance	total number of periodic test–maintenances	total operating hours	total component outage hours due to periodic test–maintenance	total number of periodic test–maintenances	total operating hours	total component outage hours due to periodic test–maintenance	1989	1990	1991					YEAR	CUMULATIVE VALUE	UNAVAILABILITY DUE TO TESTING AND MAINTENANCE

3. Unavailability due to testing and maintenance is obtained by dividing cumulative component outage hours due to periodic test–maintenance by the cumulative operating hours.

Data from plant-specific data bases are more reliable because such data bases may have more details about operating conditions, failure modes, failure causes, and corrective actions. Information about the component pedigree is also readily available. The shortcoming of plant-specific data bases is that the amount of failure data is usually very limited unless the plant has been operating for many years or there are a large number of similar components in the plant. If the failure data for a component consist of very few or no failures, we can have very little confidence in the computed failure rate or failure-on-demand probability. Generally, the more the number of failures in the data base, the more confidence we can have in the computed failure rates and probabilities.

The reliability analyst will turn to generic data bases if the number of failures in the plant-specific data base is very low (insufficient to provide reasonably accurate failure statistics). Failure rates and probabilities from generic data bases may be used to supplement plant-specific data bases or used in lieu of plant-specific data bases.

5.2.5 Generic Data Bases

Appendix II gives a list of generic data bases (generic data sources).

Generic data bases contain failure information from nationwide or worldwide experience in a particular industry or a variety of industries. For example, the Nuclear Plant Reliability Data System (NPRDS) contains information exclusively from nuclear power plant operations in the United States. The *Military Standardization Handbook* [Department of Defense (1974)] contains information on the failure statistics of a variety of components used in the defense industry but the statistics are based on operational experience from both military and civilian applications. Although most data bases contain failure data on a variety of components, there are some data bases that are specific to a few selected components. For example, the Pressure Vessel and Piping Data Base [Bush (1985)] contains information on failures of pressure vessels and piping only.

We may classify generic data bases into four groups, depending on the type of data they contain:

1. Raw data
2. Primary, consolidated data
3. Secondary, consolidated data
4. Expert opinion data

5.2.5.1 Raw data bases. These data bases provide the number of failures and the total number of operating hours or the total number of demands for different types of components (valves, switches, pumps, motors, etc.); most such data bases provide the information on a year-by-year basis. Users can compute the failure statistics from this raw information. Some raw data sources also include details on failure modes, corrective actions, and repair time.

Methods of calculating the reliability parameters (failure rates, failure-on-demand probabilities, etc.) are essentially the same as those discussed for plant-specific data bases in Section 5.2.4.

The NPRDS is an example of a raw data base (see Appendix II).

5.2.5.2 Primary, consolidated data bases. Primary, consolidated data bases provide failure rates and failure-on-demand probabilities that are derived from failure data compiled from different plants over a period of many years. Some primary, consolidated data bases are limited to a single industry whereas others are based on data from a variety of industries.

The failure statistics are provided either as a *point estimate* or as a *range* and a point estimate. The point estimate is usually the mean or median, and the range is provided as an upper and a lower bound. The upper and lower bounds are usually (not always) the 90% *confidence interval*. A confidence interval is also called a *confidence range*, *confidence band*, or *probability range*.

Let us consider an example. The mean value of the failure rate of a component is 9×10^{-5} failures per hour and the 90% confidence interval (lower and upper bounds) is 1×10^{-5} and 7×10^{-4} failures per hour. What do we mean by "the 90% confidence interval"?

Failure rates and other reliability parameters are not known exactly. If we compute the failure rate of a component (say, a pump) from the actual operational history of such components in plant 1 and plant 2, the computed failure rate from plant-1 data will seldom be identical to the computed failure rate value from plant-2 data. If we compute the failure rates separately from operational data from plants $1, 2, \ldots, N$, the N values will differ from each other. Suppose we are interested in the failure rate of the component (pump) for the reliability analysis of a plant we are building. Which of the N failure rates shall we use? The mean or median of these N values provides a point estimate for the failure rate, but how good (or accurate) is this value? How close is this point estimate to the actual value? The confidence interval provides some additional information. In the preceding example, the 90% confidence interval is 1×10^{-5} and 7×10^{-4} failures per hour. It means that there is a 90% probability (that is, there is a 90% chance) that the actual value of the failure rate lies between 1×10^{-5} and 7×10^{-4} failures per hour; it means that there is a

5% probability that the actual value will be less than 1×10^{-5} failures per hour and there is a 5% probability that the actual value will be higher than 7×10^{-4} failures per hour. Thus the confidence interval (if available) provides us some idea about the accuracy of the data.

Confidence intervals for component reliability parameters are needed if we are interested in a quantitative uncertainty analysis of the system reliability (see Section 10.8). As will be noted in Section 10.8, uncertainty analysis of system reliability is performed only in very critical projects such as nuclear power plants. In most other projects, uncertainty analyses are seldom carried out, so confidence intervals for component reliability parameters are not needed. Even in such projects, confidence intervals, if available, give us some idea about the accuracy of component failure data.

Readers interested in additional information on confidence intervals are referred to the *Reactor Safety Study* [Nuclear Regulatory Commission (1975)].

FARADA is an example of a primary, consolidated data base (see Appendix II).

5.2.5.3 Secondary, consolidated data bases.   Secondary, consolidated data bases combine the failure rates on failure-on-demand probabilities contained in a number of primary, consolidated data bases and provide users with point estimates and ranges. The method of combination (consolidation) is discussed in Section 5.2.6.

The data summarized in the *Reactor Safety Study* [Nuclear Regulatory Commission (1975)] is an example of a secondary, consolidated data base.

5.2.5.4 Expert opinion data bases.   Expert opinion data bases contain failure rates and failure-on-demand probabilities estimated by experts. The experts provide their opinion on the basis of their experience with the operation and failure of components in various applications and any available actual failure data. Thus the failure rates and failure-on-demand probabilities are based on a combination of available actual data and the experts' judgement.

The IEEE Data [Institute of Electrical and Electronics Engineers (1977)] is an example of an expert opinion data base.

5.2.5.5 Which data base is the best?   The choice of data bases depends on the requirements of the reliability analysis project and budget. Raw data bases usually contain more information about each component and failure (type of system or plant in which the component is used, operating and environmental conditions, failure mode, failure cause, corrective action, etc.) than consolidated data bases. Such information is very useful to a reliability analyst. The shortcoming is that the user has to go through the year-by-year record for each component, consolidate them, and compute the failure statistics. This could be a very expensive and time-consuming effort.

Primary, consolidated data bases are excellent if (i) they are based on a sufficiently large number of component failures to give reasonably accurate failure statistics and (ii) the failure information is from plants and applications that are similar to the plant or application for which the reliability analysis is to be carried out.

Secondary, consolidated data bases have the advantage that they are based on a large amount of data from a number of primary, consolidated data bases. However, the failure information is from so many diverse industries that it may not be closely related to the project for which the reliability analysis is performed. Most secondary data bases provide upper- and lower-bound values, and the user does get enough of an idea about the spread in the failure rates to make a decision whether to use the secondary data base or a more appropriate primary data base. A greater spread between the upper and lower bounds may mean that the data used are from a wide variety of applications.

Expert opinion data bases are only as good as the experience and judgement of the experts contributing to the data base. Expert opinion data bases could be an excellent data source if the experts estimate the failure rates not only on the basis of their personal experience but also on the basis of a reasonable amount of actual failure data. Such data bases could be even better than primary and secondary consolidated data bases.

There are no specific criteria to decide which one of the data bases to use. The decision has to be made by the reliability analyst on a case-by-case basis.

5.2.6 Combination of Data Bases

A plant-specific data base (if available) or a primary or a secondary, consolidated data base is used in most reliability projects. In major reliability analysis projects where sufficient funds are available to derive component reliabilities from a number of available data bases, including primary, secondary, raw, and plant-specific data bases, reliability parameters from each applicable data base are combined to derive more precise reliability parameters. One approach of combining the information from a number of data bases is provided here.

Let λ_i be the point estimate of the failure rate from the ith data base, where $i = 1, 2, \ldots, n$, in which n is the number of data bases. We can assign different weighting factors to the different data bases depending on the following:

1. The extent of data from which the failure rate is derived (the larger the amount of data, the higher is the weighting factor).
2. The relevance of the data from which the failure rate is derived (whether the data are from plants or applications that are similar to the present project; the closer the similarity, the higher is the weighting factor).

If sufficient information is not available to assign different weighting factors to the different data bases, all weighting factors may be set to unity.

Let W_i be the weighting factor for the ith data base. The mean failure rate is given by

$$\lambda = \frac{\sum_{i=1}^{n}(W_i \lambda_i)}{\sum_{i=1}^{n} W_i} \tag{5.1}$$

This mean value is more specifically referred to as the *arithmetic mean* to differentiate it from the *geometric mean*, which is discussed later. Usually, the term "mean" refers to the arithmetic mean.

Equation (5.1) can be used not only for failure rates but also for failure-on-demand probabilities, MTTR, and test–maintenance unavailabilities.

A geometric averaging technique is sometimes used for failure rates, failure-on-demand probabilities, and test–maintenance unavailabilities. The geometric mean of the failure rate is given by

$$\lambda = \left[\prod_{i=1}^{n} (\lambda_i)^{W_i} \right]^{1/W} \tag{5.2}$$

where $W = \sum_{i=1}^{n} W_i$. The notation $\prod$ denotes the "product symbol," which is defined as

$$\prod_{i=1}^{n} a_i = a_1 a_2 \cdots a_n$$

Some analysts prefer Equation (5.2) to Equation (5.1), but both are equally applicable.

A median value for the reliability parameters (failure rates, failure-on-demand probabilities, MTTR, and test–maintenance unavailabilities) may also be computed.

The following procedure is used if the weights of all the data bases are equal: The data bases are arranged such that their failure rates are in ascending order. Let there be n data bases under consideration. If n is even, then the mean failure rate is the average of the failure rates of the

$(n/2)$th and the $[(n/2) + 1]$th data bases. If n is odd, then the mean failure rate is the failure rate of the $[(n + 1)/2]$th data base. This procedure is applicable also for failure-on-demand probabilities, MTTR, and test–maintenance unavailabilities.

If the weights are not equal for all the data bases, the median failure rate is chosen such that the failure rates of the data bases with 50% of the total weight fall below the median value or the failure rates of the data bases with 50% of the total weight fall above the median value. This procedure is also applicable for failure-on-demand probabilities, MTTR, and test–maintenance unavailabilities. The procedure is illustrated in Example 5.5.

Which failure rate (mean or median) shall we use in a system reliability analysis? The median value seems to be a more appropriate choice unless there is a special reason to use the arithmetic or geometric mean [Nuclear Regulatory Commission (1975)]. Alternatively, the mean and median values may be calculated and the higher value may be used; this is a conservative approach.

Example 5.4

Failure-on-demand probabilities (P) of a component are collected from five data bases (see Table 5.3). Each data base is assumed to have unit weight. Calculate the arithmetic mean, geometric mean, and the median of the failure-on-demand probability.

Solution

First we rearrange the data bases in ascending order of the failure-on-demand probabilities (see Table 5.4).

$$\text{Total weight} = W = 5$$

$$\text{arithmetic mean for } P = \frac{\left\{\begin{array}{c}(5 \times 10^{-4}) + (8 \times 10^{-4}) + (1 \times 10^{-3}) \\ + (2 \times 10^{-3}) + (3 \times 10^{-3})\end{array}\right\}}{5}$$

$$= 1.46 \times 10^{-3} \text{ failures per demand}$$

Table 5.3.   Data for Example 5.4

	DATA BASE				
	A	B	C	D	E
P	3×10^{-3}	1×10^{-3}	5×10^{-4}	8×10^{-4}	2×10^{-3}

Table 5.4.   Failure Probabilities in Ascending Order
(Example 5.4)

SERIAL NUMBER	DATA BASE	P
1	C	5×10^{-4}
2	D	8×10^{-4}
3	B	1×10^{-3}
4	E	2×10^{-3}
5	A	3×10^{-3}

$$\text{geometric mean for } P = \left[(5 \times 10^{-4})(8 \times 10^{-4})(1 \times 10^{-3}) \right.$$
$$\left. \times (2 \times 10^{-3})(3 \times 10^{-3})\right]^{0.2}$$
$$= 1.19 \times 10^{-3} \text{ failures per demand}$$

$$\text{number of data bases} = n = 5$$
$$\frac{n+1}{2} = 3$$

So the median value is the P value of the third data base:

$$\text{median value of } P = 1 \times 10^{-3} \text{ failures per demand}$$

Example 5.5

Redo Example 5.4 with the weights as shown in Table 5.5.

Solution

First we rearrange the data bases in ascending order of the failure-on-demand probabilities (see Table 5.6).

$$\text{Total weight} = W = 2 + 1 + 3 + 4 + 2 = 12$$

arithmetic mean for P

$$= \frac{\left\{ \begin{array}{c} (2 \times 5 \times 10^{-4}) + (1 \times 8 \times 10^{-4}) + (3 \times 1 \times 10^{-3}) \\ + (4 \times 2 \times 10^{-3}) + (2 \times 3 \times 10^{-3}) \end{array} \right\}}{12}$$

$$= 1.57 \times 10^{-3} \text{ failures per demand}$$

geometric mean for P

$$= \left[(5 \times 10^{-4})^2 (8 \times 10^{-4})^1 (1 \times 10^{-3})^3 (2 \times 10^{-3})^4 (3 \times 10^{-3})^2 \right]^{1/2}$$

$$= 1.32 \times 10^{-3} \text{ failures per demand}$$

$$\text{half the total weight} = \frac{W}{2} = 6$$

Table 5.5.   Data for Example 5.5

| | DATA BASE | | | | |
	A	B	C	D	E
weight	2	3	2	1	4

Table 5.6.   Failure Probabilities in Ascending Order
(Example 5.5)

SERIAL NUMBER	DATA BASE	WEIGHT	P
1	C	2	5×10^{-4}
2	D	1	8×10^{-4}
3	B	3	1×10^{-3}
4	E	4	2×10^{-3}
5	A	2	3×10^{-3}

So a weight of 6 should fall below the median value of P. Weights of the first, second, and third data bases add to 6. Weights of the fourth and fifth data bases add to 6. So the median value should lie between the P values of the third and the fourth data bases. We may take the median value as the average of the P values from the third and fourth data bases. So,

$$\text{median value of } P = \frac{(1 \times 10^{-3}) + (2 \times 10^{-3})}{2}$$

$$= 1.5 \times 10^{-3} \text{ failures per demand}$$

Example 5.6

Calculate the median value for P for Example 5.5 with just one change: Weights are changed as shown in Table 5.7.

Solution

First we rearrange the data bases in ascending order of the failure-on-demand probabilities (see Table 5.8).

$$\text{Total weight} = 2 + 2 + 3 + 4 + 2 = 13$$

$$\text{half the total weight} = \frac{W}{2} = 6.5$$

Table 5.7.   Data for Example 5.6

| | DATA BASE | | | | |
	A	B	C	D	E
weight	2	3	2	2	4

Table 5.8.   Failure Probabilities in Ascending Order
(Example 5.6)

SERIAL NUMBER	DATA BASE	WEIGHT	P
1	C	2	5×10^{-4}
2	D	2	8×10^{-4}
3	B	3	1×10^{-3}
4	E	4	2×10^{-3}
5	A	2	3×10^{-3}

So a weight of 6.5 should fall below the median value of P. Weights of the first, second, and third data bases add to 7. Weights of the fourth and fifth data bases add to 6. There is no clear-cut value for the median. We may take the P value of the third data base as the median. So,

$$\text{median value of } P = 1 \times 10^{-3} \text{ failures per demand}$$

If needed, we may also determine confidence intervals for the reliability parameters. Because confidence intervals are needed only in some very critical projects (as discussed in Section (5.2.5.2), we shall not discuss them here. Interested readers are referred to the *Reactor Safety Study* [Nuclear Regulatory Commission (1975)].

5.2.7 Use of Generic Data Bases

"Let the user beware" is the keyword in using generic data bases. Some cautionary notes about using generic bases are given here.

1. Most generic data bases provide details about the sources of failure data, method of statistical analysis, and limitations of the data base. Users must read these details before using the data base.
2. Some data bases identify the failure modes and others do not. The latter type of data base may not be of much use if we are computing failure rates for a specific failure mode.
3. Many data bases do not differentiate between independent failures, dependent failures, primary failures, secondary failures, and common cause failures. This could be a source of error.
4. Component failures due to human errors should not be included in the failure count, but some data bases do include such failures.
5. Failures during test and maintenance should not be included in the failure count but some data bases do include such failures.

6. If we are assuming a constant failure rate, failures during break-in and wear-out periods should not be included in the failure count but some data bases do include such failures.

7. Failure data used in generic data bases are reported by different personnel and organizations, so the accuracy of the reported failure data may not be consistent (inaccurate operating hours, missing failures, etc.)

8. Failure rates could be significantly affected by the design code used, the quality control during manufacturing, and the test–maintenance procedures and schedules. Such information may not be found in the data bases. Especially if just one or two generic data bases are used, it is best to use data bases pertaining to an industry or application similar to the project under investigation. For example, components used in nuclear power plants or high-performance aerospace systems are designed, manufactured, tested, and maintained to a high standard. Failure rates of such components may be lower than those of components used in other industries.

9. Advances in technology may result in improved component reliabilities, so failure rates from older classes of components may be somewhat higher.

10. Compare the failure rates from one data base to another, and check for any unusual values. If the failure rate from a data base is significantly different from the other data bases, see if there is any specific reason for such a deviation; discard that failure rate if the difference cannot be explained satisfactorily.

The purpose of listing these cautionary notes is *not* to warn reliability analysts against using generic data bases but simply to caution them against using generic data bases indiscriminately.

The following extract from the *Military Standardization Handbook* [Department of Defense (1974)] is informatory. These remarks are applicable to most data bases for mechanical, electrical, and electronic components.

The use of failure rate data, obtained from field use of past systems, is applicable in future concepts depending on the degrees of similarity existing both in the hardware design and in the anticipated environments. Data obtained on a system used in one environment may not be applicable to use in a different environment, especially if the new environment substantially exceeds the design capabilities. Other variants that can affect the stated failure rate of a given system are: different uses, different operators, different maintenance practices, different measurement techniques or definitions of failure.

Another cautionary note from the *IEEE Guide* [Institute of Electrical and Electronic Engineers (1977)] is also a useful reminder about the limitations of generic data bases.

The published failure data values given in the manual are for items used under the conditions discussed in individual chapter prefaces. They are intended to apply only to such use. The user of these values should take appropriate precautions while applying these data under different conditions.

5.2.8 Correction Factors

As emphasized in previous sections, the failure statistics presented in most generic data bases are based on failure data from a variety of applications, and the actual operating and environmental conditions, such as temperature, humidity, and mechanical stress, may not be listed. If we are interested in the reliability of a component that is expected to operate in a particularly harsh environment, we may multiply the generic failure statistics by a correction factor.

The correction factors may be estimated through theoretical models in some cases. We present here two such models as examples.

Levenbach (1957) suggests a power-rule model to account for the effect of electric voltage on the mean life of certain dielectric capacitors:

$$\theta_i = \frac{C}{V_i^p} \tag{5.3}$$

where θ_i is the mean time to failure at a voltage of V_i. The empirical parameters C and p can be estimated if θ_i is known for different values of V_i.

Pershing and Hollingsworth (1964) recommend the following exponential rule to consider the effects of temperature on semiconductor materials:

$$\lambda_i = \exp\left(A - \frac{B}{T_i} \right) \tag{5.4}$$

where λ_i is the failure rate at a temperature T_i. A and B are empirical parameters that can be determined if we know λ_i for different values of T_i.

These rules are applicable only to the specific types of components for which they are derived and should not be applied to other types of components. Also, in order to use these relationships, we do need suffi-cient pairs of (θ_i, V_i) or (λ_i, T_i) values to estimate the empirical coeffi-cients. Such data may not always be available and component testing may be required. Also, theoretical models such as Equations (5.3) and (5.4) are not available for many types of component.

Faced with such difficulties, many system reliability analysis project managers turn to expert opinion for correction factors. Expert opinion is only as good as the knowledge of the experts about the failure behavior of the component. Because the failure rates and failure-on-demand probabil-ities from generic data bases are accurate only in an order-of-magnitude sense, we do not need very accurate correction factors either.

The *IEEE Guide* [Institute of Electrical and Electronics Engineers (1977)] provides environmental factors (correction factors) for some elec-trical and electronic components. For example, it is recommended there that failure rates for electric valve operators and actuators be multiplied by a factor of 1.57 if the environmental temperature is very high; a factor of 2.04 is recommended if the humidity is very high.

Failure rates may be reduced by a suitable correction factor if the design procedure used is very conservative or if very stringent quality control is used during manufacturing and testing. Expert opinion may be required to estimate the correction factor.

It is wise not to rely on a single expert (for example, a knowledgeable engineer, operator, or maintenance–test person); seek the opinion of at least a few experts. The experts' estimates of correction factors are averaged or weighted-averaged; Equation (5.1) or (5.2) may be used. Weighting factors are assigned on the basis of each expert's relative experience and knowledge.

Another method of consolidating expert opinion is through a round table discussion. Each expert presents his or her estimate and the justifi-cation for that estimate. The experts discuss the pool of estimates and arrive at a consensus estimate. One potential problem with this approach is that one or two individuals could dominate the discussion, impose their opinions on the others, and thus unduly influence the final consensus estimate.

The Delphi method may also be used for processing expert opinion. Interested readers are referred to Linstone and Turoff (1975) for informa-tion on the Delphi method.

5.2.9 Common Cause Failures

Section 5.2.2 discusses some basic concepts relating to common cause failures. Here we discuss the calculation of some basic probabilities relating to common cause failures. Section 10.3.9 and Chapter 11 present more detailed discussions on how to include the effects of common cause failures in system reliability analysis.

The following computational procedure is based on the *Probabilistic Risk Assessment Guide* published by the Nuclear Regulatory Commission (1983). These computations may be carried out if the necessary data are available.

Let us denote the common cause by Y. The following notation is used:

c = occurrence rate of the common cause Y

p = conditional probability that a specific component (for example, the first 8-inch valve or the 12th 8-inch valve, etc.) fails, given that the common cause Y has occurred

λ = failure rate of the components (for example, the 8-inch valves) due to causes other than the common cause Y

n = number of identical components in the system that are simultaneously susceptible to the common cause Y (for example, if fourteen 8-inch valves are simultaneously susceptible to Y, then $n = 14$); all n components should have the same p and λ

We calculate the basic probability parameters c, p, and λ from available failure data. Computation of λ is discussed in Section 5.2.4.

The rate of common cause occurrence, c, may be calculated from either historical data or deductive procedures. If the common cause is internal to the plant (for example, a fire, or an air-conditioner failure), historical data from the plant or from similar plants may be used to estimate c. Alternatively, deductive procedures such as fault tree analysis may be used to calculate the probability of the fire or air-conditioner failure. (Fault tree analysis is a system reliability analysis technique and is the subject of Chapters 8–10.) Occurrence rates c of external common causes, such as earthquakes, tornados, and floods, may also be calculated from historical data or through deductive procedures. Historical data on natural events such as earthquakes may be obtained from experts in the appropriate fields. Alternatively, occurrence rates of natural events may be deduced using mathematical models; for example, earthquake occurrence rates may be estimated by seismologists using mathematical models of the earthquake phenomenon. If necessary, specialized consultants should be hired to determine c values.

The conditional failure probability of the failure of a specific component given that the common cause Y has occurred (p) is also calculated either from historical data or through deductive procedures. For example, p for a pump due to an earthquake of magnitude 4 or higher may be obtained from historical data on pump failures during earthquakes of magnitude 4 or higher. Alternatively, this probability may be calculated by structural engineers through a deductive procedure known as seismic risk assessment. If necessary, specialized consultants should be hired to compute p. If sufficient information is not available, we may conservatively assume $p = 1$. Some reliability engineers assume $p = 1$ and perform the system reliability analysis; if the computed reliabilities are not acceptable (that is, do not meet system reliability requirements), a more accurate analysis using more accurate p values is performed.

Knowing the values of λ, c, p, and n, probabilities of single- and multiple-component failures may be computed as follows:

failure rate of a specific component failing due to the common cause Y

$$= z_1 = cp$$

failure rate of a specific component failing due to all causes including the common cause Y

$$= \lambda^* = \lambda + cp$$

rate of at least one component failing due to the common cause Y

$$= s_1 = c(1 - q^n) \quad \text{where } q = (1 - p)$$

rate at which a specific set of k components failing simultaneously due to the common cause Y

$$= z_k = cp^k \quad \text{where } n \geq k > 1$$

conditional probability that a specific set of k components failing simultaneously given that one of the components has failed

$$= G_k = \frac{z_k}{\lambda^*} \quad \text{where } n \geq k > 1$$

What is the difference between z_1 and s_1? This is best explained by considering two components ($n = 2$). Let these components be referred to as A and B. The failure rate z_1 is the rate at which a specific component (say, A) fails due to the common cause Y. When we deal with z_1, we restrict our attention to the failure of the specific component (in this case, A). Consider a time duration of T hours. Each failure of A due to Y during that duration is counted as an incident. Suppose, during the period of T hours, A fails m times due to Y. Then $z_1 = m/T$ failures per hour. Because A and B are identical components, the failure rate of B due to Y is also $z_1 = m/T$ failures per hour.

The failure rate s_1 is the rate at which at least one of the two components (either A or B or both) fails due to Y. When we deal with s_1, we look at the failures of A as well as B. Consider again a time duration of T hours. A failure of A or B or the simultaneous failure of A and B due to Y is counted as an incident. Note that the simultaneous failure of A and B shall not be counted as two incidents but as one incident. Let A fail m times and B fail m times during the duration of T hours. Let A and B fail simultaneously j times. So A fails alone $m - j$ times, B fails alone $m - j$ times and A and B fail simultaneously j times. The total count is

$$(m - j) + (m - j) + j = 2m - j$$

The corresponding failure rate $s_1 = (2m - j)/T$ failures per hour. Because j can never be greater than m, s_1 can never be less than z_2. So $z_1 \leq s_1$.

Example 5.7

Historical data for the region where an industrial plant is to be built show that there were four earthquakes of magnitude 5 or higher during the past 100 years. Calculate the rate of occurrence of earthquakes of magnitude 5 or higher at the proposed plant site.

Solution

Duration of historical record = 100 years

number of earthquakes (magnitude 5 or higher) = 4

rate of common cause occurrence = $\frac{4}{100}$ = 0.04 occurrences per year

Example 5.8

The performance of a certain type of component in four plants during four different floods has been recorded. The first plant contained 24 components, of which 2 failed during the flood. The second plant contained 42 components, of which 3 failed during the flood. The third plant contained 60 components, of which 5 failed during the flood. The fourth plant contained 21 components, of which 3 failed during the flood. Assuming that these failures during the flood are caused by the flood, calculate the conditional probability of component failure given that a flood (common cause) has occurred.

Solution

Total number of components exposed to the flood
$$= 24 + 42 + 60 + 21 = 147$$
total number of failures caused by the flood
$$= 2 + 3 + 5 + 3 = 13$$
conditional probability of component failure given that a flood has occurred
$$= \tfrac{13}{147} = 0.088$$

Example 5.9

Two identical components in a system are susceptible to failure due to excessive temperature caused by air-conditioning failures. The rate of occurrence of air-conditioning failures is 0.0005 occurrences per hour. The conditional probability of component failure given that an air-conditioning failure has occurred is 0.4. The failure rate of each component due to causes other than excessive temperature is 0.001 per hour. Calculate the conditional probability of the simultaneous failure of the two components given that one component has failed.

Solution

Excessive temperature (or air-conditioning failure) is the common cause. Using the notation introduced in the text, we have
$$n = 2 \qquad c = 0.0005 \qquad \lambda = 0.001 \qquad p = 0.4$$
$$q = 1 - p = 0.6$$
rate of failure of one specific component due to excessive temperature
$$= z_1 = cp = 0.0005 \times 0.4$$
$$= 0.0002 \text{ failures per hour}$$
rate of failure of at least one component failing (either of the two components or both the components failing) due to excessive temperature
$$= s_1 = c(1 - q^n) = 0.0005(1 - 0.6^2)$$
$$= 0.00032 \text{ failure per hour}$$
failure rate of one specific component failing due to all causes including excessive temperature
$$= \lambda^* = \lambda + cp$$
$$= 0.001 + (0.0005 \times 0.4) = 0.0012 \text{ failures per hour}$$

rate at which the two components fail simultaneously due to excessive temperature

$$= z_2 = cp^2 = 0.0005 \times 0.4^2$$

$$= 0.00008 \text{ failures per hour}$$

conditional probability of the two components failing simultaneously due to the excessive temperature given that one component has failed

$$= G_2 = \frac{z_2}{\lambda^*} = \frac{0.00008}{0.0012} = 0.0667$$

Example 5.10

Three identical components in a system are susceptible to failure due to dust. The three components are denoted by A, B, and C. The rate of occurrence of dust in the vicinity of all three components at the same time is 0.001 occurrences per hour. Conditional probability of a component failure given that the component is exposed to dust is 0.7. The failure rate of each component due to causes other than exposure to dust is 0.003 failures per hour. Calculate the conditional probability of the simultaneous failure of all three components given that one component has failed. Also, calculate the conditional probability of the simultaneous failure of A and B given that one of the components has failed.

Solution

Dust is the common cause. Using the notation introduced in the text, we have

$$n = 3 \qquad c = 0.001 \qquad \lambda = 0.003 \qquad p = 0.7$$

$$q = 1 - p = 0.3$$

rate of failure of one specific component due to dust

$$= z_1 = cp = 0.001 \times 0.7$$

$$= 0.0007 \text{ failures per hour}$$

rate of failure of at least one component failing due to dust

$$= s_1 = c(1 - q^n)$$

$$= 0.001(1 - 0.3^3)$$

$$= 0.000973 \text{ failures per hour}$$

failure rate of a specific component due to all causes, including dust

$$= \lambda^* = \lambda + cp = 0.003 + (0.001 \times 0.7)$$
$$= 0.0037 \text{ failures per hour}$$

rate at which two specific components (A and B) fail simultaneously due to dust

$$= z_2 = cp^2 = 0.001 \times 0.7^2$$
$$= 0.00049 \text{ failures per hour}$$

Because all three components are identical, the same failure rate is applicable to the simultaneous failure (due to dust) of B and C or of C and A also:

rate at which two specific components (B and C) fail simultaneously due to dust

$$= z_2 = 0.00049 \text{ failures per hour}$$

rate at which two specific components (A and C) fail simultaneously due to dust

$$= z_2 = 0.00049 \text{ failiures per hour}$$

conditional probability of two specific components (A and B) failing simultaneously due to dust given that one component has failed

$$= G_2 = \frac{z_2}{\lambda^*} = \frac{0.00049}{0.0037} = 0.1324$$

Because all three components are identical, the same conditional probability is applicable to the simultaneous failure (due to dust) of B and C, or of C and A also.

rate at which all three components fail simultaneously due to dust

$$= z_3 = cp^3 = 0.001 \times 0.7^3$$
$$= 0.000343 \text{ failures per hour}$$

conditional probability of all three components failing simultaneously due to dust given that one component has failed

$$= G_3 = \frac{z_3}{\lambda^*} = \frac{0.000343}{0.0037} = 0.0927$$

Example 5.11

Redo Example 5.10, with just one change: The conditional probability of a component failure given that the component is exposed to dust is 0.2.

Solution

Using the notation introduced in the text, we have

$$n = 3 \qquad c = 0.001 \qquad \lambda = 0.003 \qquad p = 0.2$$

$$q = 1 - p = 0.8$$

rate of failure of one specific component due to dust

$$= z_1 = cp = 0.001 \times 0.2$$

$$= 0.0002 \text{ failures per hour}$$

rate of failure of at least one component failing due to dust

$$= s_1 = c(1 - q^n) = 0.001(1 - 0.8^3)$$

$$= 0.000488 \text{ failures per hour}$$

failure rate of a specific component due to all causes including dust

$$= \lambda^* = \lambda + cp = 0.003 + (0.001 \times 0.2)$$

$$= 0.0032 \text{ failures per hour}$$

rate at which two specific components (A and B) fail simultaneously due to dust

$$= z_2 = cp^2 = 0.001 \times 0.2^2$$

$$= 0.00004 \text{ failures per hour}$$

Because all three components are identical, the same failure rate is applicable to the simultaneous failure (due to dust) of B and C or of C and A also:

rate at which two specific components (B and C) fail simultaneously due to dust

$$= z_2 = 0.00004 \text{ failures per hour}$$

rate at which two specific components (A and C) fail simultaneously due to dust

$$= z_2 = 0.00004 \text{ failures per hour}$$

conditional probability of two specific components (A and B) failing simultaneously due to dust given that one component has failed

$$= G_2 = \frac{z_2}{\lambda^*} = \frac{0.00004}{0.0032}$$

$$= 0.0125$$

Because all three components are identical, the same conditional probability is applicable to the simultaneous failure (due to dust) of B and C or of C and A also.

rate at which all three components fail simultaneously due to dust

$$= z_3 = cp^3 = 0.001 \times 0.2^3$$

$$= 0.000008 \text{ failures per hour}$$

conditional probability of all three components failing simultaneously due to dust given that one component has failed

$$= G_3 = \frac{z_3}{\lambda^*} = \frac{0.000008}{0.0032} = 0.0025$$

Compared to Example 5.10, G_2 and G_3 have decreased significantly. As a general rule, if p is decreased and all other data remain unchanged, then G_k will also decrease for $n \geq k > 1$.

Example 5.12

Redo Example 5.10 with just one change: The rate of occurrence of dust in the vicinity of all three components at the same time is 0.0001 occurrences per hour.

Solution

Using the notation introduced in the text, we have

$$n = 3 \qquad c = 0.0001 \qquad \lambda = 0.003 \qquad p = 0.7$$

$$q = 1 - p = 0.3$$

rate of failure of one specific component due to dust

$$= z_1 = cp = 0.0001 \times 0.7$$

$$= 0.00007 \text{ failures per hour}$$

rate of failure of at least one component failing due to dust

$$= s_1 = c(1 - q^n) = 0.0001(1 - 0.3^3)$$
$$= 0.0000973 \text{ failures per hour}$$

failure rate of a specific component due to all causes including dust

$$= \lambda^* = \lambda + cp = 0.003 + (0.0001 \times 0.7)$$
$$= 0.00307 \text{ failures per hour}$$

rate at which two specific components (A and B) fail simultaneously due to dust

$$= z_2 = cp^2 = 0.0001 \times 0.7^2$$
$$= 0.00004 \text{ failures per hour}$$

Because all three components are identical, the same failure rate is applicable to the simultaneous failure (due to dust) of B and C, or of C and A also:

rate at which two specific components (B and C) fail simultaneously due to dust)

$$= z_2 = 0.000049 \text{ failures per hour}$$

rate at which two specific components (A and C) fail simultaneously due to dust

$$= z_2 = 0.000049 \text{ failures per hour}$$

conditional probability of two specific components (A and B) failing simultaneously due to dust given that one component has failed

$$= G_2 = \frac{z_2}{\lambda^*} = \frac{0.000049}{0.00307}$$
$$= 0.016$$

Because all three components are identical, the same conditional probability is applicable to the simultaneous failure (due to dust) of B and C, or of C and A also.

rate at which all three components fail simultaneously due to dust

$$= z_3 = cp^3 = 0.0001 \times 0.7^3$$
$$= 0.0000343 \text{ failures per hour}$$

conditional probability of all three components failing simultaneously due to dust given that one component has failed

$$= G_3 = \frac{z_3}{\lambda^*} = \frac{0.0000343}{0.00307} = 0.011$$

A decrease in the occurrence rate of the common cause, with all other data remaining the same, results in a decrease in G_k for $n \geq k > 1$.

5.3 SOFTWARE ERROR DATA

Computers are increasingly used on a real-time basis to control the operation of engineering systems and/or to display vital information on display screens. Computer-controlled process systems in chemical plants are examples of real-time computer control. Many chemical and power plants also utilize computers to display system parameters (temperatures, pressures, chemical concentrations, etc.) in control rooms where human operators can "keep an eye" on plant operation status.

Computers perform their control or display functions as "instructed" by the computer software. Errors in *control software* could result in undesired events such as the production of off-specification products, system failure, and accidents. Errors in *display software* would result in the display of incorrect information in the control room, which could trigger actions by human operators that could lead to off-specification products, system failure, or accidents.

Software errors by themselves could cause an undesired event, or a combination of software, hardware, and human errors could cause an undesired event. In either case, we need to know the probability of software errors in order to compute the probability of the undesired event. Software error probabilities may be specified in terms of error rates or errors per mission (for example, 10^{-3} errors per year or 10^{-4} errors per mission).

Software reliability is a relatively new branch of reliability engineering. Some mathematical models have been developed to compute software error probabilities as a function of the number of errors detected and corrected during debugging [Musa, Iannino, and Okumoto (1987)]. There is no generic data base available for software error probabilities.

If the software vendor or developer assures that very few errors can be expected in the software (based on past experience with the same or similar software and on the number of errors detected and corrected during debugging) and the effect of software errors on system operations are not expected to be severe (for example, no catastrophic accidents), we may assume that the software error probability is zero. If not, software error probability has to be estimated for use in the system reliability analysis. Software reliability is a rather specialized area that the best approach is either to request the software vendor to supply the software error probabilities or to hire a consultant to estimate them.

Readers interested in additional information are referred to the state-of-the-art review paper by Bastani and Ramamoorthy (1988). Books entirely devoted to software reliability are also available [Musa, Iannino, and Okumoto (1987)].

(Note: Errors in computer software used in component design are seldom, if ever, considered explicitly during the system reliability analysis. Effects of such errors are implicitly accounted for in component failure probabilities, in the same way as design errors and manufacturing errors are implicitly included in component failure probabilities.)

5.4 HUMAN ERROR DATA

Maintenance, test, and operations personnel are essential to the proper functioning of engineering systems. Errors on their part could contribute to component and system failures. So human error data are necessary for quantitative system reliability analysis. Sources for human error probabilities include Pontecorro (1965), Meister (1966), Peters (1966), Recht (1966), *Reactor Safety Study* [Nuclear Regulatory Commission (1975)], Meister (1978), Joos, Sabri, and Husseiny (1979), and Swain and Guttmann (1983).

Human error probabilities are estimated by three different means: actual operating experience; simulator testing; and expert opinion. Data derived from actual operating experience is the best but sufficient data may not be available for some types of actions. Simulator data are not as good as operating experience data but are acceptable. Expert opinion is acceptable and useful if the experts are experienced and knowledgeable. Unless otherwise stated, human error probabilities available in the sources cited in the previous paragraph may be based on expert opinion, operating experience, and simulator data. Expert opinion could be the dominant basis when operating experience and simulator data are sparse.

The operating experience used in many of the generic data sources comes from a variety of plants; the maintenance–test–operations personnel may have different levels of training and experience; they may be working with or without clearly written procedures, under different work environments (day or night shift, regular or overtime hours, and varying conditions of light, temperature, and humidity); errors might have been committed during normal or stressful conditions (such as accidents, fires, earthquakes, etc.). So, what we may find in data sources is an average human error probability covering a wide range of situations. These probabilities may be used as such or increased or decreased, depending on whether we believe that our situation is an average situation or is significantly different from the average situation. For example, if our interest is in operator action during or soon after an accident, the human error probability we use in the system reliability analysis should be higher than the average values found in generic data sources; a 25% increase is used in some nuclear reactor risk assessment studies.

The manner in which controls or components are labelled may affect human error probabilities: According to the data provided by Swain and Guttmann (1983), the probability of an operator closing or opening a wrong valve is reduced from 0.005 to 0.001 if the valves are labelled clearly.

Data provided in some sources, such as Swain and Guttmann (1983), give human error probabilities under "good" versus "bad" scenarios ("clear labels" versus "unclear labels," "functional grouping of controls" versus "no such grouping," "densely packed controls" versus "well-spaced controls," etc.).

Swain and Guttman (1983) provide not only point estimates but also confidence intervals for human error probabilities. The point estimates may be used if we believe our operating conditions to be "average"; a higher value closer to the upper end of the confidence interval may be used if the operating conditions are substantially "worse" than average, and a lower value closer to the lower end of the confidence interval may be used if the operating conditions are substantially "better" than average. We should keep in mind that most human error probability estimates are approximate (this is true for most hardware failure and software error probabilities also). So undue concern need not be placed on estimating very accurate multiplication factors.

The information provided in this section, together with the data contained in the data sources cited earlier, is sufficient for most system reliability analysis projects. Readers interested in learning more about human reliability are referred to books exclusively devoted to the subject [Dhillon (1986), Park (1987), and Dougherty and Fragola (1988)].

(Note: Human errors in component design and manufacture are seldom, if ever, considered explicitly in system reliability analysis. Effects of such errors are implicitly included in component failure probabilities.)

5.5 COMPARISON OF HARDWARE AND HUMAN FAILURE PROBABILITIES

Lambert (1975) provides some approximate comparisons between the probabilities of different types of hardware failures (component failures) and human errors.

Component failures may be classified as *dynamic failures* and *quasistatic failures*. Failures like a valve failing to close, a switch failing to open, and a pump failing to start are called dynamic failures because they relate to operations that involve a change in the state of the component (open state to closed state, closed state to open state, idle state to active state, etc.). Failures like pipe rupture, wire break, and pipe-support collapse are called quasistatic failures. According to Lambert (1975), the probabilities of dynamic failures are in general one to three orders of magnitude higher than the probabilities of quasistatic failure.

Human error probabilities are in general one to three orders of magnitude higher than the probabilities of dynamic failures [Lambert (1975)].

5.6 DOCUMENTATION

Failure data documentation should contain component failure probabilities, software error probabilities, and human error probabilities, and the identification of data bases or documents from which the information is obtained. If failure statistics from a number of data bases are combined, as described in Section 5.2.6, description of the combination procedure and the weights assigned to the different data bases should be included in the documentation. If expert opinion is used, names and affiliation of the experts, a brief summary of their experience, the individual estimates, weighting factors assigned to each expert, and the method used to consolidate the individual estimates should be included in the documentation.

All the failure data may be summarized as shown in Table 5.9. (All the columns indicated in that table may not be applicable or needed in all cases; such columns may be omitted or left blank.) Whether the value

Table 5.9.   Summary of Failure Data Used in a Fault Tree

(1)	(2)	(3)	(4)	(5)	(6)	(7)	(8)
NO.	COMPONENT, SOFTWARE, OR OPERATOR	FAILURE MODE	λ OR q_D	MTTR	q_{TM}	REFERENCES	NOTES

Note: Failure rates (λ) are entered as "... /H" (failures per hour) and probabilities of failure on demand (q_D) are entered as "... /D" (failures per demand).

tabulated for each failure rate, failure-on-demand probability and test–maintenance unavailability is the mean, geometric mean, or median should be clearly stated in the table or in a footnote to the table. We may choose to provide all the three values in the table; ranges (confidence intervals) may also be included in the table, if available. All documents and data bases used in developing the table may be listed at the end of the report and identified by a reference number in column 7. The eighth column ("notes") may include notes relating to the combination of data bases, if applicable, or other relevant information. If such information is presented elsewhere in the report, reference may be made to the appropriate section, chapter, or appendix.

Some general guidelines on documentation may be found in Section 14.10.

EXERCISE PROBLEMS

5.1. Failure data for pumps in a petrochemical plant for the years 1985–1990 are summarized in Table 5.10. The failures summarized in that table are for the failure mode 'pump fails to run after having started on demand.' Assuming a constant failure rate and constant repair rate, compute the failure rate and mean time to repair.

Table 5.10. Data for Problem 5.1

	1985	1986	1987	1988	1989	1990
total number of failures	2	1	4	3	1	2
total operating hours	60,000	60,000	60,000	60,000	60,000	60,000
total repair time (hours)	180	110	370	320	80	190

5.2. Failure data for check valves in a number of power plants are summarized in Table 5.11, for the failure mode 'valve fails to open on demand.' Compute the probability of failure on demand and the mean time to repair.

5.3. Failure data for a certain type of component from five plants owned by an oil company are summarized in Table 5.12. Assuming constant failure rates, compute the failure rates for primary failures and failure rates for secondary failures for each plant separately and for all plants combined. (*Note*: Each

Table 5.11. Data for Problem 5.2

	1971–1975	1976–1980	1981–1985	1986–1990
total number of failures	1	2	0	1
total number of demands	9500	10,800	9700	9900
total repair time (hours)	70	135	0	60

plant has dozens of these components. The operating hours and number of failures given in Table 5.12 are the sum total for all components in each plant.)

Table 5.12. Data for Problem 5.3

PLANT	OPERATING HOURS	NUMBER OF PRIMARY FAILURES	NUMBER OF SECONDARY FAILURES
plant 1	104,000	2	1
plant 2	320,000	5	2
plant 3	150,000	4	2
plant 4	308,000	6	1
plant 5	160,000	4	4

5.4. Eighty failures of component A and 50 failures of component B were observed during a period of 40,000 hours. Among the 80 component-A failures, 50 occurred simultaneously with component-B failures. What is the failure rate of component A? What is the failure rate of component B? What is the conditional probability of component-A failure given that component B has failed? What is the conditional probability of component-B failure given that component A has failed?

5.5. Failure histories of circuit breakers are collected from six plants. Failure-on-demand probabilities (P) of the circuit breakers are determined separately for each plant and are summarized in Table 5.13. The failure-on-demand probability from each plant is assigned unit weight. Compute the arithmetic mean, the geometric mean, and the median of the failure-on-demand probability.

Table 5.13. Data for Problem 5.5

PLANT	P
A	5×10^{-3}
B	2×10^{-4}
C	8×10^{-4}
D	1×10^{-3}
E	3×10^{-4}
F	6×10^{-3}

5.6. Consider the data described in Problem 5.5. Based on the number of circuit breakers in each plant and the length of operation of each plant, the chief reliability engineer has assigned the following weights to the failure-on-demand probabilities from the six plants—Weights for A, B, C, D, E and F are 7, 4, 1, 8, 10 and 8, respectively. Compute the arithmetic mean, the geometric mean, and the median of the failure-on-demand probability.

5.7. Consider Problem 5.6. If we change the weight of C to 8 and the weight of D to 1, how will the arithmetic mean, the geometric mean, and the median of the failure-on-demand probability change from the results of Problem 5.6?

5.8. Redo Problem 5.6, with the weight of D equal to 7; all other data remain the same as in Problem 5.6.

5.9. The historical records of a coastal town show that there were three hurricanes during the past 40 years. Calculate the rate of occurrence of hurricanes in that town.

5.10. Failures of a certain type of electronic component in three different plants due to excessive temperatures during air-conditioning system failures have been recorded. Each plant had 140 of these components. Fifty-two out of the 140 components failed in the first plant. Forty-one out of the 140 components failed in the second plant. Forty-three out of the 140 components failed in the third plant. Compute the conditional probability of component failure given that an air-conditioning system failure has occurred.

5.11. There are three identical pumps in a waste-treatment facility. These pumps are susceptible to major earthquakes. The rate of occurrence of major earthquakes in that region is 10^{-3} per year. The conditional probability of pump failure given that a major earthquake has occurred is 0.6. The failure rate of each of those pumps due to causes other than earthquakes is 3×10^{-5} failures per hour. Compute the conditional probability of the simultaneous failure of the three pumps given that one pump has failed during an earthquake. Also, compute the conditional probability of the simultaneous failure of any two pumps given that one pump has failed during an earthquake.

5.12. Consider Problem 5.11 with just one change. There are two identical pumps in the waste-treatment facility (instead of three). Compute the conditional probability of the simultaneous failure of both pumps given that one pump has failed during an earthquake.

5.13. There are three identical electronic components in a system. These components are susceptible to excessive temperatures during air-conditioning system failures. The rate of occurrence of air-conditioning failures is 1 failure per year. The conditional probability of component failure given that the air-conditioning system has failed is 0.6. The failure rate of each of these electronic components due to causes other than air-conditioning system failure is 3×10^{-5} failures per hour. Compute the conditional probability of

4

the simultaneous failure of the three components given that one component has failed during an air-conditioning system failure. Also, compute the conditional probability of the simultaneous failure of any two components given that one component has failed during an air-conditioning system failure. (Note the similarity between Problems 5.11 and 5.13.)

REFERENCES

Bastani, F. B. and C. V. Ramamoorthy (1988). Software reliability. In *Handbook of Statistics*, vol. 7, P. R. Krishnaiah and C. R. Rao, eds. Elsevier Science Publishers, Amsterdam, pp. 7–25.

Bush, S. H. (1985). Statistics of pressure vessel and piping failures. In *Pressure Vessel and Piping Technology—A Decade of Progress*, C. Sundararajan, ed. American Society of Mechanical Engineers, New York, pp. 875–894.

Department of Defense. (1974). *Military Standardization Handbook—Reliability Prediction of Electronic Equipment* (MIL-HDBK-217C. Department of Defense, Washington, DC.

Dhillon, B. S. (1986). *Human Reliability: With Human Factors*. Pergamon Press, New York.

Dougherty, E. M. and J. R. Fragola (1988). *Human Reliability Analysis*. John Wiley and Sons, New York.

Haugen, E. B. (1980). *Probabilistic Mechanical Design*. John Wiley and Sons, New York.

Institute of Electrical and Electronic Engineers (1977). *IEEE Guide to the Collection and Presentation of Electrical, Electronic and Sensing Component Reliability Data for Nuclear Power Generating Stations*. Institute of Electrical and Electronic Engineers, New York.

Joos, D. W., Z. A. Sabri, and A. A. Husseiny (1979). Analysis of gross error rates in operation of commercial nuclear power stations. *Nuclear Engineering and Design* 53 265–300.

Lambert, H. E. (1975). *Fault Trees for Decision Making Systems Analysis* (Report No. UCRL-51829). Lawrence Livermore National Laboratory, Livermore, CA.

Levenbach, G. J. (1957). Accelerated life testing of capacitors. *IRE Transactions* 10 9–20.

Linstone, H. A. and M. Turoff (1975). *The Delphi Method: Techniques and Applications*. Addison-Wesley, Reading, MA.

Mann, N. R., R. E. Schafer, and N. D. Singhpurwalla (1974). *Methods for Statistical Analysis of Reliability and Life Data*. John Wiley and Sons, New York.

Meister, D. (1966). Human factors in reliability. In *Reliability Handbook*, W. G. Ireson, ed. McGraw-Hill Book Company, New York, pp. 12.2–12.37.

Meister, D. (1978). Subjective data in human reliability estimates. In *Proceedings of the Annual Reliability and Maintainability Symposium*, Institute of Electrical and Electronics Engineers, New York, pp. 380–384.

Musa, J. D., A. Iannino, and K. Okumoto (1987). *Software Reliability: Measurement, Prediction and Applications*. McGraw-Hill Book Company, New York.

Nuclear Regulatory Commission (1975). *Reactor Safety Study: An Assessment of Accident Risks in U.S. Commercial Nuclear Power Plants* (WASH-1400). Nuclear Regulatory Commission, Washington, DC.

Nuclear Regulatory Commission (1983). *PRA Procedures Guide* (NUREG/CR-2300). Nuclear Regulatory Commission, Washington, DC.

Park, K. S. (1987). *Human Reliability: Analysis, Prediction and Prevention of Human Errors*. Elsevier Publishing Company, Amsterdam.

Pershing, A. V. and G. E. Hollingsworth (1964). Derivation of Delbruck's model for random failure of semiconductor material. In *Physics of Failures in Electronics*, vol. 2, M. F. Goldberg and J. Vaccaro, eds. pp. 61–67.

Peters, G. A. (1966). Human error: Analysis and control. *American Society of Safety Engineers Journal* **11** 9–15.

Pontecorro, A. (1965). A method for predicting human reliability. In *Proceedings of the Annual Reliability and Maintainability Symposium*, Institute of Electrical and Electronics Engineers, New York, pp. 337–342.

Recht, J. L. (1966). Systems safety analysis: Error rates and costs. *National Safety News* 20–23.

Sundararajan, C. (1986). Probabilistic assessment of pressure vessel and piping reliability. *Journal of Pressure Vessel Technology* **108**(1) 1–13.

Swain, A. D. and H. E. Guttmann (1983). *Handbook of Human Reliability Analysis with Emphasis on Nuclear Power Plant Applications*. Nuclear Regulatory Commission, Washington, DC.

Chapter 6
Preliminary Hazard Analysis

6.1 INTRODUCTION

Preliminary hazard analysis (PHA) is conducted at the conceptual design stages of a system (or plant) to identify the hazardous elements in the system and to assess their safety implications. The preliminary hazard analysis is a global type of analysis that looks at the big picture of the system as a whole from the point of view of safety. More detailed analyses of components and systems may follow a preliminary hazard analysis, for example, a failure modes and effects analysis (Chapter 7) and/or a fault tree analysis (Chapters 8–10). Some reliability analysis projects do end after the preliminary hazard analysis and no other analysis is carried out; this depends on the scope and purpose of the reliability project.

Findings and recommendations from the analysis are transmitted to the design team and appropriate changes are made in the design to reduce accidents and minimize their consequences. If necessary, another preliminary hazard analysis may be conducted after design changes are made. Preliminary hazard analyses may also be conducted for operating plants if no preliminary hazard analysis was conducted during plant design.

6.2 ANALYSIS PROCEDURE

A preliminary hazard analysis consists of the following steps:

1. Data collection
2. Identification of hazardous sources
3. Preparation of hazards and effects table

Each of these steps is described in the following subsections.

6.2.1 Data Collection

The team performing a preliminary hazard analysis should examine all available information about the hardware, materials, processes, tests, and operations relating to the system under study. Because the preliminary hazard analysis is normally performed during the conceptual design stages, detailed drawings and data may not be available. System layout drawings showing the relative placement of hardware, storage facilities, operator stations, and process flows should be examined.

6.2.2 Identification of Hazardous Sources

Any hardware, material, process, or operation that could, under certain conditions, cause loss of life, personnel injury, or property damage is considered a *hazardous source* (*hazardous element*). Under normal conditions, these hazardous sources are harmless but certain abnormal conditions can trigger the hazardous sources to cause accidents. Consider, for example, a poisonous gas produced as a by-product of a chemical process. Under normal conditions, plant personnel and the public are protected by containing the gas in a pressure vessel. Here the hazardous element (poisonous gas) does not pose a threat to plant personnel or the public. A leak in the pressure vessel is a *triggering event* that could lead to an accident. A leak in the pressure vessel would release the gas to the environment and affect plant personnel. A more serious triggering event might be an explosion in the pressure vessel that could release enormous quantities of the gas, and both plant personnel and the general public in the neighborhood would be at risk.

All hazardous sources, whether they are hardware, materials, processes, or operations, should be identified and listed. A number of hazardous elements are listed here:

1. Explosive charges
2. Flammable materials
3. Excessively high or low pressure
4. Excessively high or low temperature
5. High voltage
6. Radioactive materials
7. Poisonous materials
8. Reciprocating equipment
9. Rotating equipment
10. Spring-loaded devices
11. Falling objects
12. Suspended objects

It is important that hazardous elements encountered not only during the operation of the system but also during installation, testing, maintenance, and decommissioning be considered.

A chemical may not be dangerous as such but it could produce a dangerous situation if it reacts with another chemical. All chemicals and materials should be viewed in such a context. A piece of equipment of a component suspended from the ceiling of a work space may pose a safety hazard because it could injure workers if the suspension system fails. Also, equipment and components fixed at some elevation from the floor could become a safety hazard if the fixture fails.

Secondary events should also be considered. For example, a high-mounted piece of equipment may be located further away from work places, thus posing no hazard to workers. However, should the fixture fail and the equipment fall on another piece of equipment, say on a storage tank, the latter may fail due to the impact of the falling object and release a toxic chemical to the environment. This is called a *secondary accident*. Thus a seemingly harmless piece of equipment, fixed at an elevation, could pose a safety hazard if its location is such that it could interact with other hazardous sources.

The preliminary hazard analysis may be stopped here if no hazardous sources are found.

Table 6.1.   Sample Hazards and Effects Table

	(1)	(2)
1. hazardous source	gas supply	fire in the burner
2. location and identification of the hazardous source	—	—
3. trigger	(i) stop valve open or leak in the system; (ii) pilot light out	(i) burner ignites "explosively," causing flames to escape through air inlet; (ii) flammable materials close to the burner
4. accident	gas collects within the room; may spread to adjacent rooms	fire in the room; may spread to adjacent rooms and areas
5. effect of the accident	gas poisoning of people in gas-filled rooms; possible health hazard and death	possible death, injury and property damage
6. warnings	none	none
7. safeguards (accident prevention)	none	none
8. safeguards (mitigation of effects)	none	none
9. accident frequency	—	—
10. criticality of the accident	critical (III)	critical (III)
11. remarks	recommendation: provide a gas detector and alarm	recommendations: (i) provide inflammable barriers around the burner; (ii) provide fire alarm

6.2.3 Preparation of Hazards and Effects Tables

Table 6.1 shows a suggested format for hazards and effects tables. There are 10 items to be entered. Some of the items may be omitted or left blank, depending on the scope of the analysis. Each item is explained in the following paragraphs. (*Note:* We have listed each item in a row. Some analysts prefer to list them columnwise. Either practice is acceptable.)

1. *Hazardous Source:* The hazardous sources identified in step 2 of the analysis (see Section 6.2.2) are entered in this column.

2. *Location and Identification of the Hazardous Source:* The identifying name or number of the hazardous source, if any (as referred to in drawings or other documents), and its location are entered here. The location of a hazardous source may be identified by building name, by a grid number used in drawings, or by other means. Location is important because accidents may be reduced or avoided, in some cases, by locating the hazardous source away from the trigger mechanism. If the system is a relatively simple one involving only a few components and is located in a narrow area, this item may be omitted or left blank. If the system contains a large number of components or is spread over a wide area covering a number of rooms or floors, location and identification of the hazardous source will be useful.

3. *Trigger Mechanisms:* As discussed earlier, a hazardous element may not pose a danger under normal conditions. It becomes a safety problem only when some operation, event, or process triggers it toward an accident. (More than one trigger mechanism may be needed to trigger an accident in some cases.) The following is a partial list of trigger mechanisms:

1. Chemical reaction
2. Corrosion
3. Explosion
4. Fire
5. Leak
6. Overloading (electrical)
7. Overloading (mechanical)
8. Vibration
9. Material degradation
10. Excessively high voltage
11. Excessively high or low temperature
12. Excessively high or low pressure
13. Mechanical shock
14. Equipment failure
15. Structural failure
16. Power failure
17. Operator error
18. Sabotage

What is a trigger mechanism for one hazardous source may not necessarily be a trigger mechanism in relation to another hazardous source. Also, whether a particular piece of hardware, material, or chemical is a trigger may depend on its proximity to the hazardous source. Even if the hazardous source and the trigger are located at a reasonable distance, the hazards and effects table may include a cautionary note that no design changes shall be made to bring the two closer.

In some instances, whether a candidate piece of hardware, material, or chemical triggers an accident may depend on the environmental conditions (temperature, pressure, etc.). This should be considered and, if appropriate, a cautionary note may be added about temperature and pressure limits.

It is important that not only trigger mechanisms encountered during the operation of the system but also those encountered during installation, testing, maintenance, and decommissioning by included.

4. *Accident:* The interaction of the hazardous element with the trigger mechanism results in an accident. For example, corrosion (trigger) in a tank containing poisonous gas (hazardous element) may eventually result in gas leakage (accident). A strong mechanical shock (trigger) trips a piece of equipment located at an elevation (hazardous element) and the resulting fall of the piece of equipment is the accident.

5. *Effect of Accident:* The effect of an accident could be death, sickness, injury, and/or property damage. The effect could also be a secondary accident (see Section 6.2.2).

6. *Warnings:* Some accidents are not of a sudden or catastrophic nature and the effects of the accident may be mitigated if there is an early warning system. Fire alarms and leak detectors are examples of such warning systems. If a warning system is present, it should be noted here. If not, state "none."

This item is omitted in some hazards and effects tables.

7. *Safeguards (Accident Prevention):* Accident prevention measures may not completely eliminate an accident but could significantly reduce the accident probability. The safeguard could be a piece of hardware or an operating–maintenance–test procedure. Physical separation of the hazardous source from triggers is also a possible safeguard against accidents.

Consider a potential explosion (accident) of a pressure vessel. Pressure inside the vessel may be continually monitored by a pressure gauge and the operator may be alerted to any abnormal pressure increases by alarms, flashing lights, or other means. Alternatively, automatic pressure relief valves may be added to the vessel. Such preventive measures could significantly reduce the probability of pressure vessel explosions.

Consider the example of a turbine blade rupture when the turbine is rotating. Large fragments of the blade will be ejected and could kill and injure personnel and damage equipment. One cause for blade rupture is cracks in the blades. Frequent inspections of the blades for cracks and the repair of large cracks could significantly reduce the probability of turbine blade ruptures.

The frequency of accidents and their consequences may also be reduced by derating the system. That is, accident frequencies and consequences may be reduced by reducing operating pressures, operating temperatures, flow rates, etc. that may have the potential to be triggering mechanisms.

If a safeguard is built into the system, it should be noted here. If not, state "none."

This item is omitted in some hazards and effects tables.

8. *Safeguards (Mitigation of Effects):* Adverse effects of the accident could be mitigated by installing appropriate hardware or through appropriate operating procedures. Fire extinguishers are examples of this kind of safeguard.

Consider the turbine blade failure accident. The possibility of "flying" turbine-blade fragments hurting workers can be minimized by locating work spaces away from the range of the fragments. Expensive equipment can also be protected by locating away from the range of the fragments. If it is necessary to locate work spaces and equipment close by, safety enclosures may be provided to safeguard workers and equipment.

Emergency plans for the evacuation of workers and local population in case of poisonous gas leaks also fall under the category of mitigation of accident consequences.

If a safeguard is built into the system, it should be noted here. If not, state "none."

This item is omitted in some hazards and effects tables.

9. *Accident Frequency:* Expected accident frequency (accident probability) is seldom known at the time of a preliminary hazard analysis. If rough estimates can be made, the estimate is entered in this column. Warning systems and preventive safeguards could reduce accident probability, and this should be considered in the estimate. Instead of a numerical estimate of the accident probability, a *frequency classification* (I, II, III, or IV, as defined next) is used in some preliminary hazard analyses. A sample frequency classification is as follows:

I: Extremely remote; accident probability is less than 10^{-6} per hour.
II: Remote; accident probability is between 10^{-5} and 10^{-6} per hour.
III: Possible; accident probability is between 10^{-4} and 10^{-5} per hour.
IV: Probable; accident probability is greater than 10^{-4} per hour.

This is not the only possible classification. One may define and use other systems of frequency classification.

This item may be left blank if an estimate of the accident frequency is not available. Some hazards and effects tables simply omit this item.

10. *Criticality of the Accident:* The effect of an accident is ranked according to its criticality (I, II, III, or IV, as defined next). Warning systems and mitigation safeguards could reduce the criticality of an accident, and this should be considered. A sample ranking is as follows:

I: Insignificant; accident will cause very limited property damage.
II: Minor; accident will cause damage to the system, requiring repair, but death, sickness, injury, or extensive property damage is not possible; only moderate property damage.
III: Critical; accident may cause death, sickness, injury, or severe property damage but there is time for the plant operators to prevent most of the potential damage.
IV: Catastrophic; accident will cause death, sickness, injury, or severe property damage immediately (no time to react).

This is not the only possible system of ranking accidents. One may define and use other systems of criticality ranking.

Criticality ranking is based on judgement about the effects of the failure. It may be left blank in the hazards and effects table if sufficient information is not available to estimate the criticality. Some hazards and effects tables omit this item.

11. *Remarks:* Any comments relevant to the safety of the system may be entered here. Recommendations for warning systems, safeguards, and changes in hardware or operating procedures may also be entered here.

This item may be left blank if there is nothing to remark about or to recommend.

6.3 BENEFITS AND LIMITATIONS

A preliminary hazard analysis looks at the system from the perspective of safety. All hazardous sources are identified. All hardware, materials, chemicals, and operations that can interact with the hazardous source and trigger an accident are also identified. These findings can be used to improve system safety. A preliminary hazard analysis also identifies areas requiring further study through more rigorous reliability analysis techniques, such as failure modes and effects analysis (Chapter 7) and/or fault tree analysis (Chapter 8–10).

Findings of the preliminary hazard analysis are also useful in developing training procedures, test–maintenance schedules, and, if necessary, emergency plans. Preliminary hazard analysis results can also be used to decide on optimal allocation of resources to get maximum possible safety. It is possible to decide which potential accidents require priority consideration.

Accidents with both high criticality ranking and high frequency classification are of a more serious nature and should receive more attention. What if "Accident A" has a higher criticality than "Accident B" but is lower in frequency classification? Which accident should receive more attention? The answer is not obvious at the outset. The decision has to be made on the basis of engineering judgement. If appropriate, the decision may be made after quantitative information about the accident frequency and criticality (number of deaths, sickness and injury, value of property damage) are obtained through detailed reliability and risk assessment.

The preliminary hazard analysis is best-suited to identify accidents involving one hazard source and one or two triggering mechanisms. More complex situations involving one or more hazard sources and a number of triggering mechanisms are difficult, if not impossible, to analyze through a preliminary hazard analysis. Deductive techniques such as the fault tree analysis are better-suited for such situations.

Another limitation of the preliminary hazard analysis is that its findings are essentially qualitative in natural though some rough estimates of the frequency of accidents may be included. Fault tree analysis has the ability to compute accident frequencies in a more rigorous and rational manner.

In spite of the preceding limitations, preliminary hazard analysis is a valuable aid in identifying potential safety problems during the conceptual design stage, at which time sufficient information may not be available for more rigorous analyses. Results of a preliminary hazard analysis enables system designers to avoid many potential safety problems at an early stage of the design process.

6.4 DOCUMENTATION

The primary documentation of a preliminary hazard analysis is the hazards and effects table (Table 6.1). It is useful to include an appendix that lists all the documents used during the analysis (system descriptions, plant layout drawings, parts lists, etc.; revision numbers or revision dates of each such document should be noted.

Some general guidelines on documentation may be found in Section 14.10.

6.5 EXAMPLE PROBLEM[1]

Consider the domestic hot water system shown in Figure 6.1. The system supplies hot water at a preset temperature range (say, from 180 to 240°F). The water is heated by gas. When the water temperature is below a

---

[1]This example problem is based on Lambert (1973).

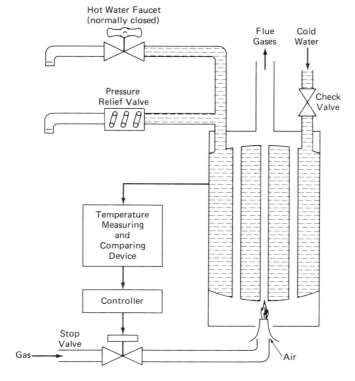

Figure 6.1. Domestic hot water system. [*Source*: Lambert (1973).]

present level (say, 180°F), the temperature measuring and comparing device (TMCD) sends a signal to the controller to open the stop valve (gas valve) and to turn on the gas burner. The controller does so and the water heats up. As soon as the water temperature reaches the present upper level (say, 240°F), the TMCD sends a signal to the controller to close the stop valve and turn off the gas burner. The controller does so and water heating stops. The water starts cooling down and as soon as the water temperature reaches the preset lower level (180°F), the TMCD signals the controller to open the stop valve and turn on the gas burner. This cycle is repeated again and again.

The check valve in the cold water inlet prevents reverse flow due to overpressure in the hot water system. The pressure relief valve (PRV) is set to open when pressure in the hot water system exceeds 100 psi; that will relieve the pressure.

The hazards and effects table for the domestic hot water system is given in Table 6.1. In Table 6.1, we have made entries for only two hazardous sources, namely, "gas supply" and "fire in the burner." There are other hazardous sources, such as "excess water pressure," "combustion products," etc. All these hazardous sources would be included in an actual hazards and effects table.

The entry for "location and identification of the hazardous source" is left blank because the domestic hot water system is a rather simple system located in just one room.

The entry for "accident frequency" is left blank because sufficient information to estimate the accident frequency is not available at the time of the preliminary hazard analysis. An entry may be made at a later date if and when sufficient information becomes available.

EXERCISE PROBLEM

6.1. Consider the domestic hot water system described in Section 6.5. A hazards and effects table with entries for just two hazardous sources is given in Table 6.1. Identify all potential hazardous sources in the domestic hot water system and prepare a complete hazards and effects table.

REFERENCE

Lambert, H. E. (1973). *System Safety Analysis and Fault Tree Analysis*. Report No. UCID-16238. Lawrence Livermore National Laboratory, Livermore, CA.

Chapter 7
Failure Modes and Effects Analysis

7.1 INTRODUCTION

Failure modes and effects analysis (FMEA) is a *qualitative* procedure that identifies potential component failures and assesses their effects on the system. If the criticality of the effects is also considered in the analysis, the analysis is referred to as the failure modes, effects, and criticality analysis (FMECA).

During a failure modes and effects analysis, the following questions are answered for each and every component of the system.

1. How can the component fail? (There could be more than one mode of failure.)
2. What are the consequences (effects) of the failure?
3. How critical are the consequences?
4. How is failure detected?
5. What are the safeguards against the failure?

(Some failure modes and effects analyses consider only the first two questions, whereas others consider one or more of the other questions also. What questions are considered depends on the scope and purpose of the analysis.) In answering these questions, all significant failure modes of the different components are identified, their detection and safeguards are documented, and their effects on the system are determined.

Failure modes and effects analysis is used to do the following:

1. Ensure that all conceivable failure modes and their effects are understood.
2. Assist in the identification of design weaknesses.
3. Provide a basis for selecting design alternatives during early stages of design.
4. Provide a basis for recommending design improvements.
5. Provide a basis for corrective action priorities.
6. Provide a basis for recommending test programs.
7. Assist in troubleshooting existing systems with operating problems.

A well-organized failure modes and effects analysis can benefit the design team in a number of ways. It screens the whole system from the point of view of failures, thus catching any weak spots in design. Communication about the effects of potential failures is established between the various component design teams and system design teams.

Preliminary hazard analysis is not a prerequisite for a failure modes and effects analysis. Failure modes and effects analysis may be conducted at the conceptual design stages, preliminary design stages, or at later stages of detailed design. In fact, one may conduct progressively more detailed failure modes and effects analyses at different stages of the design process. An initial failure modes and effects analysis may be conducted at the conceptual design stage and the findings used in improving the conceptual design. Then one or more additional failure modes and effects analyses may be conducted during the preliminary and/or detailed design stages to improve the design further.

Some reliability analysis projects may end at the completion of the failure modes and effects analysis whereas other projects may proceed to conduct more rigorous analyses such as the fault tree analysis, depending on the purpose and scope of the project.

Failure modes and effects analysis may also be used for identifying failure causes in existing systems with operating problems. If a failure modes and effects analysis was conducted during the design of the system, the information from that FMEA may be used as the basis for troubleshooting; if necessary, the failure modes and effects analysis sheets from the old analysis may be updated, using any newly available information.

7.2 ANALYSIS PROCEDURE

Failure modes and effects analysis consists of four steps:

1. Establishment of the scope of the analysis
2. Data collection
3. Preparation of components list
4. Preparation of the failure modes and effects sheets (FMEA sheets)

Each of these steps is described in the following subsections.

7.2.1 Establishment of the Scope

The scope of the failure modes and effects analysis should clearly identify the following:

1. *System Boundaries*: System boundaries should be specified so that no component is left out of consideration.

2. *Extent of the Analysis*: FMEA sheets may include the following information about each potential component failure:

- Underlying causes of the failure
- Possible effects of the failure
- Failure detection
- Safeguards
- Frequency of the failure
- Criticality of the effects of the failure

All failure modes and effects analyses include information about the first two items in the preceding list. Information about the other items may or may not be included, depending on the scope of the analysis. What information is to be included in the FMEA sheets should be specified in the scope of the FMEA.

The extent of a failure modes and effects analysis may depend on when it is performed. If two failure modes and effects analyses are performed for the same system, one at the conceptual design stage and another at the detailed design stage, the extent of the latter analysis may be broader than the former. The extent of a failure modes and effects analysis should be decided on a case-by-case basis depending on the purpose of the analysis.

7.2.2 Data Collection

The team performing the analysis should have access to all pertinent documents relating to system configurations, designs, specifications, and operating procedures. All these documents may not be available at the time of a failure modes and effects analysis. Whatever is available should be used.

If possible, the analysis team should also interview (in person or by telephone) design personnel, operations, testing, and maintenance personnel, component vendors, and outside experts to gather as much information as possible and necessary. Sometimes questionnaires may be sent to these personnel in lieu of interviews.

7.2.3 Preparation of Components List

A list of all components in the system is prepared before examining the potential failure modes of each of those components. Functions, operating conditions (temperature, loads, pressure, etc.), and environmental conditions of each component may be included in the components list. Such a list will be very useful during the preparation of FMEA sheets.

7.2.4 Preparation of FMEA Sheets

Findings of the failure modes and effects analysis are recorded in a tabular format in FMEA sheets. Table 7.1 shows sample FMEA sheet. There are 10 items to be entered; some of the items may be omitted or left blank, depending on the scope of the analysis and the information available at the time the FMEA sheets are prepared. (*Note:* We have listed each item in a row. Some analysts prefer to list them columnwise. Either practice is acceptable.)

1. *Component:* The unique identifying name or code of the component is entered here. It is not sufficient to enter "valve"; it is prudent to enter the identifier name or code, for example, valve 1 or valve B2K. This name should correspond to the names used in system drawings, design drawings, or other pertinent documents. If it is a part of a component, it may be listed as, for example, impeller blades of pump 17J.

Table 7.1. Sample FMEA Sheets

	(1)	(2)
1. component	stop valve	stop valve
2. function	controls the flow of gas	controls the flow of gas
3. failure mode	fails to close	fails to open
4. causes of failure	corrosion	corrosion
5. effects of failure	abnormal increase in water pressure and temperature	failure to produce hot water
6. failure detection	(i) water temperature too hot; (ii) SRV opens (if the SRV fails to open, water tank may rupture)	water temperature too low
7. safety features	SRV opens and excess pressure in water tank is relieved	none
8. failure frequency	possible (III)	possible (III)
9. criticality of effects	critical (IV) (*Note:* "critical" only if SRV fails to open)	major (III)
10. remarks	none	none

2. *Function:* A brief statement of the intended function of the component in different modes of operation is entered here. Some failure modes and effects analyses (FMEAs) omit this item.

3. *Failure Modes:* Possible ways in which the component can fail to function as intended are listed here. Failures due to degradation with age (corrosion, fatigue, etc.) should be considered. Possible failure modes in all operating modes of interest (automatic, manual, test, bypass, etc.) should be considered. Operation or shutdown of the system under all possible environmental conditions (earthquake, tornado, flood, etc.) should be considered, as applicable; failure to shut down, when necessary, could cause property damage and personnel injury. Premature operation (for example, a valve that is expected to close at a certain time closes prematurely), failure to operate at the prescribed time (a valve fails to close when required), and failure during operation should be included. Failure modes such as excessive deformations of components (for example, excessive deflection of a spring) and structural failures (for example, collapse of a pipe) should also be included here.

4. *Causes of Failure:* All possible causes of the failure are noted here.

5. *Effects of Failure:* All possible effects of the failure are listed here.

6. *Failure Detection:* How the failure will first become apparent to operating personnel is noted here. Failures may be initially detected because of an alarm, noise, meter reading, cessation of function, etc. Some failures may be detected only during maintenance or testing. This item is omitted in some FMEAs.

7. *Safety Features:* Provisions built into the system that will reduce the failure probability or mitigate the effects of the failure are listed here. This item is omitted in some FMEAs.

8. *Failure Frequency:* If the failure probability (failure frequency) is known, it may be noted here.

In some FMEAs, instead of noting a numerical estimate of the failure probability, a *frequency classification* is noted. The frequency classification system described in Section 6.2.3 may be used.

This item is omitted in some FMEAs. This item may be left blank if sufficient information is not available to estimate the failure frequency at the time the FMEA is conducted.

9. *Criticality of Effects:* Effect of the failure is ranked according to its criticality. The following is a sample ranking system:

I: Insignificant; not a safety hazard and will have very little effect on reliability and availability.

II: Minor; not a safety hazard but will affect reliability and availability somewhat.

III: Major; not a safety hazard but will affect reliability and availability significantly.
IV: Critical; potential safety hazard.

One may establish other systems of criticality ranking but whatever system is used, it should be used consistently throughout the FMEA.

The criticality ranking is based on engineering judgement. It is omitted in some FMEAs.

10. *Remarks:* Any comments relevant to the failure modes and effects are entered here. Recommendations for system changes may also be made here.

7.3 BENEFITS AND LIMITATIONS

Failure modes and effects analysis concentrates on identifying possible component failures and their effects on the system. Design deficiencies are thus identified and improvements can be made. Because potential failures are identified, effective test programs can be recommended. Failure modes may be prioritized according to their frequency and criticality so that more effort can be spent in fixing the higher-priority failure modes. (The prioritizing procedure is similar to that discussed in Section 6.3.)

A limitation of the failure modes and effects analysis is that it is a "single-failure analysis." That is, each failure mode is considered individually. If a failure mode, all by itself, affects the system performance, this is identified by the failure mode and effects analysis. There are many cases, particularly in complex systems, where a single failure may not adversely affect the system, but two or more failures together may adversely affect the system. Failure modes and effects analysis is not well-suited for assessing the combined effects of two or more failures. Deductive methods such as the fault tree analysis (Chapters 8–10) are better-suited for this purpose.

Although a failure modes and effects analysis does not provide all the answers to reliability and safety problems, it does provide much useful information at an early stage in the design process.

7.4 DOCUMENTATION

Final documentation may include the following:

1. Scope of the analysis
2. Components list
3. FMEA sheets

References (drawings, specifications, reports, vendor data, etc.) from which information is obtained should be clearly identified and the dates or version numbers of these references should be indicated.

If more than one failure modes and effects analysis is conducted at different stages of the design process (during conceptual design stage, preliminary design stage, final design stage, etc.), then the FMEA sheets should identify at what stage of the design process the particular sheets were prepared. A progressively increasing version number may be assigned (version 1, 2, ...) and noted on the FMEA sheets.

Some general guidelines on documentation may be found in Section 14.10.

7.5 EXAMPLE PROBLEM[1]

Consider the domestic hot water system described in Section 6.5 and shown in Figure 6.1.

Sample FMEA sheets for this system are given in Table 7.1, where we have made entries for just one component, namely, the stop valve (gas valve). There are other components, for example, the pressure relief valve, the temperature measuring and comparing device, and the check valve. All these components would be included in an actual analysis.

There are two failure modes associated with the stop valve and both are considered in Table 7.1.

EXERCISE PROBLEM

7.1. Consider the domestic hot water system described in Section 6.5. FMEA sheets with entries for just one component are given in Table 7.1. Perform a complete failure modes and effects analysis for the domestic hot water system, considering all the components.

REFERENCE

Lambert, H. E. (1973). *System Safety Analysis and Fault Tree Analysis*. Report No. UCID-16238. Lawrence Livermore National Laboratory, Livermore, CA.

---

[1]This example is based on Lambert (1973).

Chapter 8
Fault Tree Construction

8.1 INTRODUCTION

Fault tree analysis is one of the most widely used methods for system reliability analysis. It is a formal deductive procedure for determining the various combinations of component-level failures[1] that could result in the occurrence of specified "undesired events" at the system level. The undesired event could be an accident or some other system-level failure. Fault trees may also be used to compute the probability of the undesired event as a function of the probabilities of component failures. We may construct a number of fault trees for the same system if we are interested in studying a number of undesired events relating to that system. Each undesired event will have a different fault tree.

Fault trees may be constructed at any stage of system design. If a fault tree is constructed at an early stage of system design, it may be updated as more precise information becomes available or as changes are made in the design. Fault trees may also be constructed for operating systems to identify the root causes of system-level failures encountered during operation.

A fault tree analysis may follow a preliminary hazard analysis or a failure modes and effects analysis, although neither a preliminary hazard analysis nor a failure modes and effects analysis is a prerequisite.

A full fault tree analysis consists of the following steps:

1. Fault tree construction.
2. Qualitative fault tree analysis.
3. Quantitative fault tree analysis (a quantitative analysis may be performed without first performing a qualitative analysis).

Depending on the scope of the reliability project, either steps 1 and 2, steps 1 and 3, or steps 1, 2, and 3 may be performed.

---

[1]Not only hardware (component) failures but also human errors and software errors are considered "component-level failures" within the context of fault tree construction and analysis. We use the term "failure" to refer to both failures and faults (see Section 3.2 for distinctions between failures and faults).

We discuss fault tree construction in this chapter. Qualitative analysis and quantitative analysis are discussed in Chapters 9 and 10, respectively.

Before we go into fault tree construction, a word of caution is in order. If two reliability analysts develop fault trees for the same undesired event of a system, the two fault trees may not necessarily be identical; they may seem different at the outset, particularly if the system is complex. This is because the way in which system logic is modelled by the two analysts may differ. However, the two fault trees *must* provide the same results when a qualitative and/or quantitative analysis is carried out. Both trees should provide identical minimal cut sets and identical probability for the undesired event. (Minimal cut set is defined in Chapter 9.)

8.2 AN EXAMPLE FAULT TREE

A *fault tree* is a diagrammatic representation of the relationship between component-level failures and a system-level undesired event. It depicts how component-level failures propagate through the system to cause a system-level failure (system-level undesired event). The component-level failures are called the *terminal events*, primary events, or end events of the fault tree; we will mostly use the term "terminal event" in this book. The system-level undesired event is called the *top event* of the fault tree.

The concept of fault trees is best explained through a simple illustration. We have purposely chosen a very simple example; fault trees encountered in practice could be tens or hundreds of times larger than this simple illustration.

A system diagram for the operation of an electric motor is shown in Figure 8.1. We are interested in developing a fault tree for the undesired event 'motor overheats.' The fault tree is shown in Figure 8.2. Let us discuss how we draw that fault tree.

First, we place the undesired event at the top of the tree, within a *rectangle*. The undesired event, in this case, is 'motor overheats.' For the

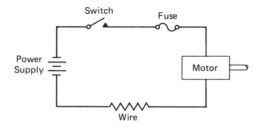

Figure 8.1. System diagram for the operation of an electric motor.

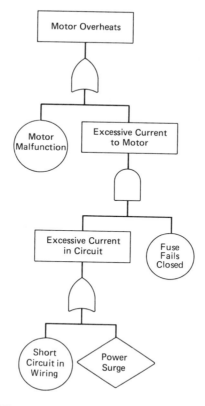

Figure 8.2. Fault tree for 'motor overheats.'

sake of simplicity in further discussions, let us denote it by the symbol A. Next, we ask the question, "How can the motor overheat?" (or, "What is necessary and sufficient for the motor to overheat?"). Motor overheating can happen due to (i) an internal malfunction of the motor itself *or* (ii) excessive current supplied to the motor. So we place the events 'motor malfunction' and 'excessive current to motor' below the top event (the undesired event). Let us denote these events by $B1$ and $B2$. Because A is possible either by $B1$ *or* by $B2$, we connect A to $B1$ and $B2$ by an *OR gate*. At this point, we may either explore the reasons for 'motor malfunction' ($B1$) by examining the failures of the various parts of the motor or treat 'motor malfunction' as a terminal event (component-level failure). In this example, we choose to do the latter. So we place the event 'motor malfunction' within a *circle* to indicate that it is a terminal event. (Criteria for deciding whether to treat an event as a terminal event are discussed in

Section 8.6.3.) We decide to resolve further the event 'excessive current to motor' ($B2$). So this event is called an *intermediate event* and is placed within a rectangle. (Top events and intermediate events are placed within rectangles.) We now ask the question, "How can there be excessive current to the motor?" (or, "What is necessary and sufficient to cause excessive current to the motor?"). Excessive current to the motor ($B2$) can happen only if (i) there is 'excessive current in the circuit' ($C1$) *and* (ii) the 'fuse fails closed' ($C2$). Because both $C1$ *and* $C2$ have to happen, we connect $B2$ to $C1$ and $C2$ by an *AND gate*. We do not plan to resolve the event 'fuse fails closed' ($C2$) further, so it is a terminal event and is placed within a circle. We do plan to resolve the event 'excessive current in the circuit' ($C1$) further, so it is an intermediate event to be placed within a rectangle. We ask the question, "How can there be excessive current in the circuit?" (or, "What is necessary and sufficient to cause excessive current in the circuit?"). Either (i) 'short circuit in the wiring' ($D1$) *or* (ii) a 'power surge' ($D2$) could produce excessive current in the circuit. Because either $D1$ or $D2$ can produce $C1$, we connect $C1$ to $D1$ and $D2$ by an OR gate. We do not plan to resolve the events 'short circuit in the wiring' ($D1$) or 'power surge' ($D2$) further, so these events are treated as terminal events. $D1$ is placed within a circle to indicate that it is a terminal event, and $D2$ is placed within a *diamond* (which represents a special type of terminal event that is discussed in Section 8.6.3).

The preceding example illustrates the basic concepts and procedure of constructing a fault tree and introduces (Figure 8.2) some symbols used in fault trees, such as OR gate, AND gate, circle, and rectangle. It also defines some terms used in the fault tree analysis, such as top event, intermediate event, terminal event, OR gate, and AND gate. A number of other symbols and terms are defined in later sections of this chapter. A fault tree for a more complex system is described in Section 8.17.

8.3 SYSTEM DESCRIPTION

Within the context of fault tree analysis, one may consider a complete plant as a system or some distinct parts of a plant as a system (for example, a steam generation system or a distillation system), or even a piece of equipment may be treated as a system.

The first step in any reliability analysis, including fault tree analysis, is understanding the system well. A clear understanding of the system, as it pertains to the fault tree, is necessary. Detailed information about each and every component in the system, the functional and physical interconnections between components, and the normal and abnormal environments experienced by each component are gathered from many sources: drawings; schematic diagrams; block diagrams; logic diagrams; piping and instrumentation diagrams; process flow sheets; installation diagrams; parts lists; operating procedures; maintenance–test procedures; discussions with plant personnel, etc. In addition, interfaces of the system with other systems (for example, with a power supply system or with a cooling-water supply system) should be identified.

The reliability analyst may prepare a *system description*, which includes the following information:

1. A list of components (note that some components may be part of more than one system because system boundaries may overlap).
2. Physical interconnections and proximity of components.
3. Physical interfaces with other systems.

The system may function in more than one mode; for example, normal operating mode, emergency operating mode, start-up mode, shutdown mode, preoperational test mode, etc. Only a few of the functional modes may pertain to a particular fault tree. A *functional description* may be prepared for each functional mode of interest. Functional descriptions should include the following information:

1. The function of each component.
2. The initial conditions of each component (valve is closed or open, tank is empty or full, motor is operating or idle, etc.).
3. Functional interfaces between components.
4. Normal operating and environmental conditions of each component (temperature, pressure, mechanical stress, vibrations, etc.).
5. Abnormal operating and environmental conditions of each component during emergencies or accidents.
6. Failure modes of each component.
7. Dependence between component failure modes.
8. Functional interfaces with other systems (supply of power, water, fuel, etc.).
9. Function of operators (human).
10. Computer controls, if applicable.
11. Operation–maintenance–test procedures relevant to the fault tree (for example, motor A and motor B should not be undergoing maintenance at the same time, etc.).

Preparation of system descriptions and functional descriptions is not mandatory for fault tree construction, but they will be immensely helpful in constructing, reviewing, and updating fault trees, particularly for complex systems.

8.4 SYMBOLS

Symbols used to represent gates, events, and transfers in fault trees are summarized in Figures 8.3, 8.4, and 8.5, respectively. Detailed description of gates, events, and transfers are provided in Sections 8.5, 8.6, and 8.7, respectively.

8.5 GATES

We have already mentioned OR gates and AND gates in Section 8.2. There are other types of gates, such as the tabular OR gates, tabular AND gates, m-out-of-n gates, exclusive OR gates, priority AND gates, inhibit gates, and AND–NOT gates. OR gates and AND gates are the most widely used gates, and some fault tree analysis computer programs may accept only these two gates. As is shown later in this section, virtually all types of gates can be reduced to OR gates and AND gates through the proper transformations.

8.5.1 OR Gate

The *OR gate* is used to indicate that the output event occurs if and only if at least one of the input events occurs (Figure 8.6). The output event could be the top event or an intermediate event. The input events could be terminal events or intermediate events or a combination of both. There should be at least two input events to an OR gate. (A vertical line followed by a few dots indicates that the part of the fault tree below that particular vertical line is not shown in the figure because that part of the tree is not particularly relevant to the discussion here. We use this convention throughout the book.)

8.5.2 AND Gate

The *AND gate* is used to indicate that the output event occurs if and only if all the input events occur (Figure 8.7). All the input events need not occur simultaneously; they may occur at different times. The output event will occur when all the input events are present. The output event could be the top event event or an intermediate event. The input events could be terminal events, intermediate events, or a combination of both. There should be at least two input events to an AND gate.

	Symbol	Description

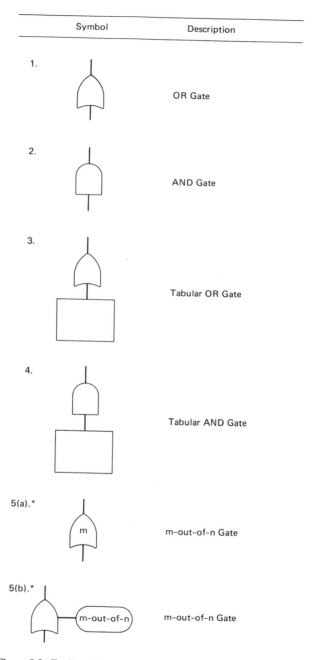

1. OR Gate

2. AND Gate

3. Tabular OR Gate

4. Tabular AND Gate

5(a).* m-out-of-n Gate

5(b).* m-out-of-n Gate

Figure 8.3. Fault tree symbols for gates (continued on next page).

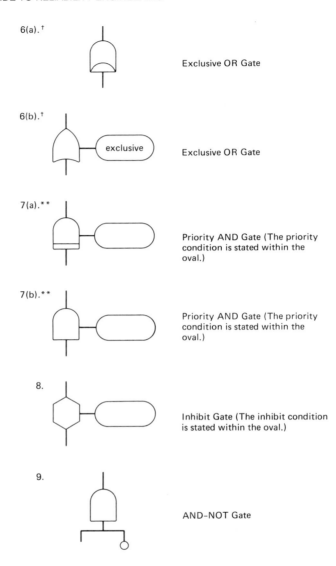

6(a).[†]   Exclusive OR Gate

6(b).[†]   exclusive   Exclusive OR Gate

7(a).**   Priority AND Gate (The priority condition is stated within the oval.)

7(b).**   Priority AND Gate (The priority condition is stated within the oval.)

8.   Inhibit Gate (The inhibit condition is stated within the oval.)

9.   AND–NOT Gate

*Either 5(a) or 5(b) may be used to denote "m-out-of-n Gates".
[†] Either 6(a) or 6(b) may be used to denote "Exclusive OR Gates".
**Either 7(a) or 7(b) may be used to denote "Priority AND Gates".

Figure 8.3. (Continued).

Symbol	Description
1.	 Circle — Basic Event
2.	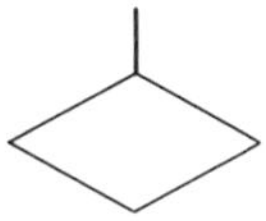 Diamond — Undeveloped Event
3.	 Double Diamond — Undeveloped Event (to be developed at a later stage. *Note*: some analysts use a single diamond for this also.)
4.	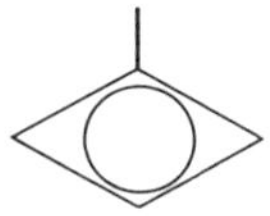 Circle Within a Diamond — Undeveloped Event (separate subtree is to be constructed. *Note*: some analysts use a single diamond for this also.)
5.	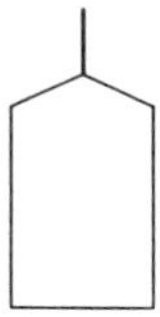 House — House Event (also known as switching event and trigger event.)
6.	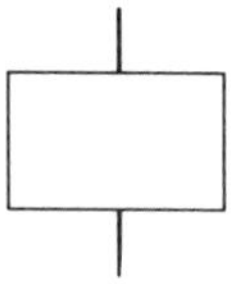 Rectangle — Intermediate Event
7.	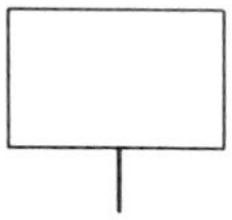 Rectangle — Top Event

Figure 8.4. Fault tree symbols for events.

	Symbol	Description
1(a).*		Transfer-in
1(b).*	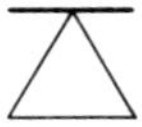	Transfer-in
2(a).†		Transfer-out
2(b).†	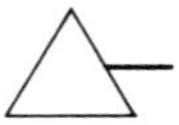	Transfer-out
2(c).†		Transfer-out

*Either 1(a) or 1(b) may be used to denote a "transfer-in."
† 2(a), 2(b), or 2(c) may be used to denote a "transfer-out."

Figure 8.5. Fault tree symbols for transfers.

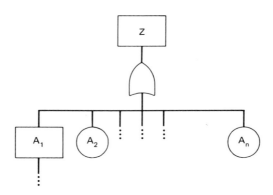

Figure 8.6. OR gate.

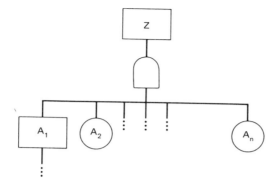

Figure 8.7. AND gate.

8.5.3 Tabular OR Gate

When there are a large number of terminal events as inputs to an OR gate, the OR gate may be represented by a *tabular OR gate* (Figure 8.8). All the input events should be terminal events. The output event could be the top event or an intermediate event.

Tabular OR gates are used as a convenience, because a tabular OR gate is more compact than the corresponding OR gate with a large number of input events.

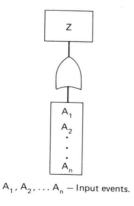

$A_1, A_2, \ldots A_n$ — Input events.

Figure 8.8. Tabular OR gate.

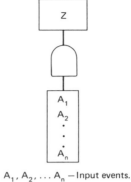

$A_1, A_2, \ldots A_n$ — Input events.

Figure 8.9. Tabular AND gate.

8.5.4 Tabular AND Gate

When there are a large number of terminal events as inputs to an AND gate, the AND gate may be represented by a *tabular AND gate* (Figure 8.9). All the input events should be terminal events. The output event could be the top event or an intermediate event.

Tabular AND gates are used as a convenience, because a tabular AND gate is more compact than the corresponding AND gate with a large number of input events.

8.5.5 *m*-out-of-*n* Gate

The *m-out-of-n gate* is used to indicate that the output event occurs if and only if at least *m* input events out of the *n* input events occur (Figure 8.10a). The *m* input events need not occur simultaneously; they may occur at different times. The output event will occur when at least *m* input events are present. When $m = 1$, the *m*-out-of-*n* gate reduces to an OR gate. An *m*-out-of-*n* gate can be replaced by a combination of OR gates and AND gates, as shown in Figure 8.10b. Note that we have introduced three intermediate events ($D1$, $D2$, and $D3$) in Figure 8.10b. $D1$ represents the intermediate event 'A and B occur,' $D2$ represents the intermediate event 'B and C occur,' and $D3$ represents the intermediate event 'C and A occur.'

The output event of an *m*-out-of-*n* gate could be the top event or an intermediate event. The input events could be either terminal events or intermediate events or could be a combination of both.

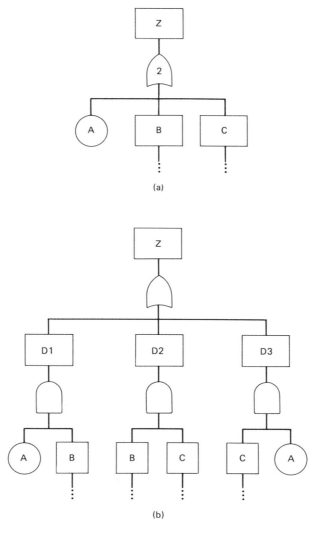

Figure 8.10. *m*-out-of-*n* gate for *m* = 2 and *n* = 3: (a) *m*-out-of-*n* gate; (b) equivalent representation.

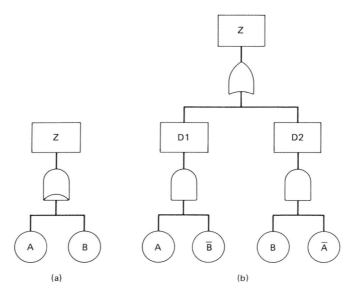

Figure 8.11. Exclusive OR gate: (a) exclusive OR gate; (b) equivalent representation.

8.5.6 Exclusive OR Gate

The *exclusive OR gate* is used to indicate that the output event occurs if only one of the input events occurs; all the other input events should be absent. If $I_1, I_2, \ldots, I_n$ are the input events, the output event Z occurs only when I_i, where $i = 1, 2, \ldots$, or n, occurs and all the other input events I_j, where $j \neq i$, are absent. Suppose there are three input events, A, B, and C, and the output event is Z. Then Z occurs if (i) A occurs and B and C are absent or (ii) B occurs and A and C are absent or (iii) C occurs and A and B are absent.

The output event of an exclusive OR gate could be the top event or an intermediate event. The input events could be either terminal events or intermediate events or could be a combination of both.

Figure 8.11a illustrates the use of an exclusive OR gate. The exclusive OR gate can be replaced by a combination of OR gates and AND gates, as shown in Figure 8.11b; $\overline{A}$ and $\overline{B}$ are the complements of A and B, respectively. $\overline{A}$ represents the nonoccurrence of A, and $\overline{B}$ represents the nonoccurrence of B. If A is 'failure state of component X,' then $\overline{A}$ is 'success state (normal state) of component X.'

The use of "success events" is not advisable in quantitative fault tree analysis because success events usually have a high probability of occurrence and quantitative analysis is difficult if there are high-probability

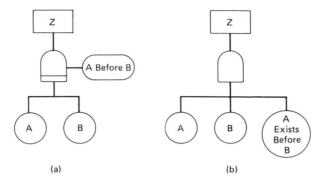

Figure 8.12. Priority AND gate: (a) priority AND gate; (b) equivalent representation.

terminal events in the fault tree. A conservative approximation is to replace an exclusive OR gate by a conventional OR gate during quantitative analysis, provided that the probabilities of the input events are small (say, less than 0.1); such an approximation will slightly increase the output event probability.

8.5.7 Priority AND Gate

The *priority AND gate* is used to indicate that the output event occurs when both of the input events are present but one of the input events should precede the other, as specified by the priority condition (Figure 8.12a). A priority AND gate can be replaced by a conventional AND gate as shown in Figure 8.12b.

Consider the situation where a fire starts in a factory and the fire alarm fails to alert the fire department; consequently there occurs extensive damage to the factory. In this case, the output event 'extensive damage to the factory' occurs only if the fire alarm failure precedes the fire. If the fire precedes the firm alarm failure, the fire starts first, the firm alarm alerts the fire department and then fails, so the output event 'extensive damage to the factory' does not occur. So, in this case, it is a necessary condition that the fire alarm failure (A) precede the fire (B).

There could be more than two input events to a priority AND gate, but which event or events should precede the other event or events, and in what order, should be specified. Let there be three input events, A, B, and C. The priority condition could be C precedes A and B; in such a case, A and B may occur in any order but C should occur before both A and B. Another example of a priority condition is that C precedes B and B precedes A; in this case, C should occur first, B next, and finally A.

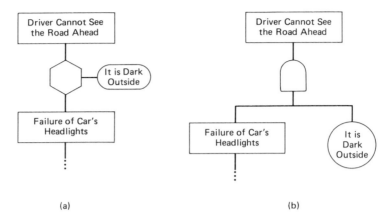

(a)   (b)

Figure 8.13. Inhibit gate: (a) inhibit gate; (b) equivalent representation.

The output event of a priority AND gate could be the top event or an intermediate event. The input events could be either terminal events or intermediate events or a combination of both.

8.5.8 Inhibit Gate

The *inhibit gate* is used to indicate that the output event occurs when the input event occurs (or is present) and the inhibit condition is satisfied. There is only one input event to an inhibit gate. The input event could be a terminal event or an intermediate event. The output event could be the top event or an intermediate event.

Figure 8.13a shows the use of an inhibit gate. The output event ('driver cannot see the road ahead') occurs because of the input event ('failure of car's head lights'), but the input event can lead to the output event only when the inhibit condition ('it is dark outside') is satisfied.

Both the input event and the inhibit condition should be present for the output event to occur. In fact, the inhibit gate is a special case of the AND gate; the only difference is that the inhibit condition is not necessarily an event but could be a condition. The inhibit gate can be replaced by an AND gate as shown in Figure 8.13b.

8.5.9 AND – NOT Gate

The *AND–NOT gate* indicates that the output event occurs if one input event occurs and the second input event does not occur, as specified (that is, one input event is present and the other is absent, as specified) (Figure

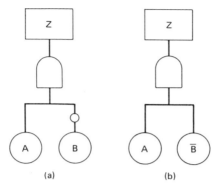

Figure 8.14. AND–NOT gate: (a) AND–NOT gate (A AND–NOT B); (b) equivalent representation.

8.14a). The fault tree in Figure 8.14a indicates that the output event Z occurs if the input event A occurs and the other input event B does not occur (the input event that should not occur is identified by a small circle between the AND gate and the input event).

An AND–NOT gate can have more than two input events. Those input events that should not occur are identified by a small circle between the event and the AND gate. Those input events that should occur will not have the small circle. The output event could be the top event or an intermediate event. The input events could be either terminal events or intermediate events or could be a combination of both.

The AND–NOT gate can be replaced by an AND gate as shown in Figure 8.14b. In that figure, the event $\bar{B}$ refers to the complement of event B. If B represents the failure of a component in a particular failure mode(s), then $\bar{B}$ represents the absence of that failure mode(s).

8.6 EVENTS

Three types of event, namely, top event, intermediate event, and terminal event, appear in fault trees. There are three types of terminal event, namely, basic event, undeveloped event, and house event. These different types of event are defined and discussed in the following subsections. The symbols used to represent these events are summarized in Figure 8.4.

8.6.1 Top Event

Top events are also referred to as undesired events in some of the literature.

The scope of a fault tree analysis is to determine the combinations of component-level failures that can cause a system-level undesired event, and/or to determine the probability of occurrence of the undesired event. There is only one undesired event (top event) in a fault tree. However, there can be more than one undesired event associated with a system, and, if necessary, we can construct more than one fault tree in relation to a system (one fault tree for each undesired event associated with the system).

The undesired event (top event) appears at the top of the fault tree and is always placed within a rectangle (see Figure 8.2). The undesired events could be 'accidental conditions' (explosion, toxic gas release, etc.), 'system shutdown,' or 'system malfunction (failure to perform as intended).'

If time is an essential factor, the time frame should be specified for the undesired event. An example is 'secondary feedwater supply system fails within 12 hours of main feedwater supply system failure.' In some cases, time duration may be important, for example, 'feedwater pump is in a failed state for 36 hours.' Here, failure of the pump in itself is not critical if it can be restored to operation within 36 hours, because there is a reserve water tank that can supply the necessary water for 36 hours.

Qualitative information may be an essential factor in some problems, for example, 'pump delivers less than 80 gallons of water per minute.'

8.6.2 Intermediate Event

Any event within the fault tree (except for the top event) that is further resolved into events that could cause it is called an intermediate event. In the example discussed in Section 8.2 (Figure 8.2), 'excessive current to motor' and 'excessive current in circuit' are the intermediate events.

Intermediate events are placed within rectangles in a fault tree (see Figure 8.2).

8.6.3 Terminal Event

Terminal events are also referred to as *end events* and *primary events* in some of the literature.

A *terminal event* is an event that is not resolved further into its causes. Terminal events are at the bottom of the fault tree and are represented by either circles or diamonds. Whether an event should be resolved further into its causes is a decision to be made by the reliability analyst. The decision depends on the availability of quantitative data for the event and the level of detail (level of resolution) the analyst wants in the fault tree. If the probability of the undesired event is to be computed (quantitative fault tree analysis), then quantitative data for the terminal events (probabilities of the terminal events) is a necessity. However, just because we know the probability of an event we need not necessarily treat it as a terminal event; we may still treat it as an intermediate event and resolve it into its causes.

Some general guidelines for deciding on the level of resolution are given next.

1. Usually, the terminal event is a component-level event, an event caused by another system, or an external event. (Component-level events include component failures, human errors, and software errors. 'Abnormally high temperature due to air-conditioning system failure' and 'no power because of external power supply failure' are examples of events caused by another system. Earthquakes and floods are examples of external events.)

2. Probabilities of the terminal events should be available if quantification of the fault tree (computation of the probability of occurrence of the undesired event and/or intermediate events) is an objective.

3. If the probability of a component-level event is not known, it may be treated as an intermediate event and further resolved to failures at the parts level (parts of the component) for which failure probabilities are available. Note that a component may have more than one failure mode and each failure mode may constitute an event; some failure modes may be treated as terminal events whereas others may best be treated as intermediate events.

4. Sometimes, even subsystem-level events are treated as terminal events if further resolution of the event will not improve our understanding of the problem. This is acceptable if (i) the probability of the subsystem-level event is available, (ii) components of the subsystem do not enter the fault tree as terminal or intermediate events elsewhere in the fault tree, and (iii) there is no statistical dependence between components of the subsystem and other events in the tree.

There are three types of terminal event:

(i) Undeveloped events
(ii) Basic events
(iii) House events

8.6.3.1 Undeveloped Event.   Undeveloped events are used in the following situations.

1. A subsystem-level event is used as an undeveloped terminal event if further resolution of that event will not improve our understanding of the problem. These events are represented by diamonds (see Figure 8.4).

2. A subsystem-level event is treated as an undeveloped terminal event in fault trees developed during early stages of the project, but will be treated as an intermediate event and resolved further at a later date. Such events are represented either by a diamond or by a double diamond (see Figure 8.4).

3. An outside event (that is, an event caused by a system or subsystem outside the boundaries of the system under analysis) is treated as an undeveloped terminal event. Such events are represented by a diamond (see Figure 8.4). There should be no common components between the system under analysis and the outside system or subsystem. Also, there should be no statistical dependence between components of the outside system (or subsystem) and events in the fault tree.

4. Sometimes a subsystem event is treated as an undeveloped terminal event in the main fault tree but a separate subtree is constructed for this event. This subsystem event will be the top event of the subtree and the probability of the top event of this subtree is transferred to the main tree as the probability of the undeveloped terminal event. It is important that the subtree be independent of the other events of the fault tree; no event from the main tree should appear on the subtree and there should be no statistical dependence between events of the main tree and the subtree. Undeveloped terminal events of this type are represented by a diamond or a circle within a diamond (see Figure 8.4).

8.6.3.2 Basic Event.   A *basic event* is either a component-level event that is not resolved further or an external event. Component-level events include component failures, human errors, and system failures. Although component-level events treated as basic events are usually failure events, success events may also enter a fault tree (see Section 8.10). External events include earthquakes, floods, tornados, etc.

A component failure is a state of the component; for example, a valve in a closed state (when it should be open), a pipe in a ruptured state, or a pump in an idle state (when it should be operating). When a component is in a failed state, it either does not perform its intended function correctly or it performs an unintended function. For example, consider a fire alarm. It is a failure if the alarm fails to sound when there is a fire; it is also a failure (fault) if the alarm sounds when there is no fire.

Degrees of component failure: There may be different *degrees* of component failure. For example, a one-gallon-per-minute (1-gpm) leak from a pipe may not have the same effect as a 10-gpm leak. Such degrees of failure may be treated in two ways. In the first approach, we set a limit below which it is not considered a failure; for example, if the leakage rate is below 0.5 gpm, it is not a failure. A corollary conservative approach is to treat any leak at all as a failure and to assume that even a small leak has the same effect as a large leak. The second approach is to draw different fault trees with different degrees of failure: one tree with a 1-gpm leak, another with a 2-gpm leak, and so forth. Such an approach could result in numerous fault trees, and thus be impractical to use, if there are a number of components, each with many different degrees of failure.

Primary, secondary, and command failures: Component failures are classified into three categories:

1. Primary failures
2. Secondary failures
3. Command failures (signal failures)

A *primary failure* is the result of a deficiency in the component, and the failure occurs even when the operating and environmental conditions are within design limits. Rupture of a pressure vessel even when the operating temperature and pressure are within design limits is an example of a primary failure.

A *secondary failure* is the result of abnormal operating and/or environmental conditions (operating and/or environmental conditions outside design limits); the component may not have any deficiency and may operate without failure under normal operating and environmental conditions. Operating or environmental conditions causing secondary failures include abnormal temperature, pressure, load, speed, vibration, electric current, moisture, dust, and chemical concentration.

The abnormal conditions could be the result of one or more of the following causes:

1. *External Events:* earthquake, flood, tornado, hurricane, external explosion, etc.
2. *Internal Events:* internal explosion, fire, etc.
3. *Malfunction of Other Components:* for example, a safety relief valve failure results in abnormal pressures in a pressure vessel.
4. *Malfunction of Other Systems:* for example, an air-conditioning system failure results in abnormal temperatures in an electronics component panel.
5. *Installation–Maintenance–Test Errors:* technicians fail to restore the component to its proper configuration; for example, technicians leave a valve open instead of closed.

These are known as *secondary causes*. Some of these secondary causes could affect more than one component simultaneously, and such sec-

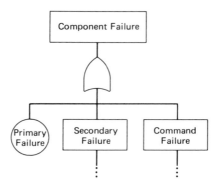

Figure 8.15. Resolution of a component failure into primary, secondary, and command failures.

ondary causes are called *common causes* (more details on common causes are provided in Section 5.2.9, Section 8.8, and Chapter 11). Common causes will appear in more than one place in the fault tree.

A *command failure* (*signal failure*) is a component failure due to an incorrect command or an incorrect signal input to the component. For example, a sprinkler system fails to extinguish a fire because the signal that is to trigger the sprinkler system does not operate.

A component failure may be resolved into a primary failure, a secondary failure, and a command failure, as necessary (Figure 8.15). All three types of failure may not be present in every situation; if only one type of failure (say, a primary failure) is present, the event 'component failure' may not have to be resolved further. If only two types of failure are present (say, a primary failure and a command failure), we will have just two events, instead of three as in Figure 8.15, under the event 'component failure.' Whether a primary, a secondary, or a command failure has the highest probability of occurrence depends on the problem. In general, there could be more than one secondary failure and/or more than one command failure. If there are m secondary failures and n command failures, then the component failure is resolved into $(m + n + 1)$ events. Any primary, secondary, or command failure that has a significantly smaller probability of occurrence (about an order of magnitude smaller than the highest-probability event among them) may be discarded, provided (i) the sum of the probabilities of the events thus discarded is at least an order of magnitude less than the probability of the highest-probability event and (ii) the discarded events or their causes are not statistically dependent with other events in the tree because of common causes or other reasons. [*Note:* Some analysts prefer to include all the primary, secondary, and command failures in the tree. The low-probability events among them that satisfy the two conditions just stated are later discarded from the tree through *fault tree reduction* (see Section 8.12)].

If possible, a number of secondary failures should be combined together as a single event, provided (i) the total probability of the combined event is known, (ii) the secondary causes of these secondary failures do not appear elsewhere in the tree, and (iii) these secondary causes are not statistically dependent with other terminal events in the tree. The same is true with command failures also.

Usually the primary failure is treated as a terminal event, unless we want to resolve it into failures at the parts level. Secondary failures and command failures are treated as terminal events or intermediate events, as appropriate.

Some secondary causes may not produce secondary failures every time they occur; instead, they may produce secondary failures only a fraction of the time. For example, an increase in temperature beyond a certain limit may induce failure of an electronic component only about 20% of the time (based on analytical studies, test data, or operational experience); that is, the possibility that high temperature induces secondary failure of the electronic component is 0.2. Such a secondary event is represented through an inhibit gate (Figure 8.16). The inhibit gate and the inhibit condition may be omitted and Figure 8.15 may be used as a conservative approximation; the higher the conditional probability, the less is the error due to such an approximation.

In cases where the probabilities of primary, secondary, and command failure are not known separately, we may not be able to resolve the event 'component failure' as primary, secondary, and command failures during the quantitative analysis; we may have to combine them as a single event during the quantitative analysis.

8.6.3.3 House event.

House events are also referred to as *trigger events* and *switching events*.

A *house event* is a special type of terminal event that can be "turned on" or "turned off" by the reliability analyst. When a house event is turned on, that event is presumed to have occurred; the probability of a turned on house event is 1. When a house event is turned off, it is presumed not to have occurred; the probability of a turned-off house event is 0.

House events are used primarily to study the failure behavior of systems under different scenarios. Consider a plant that has two cooling-water injection systems; only one is needed, the other is a back-up system. Let the two systems (system A and system B) be not identical, that is, they are

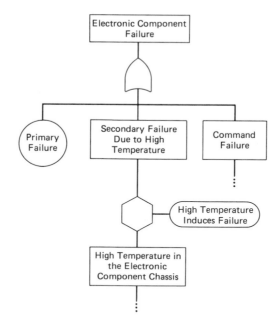

Figure 8.16. Secondary failure with inhibit gate.

of different design or manufacture (component configurations, or component failure probabilities, are different). So the failure behavior of the plant will depend on which injection system is used. We may construct two different fault trees: one to represent the plant operation with cooling-water injection through system A and another to represent the plant operation with cooling-water injection through system B. Of course, only parts of the fault trees relating to the cooling-water injection system will be different; other parts of the trees will be identical. A more efficient way is to construct a single fault tree that can be used to represent either case (use of system A or system B).

Such an application is illustrated in Figure 8.17. A house event HA is attached to the event 'failure of system A' through an AND gate, and another house event HB is attached to the event 'failure of system B' through another AND gate. (In general, the events 'failure of system A' and 'failure of system B' could be intermediate or terminal events.) By turning on the house event HA and turning off the house event HB, the scenario in which injection system A is used is represented by the fault tree. When HB is turned on and HA is turned off, the scenario in which injection system B is used is represented. The use of house events is

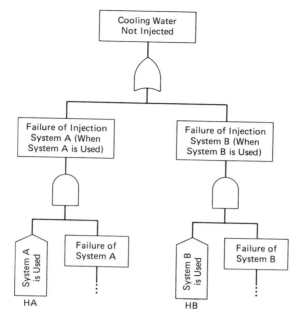

Figure 8.17. Use of house events.

particularly useful when the fault tree is analyzed using computer programs. A single tree can be input to the program and the two scenarios can be considered by simply changing just a few of the data (just change the house events from ON to OFF or vice versa).

Although we have illustrated the use of house events by considering only two scenarios, any number of scenarios can be considered using as many house events.

When using a house event, make sure that it turns off only those parts of the tree that need to be turned off. When a house event is turned off, the branch of the tree containing the house event will be effectively removed from the tree until that branch reaches an OR gate. In Figure 8.18, if the house event HS is turned off, it will remove the part of the tree enclosed within broken (dashed) lines. The effective fault tree when the house event HS is turned off is shown in Figure 8.19. The effective tree when HS is turned on is shown in Figure 8.20. When the house event HS is turned on, HS is presumed to have occurred. Therefore, the event P occurs whenever the event R occurs. So we may say that $R = P$ when HS is turned on. That is why the two events P and R are shown as '$R (= P)$' in Figure 8.20.

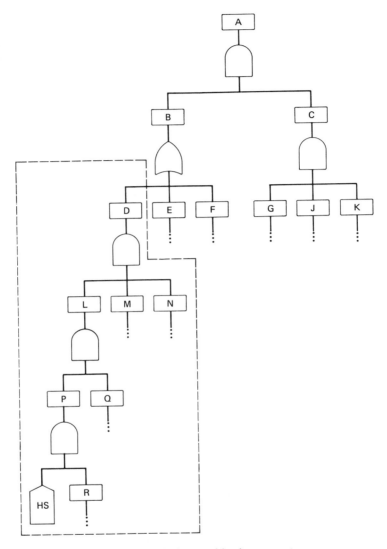

Figure 8.18. Fault tree with a house event.

If a quantitative fault tree analysis program is used, simply input whether each house event in the fault tree is ON or OFF; the program will automatically compute the top event probability of the effective fault tree. If manual calculations are carried out to compute the top event probability, either (i) construct the effective fault tree and perform a quantitative analysis of that tree or (ii) perform a quantitative analysis of the full fault

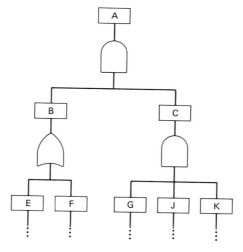

Figure 8.19. Effective fault tree when the house event *HS* in Figure 8.18 is turned off.

tree using a probability of 1.0 for the turned-on house events and a probability of 0.0 for the turned-off house events. There is no need to construct the effective fault tree if the second approach is used.

8.6.3.4 Identification of terminal events. In order to keep track of the large number of terminal events in a typical fault tree, each terminal event is identified by a unique name or code. If a terminal event appears in a fault tree in more than one place, the same name should be used in all those places. No two terminal events should have the same name.

Although we may name terminal events in any way we please, a systematic method is advisable. The *Probabilistic Risk Assessment Procedures Guide* [Nuclear Regulatory Commission (1983)] recommends an eight-character naming system. We present here a slightly different eight-character scheme. Each character can be either a letter (A through Z) or a number (0 through 9).

First character: Identifies the system to which the component belongs (see Table 8.1). Because of overlapping system boundaries, the same component may belong to two or more systems; in such situations, arbitrarily choose one of the systems for the purpose of naming the component. If more than one fault tree is developed for those systems, the same component name should be used in all those trees.

Second and third characters: Identify the component type (see Table 8.2).

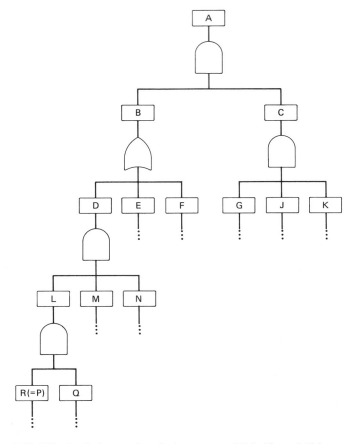

Figure 8.20. Effective fault tree when the house event *HS* in Figure 8.18 is turned on.

Table 8.1.   Sample System Identifiers

SYSTEM NAME	IDENTIFIER
system-independent (or not applicable)	0
feedwater system	1
electric power supply system	2
heating and ventilation system	A
fuel system	B

Table 8.2.   Sample Component-Type Identifiers

COMPONENT TYPE	IDENTIFIER
not applicable	00
blower	BL
current transformer	CT
external event *	EE
flow transformer	FT
human operator *	HO
limit switch	LS
nozzle	NZ
safety valve	SV
tank	TK
undeveloped event *	UE
vacuum relief valve	VV

*Human operators, external events, and undeveloped events have also been included as "component types."

Fourth character: Identifies the specific component (unique to each component within a subsystem; the same identifier may be used in another subsystem to identify another component).

Fifth character: Identifies the subsystem to which the component belongs (see Table 8.3).

Sixth character: Identifies the failure mode (see Table 8.4).

Seventh and eighth characters: Identify special situations (see Table 8.5).

Table 8.3.   Sample Subsystem Identifiers

SUBSYSTEM NAME	IDENTIFIER
subsystem-independent (or not applicable)	0
diesel generator subsystem	1
fuel storage subsystem	2
feedwater piping subsystem	A

Table 8.4.   Sample Failure-Mode Identifiers

FAILURE MODE	IDENTIFIER
not applicable	0
fails closed	C
fails open	O
leak	L
off-site power off	Z
overload	V
rupture	R

Table 8.5.   Sample Special-Situation Identifiers

SITUATION	IDENTIFIER
not applicable (no special situation)	00
common cause	CC
house event	HE
mutually exclusive events	*
statistically dependent events	SD

*Mutually exclusive events are denoted by the following scheme: If a set of n events are mutually exclusive with each other, the seventh character is the same for all the n events and the eighth character is different for the different events. For example, the seventh character is A (arbitrarily chosen by the analyst) and the eighth character is $1, 2, 3, \ldots, n$. If there is another set of m mutually exclusive events, we may use B as the seventh character and $1, 2, 3, \ldots, m$ as the eighth character.

Additional characters may be added to identify component location and other information.

In order to develop a systematic coding system, first list all the systems, subsystems, component types, components, and their failure modes that will be part of the fault tree. Next, provide a unique identifier for the different systems, subsystems, component types, components, and failure modes. Some sample identifiers are shown in Tables 8.1–8.4. Also list all possible special situations and provide unique identifiers to each of them; some sample identifiers are given in Table 8.5. It is not necessary to use the identifiers given in Tables 8.1–8.5. The reliability analyst may make his or her own tables of identifiers.

Let us go over an exercise of naming the 'fails closed' event (failure mode) of an 8-inch safety valve used in the fuel storage subsystem of the fuel system. Tables 8.1–8.5 are used as the basis for naming the event. The first character should be B because the component is part of the fuel system (per Table 8.1). The second and third characters should be SV because the component type is safety valve (per Table 8.2). The fourth character is an identifier for this specific component; we arbitrarily assign 8 as the component identifier. The fifth character should be 2 because the component is part of the fuel storage subsystem (per Table 8.3). The sixth character should be C because the failure mode is 'Fails closed' (per Table 8.4). The seventh and eighth characters are J1, to indicate that it is mutually exclusive with another event (or other events) whose seventh character is J and whose eighth character would be any number or letter other than 1, which is used here (per Table 8.5). So the eight-character name of the event is BSV82CJ1.

The eight-character name for the 'fails open' failure mode of the same valve would be BSV82OJ2. The only changes are in the sixth and eighth characters. The sixth character is O because the failure mode is 'fails open' (per Table 8.4). The seventh and eighth characters are J2, which signifies that this event is mutually exclusive with BSV82CJ1 because 'fails open' is mutually exclusive with 'fails closed' (the valve cannot be open and closed at the same time).

Eight-character names may be used for external events also. We propose the following scheme. Because external events such as earthquakes, lightning, floods, tornados, etc. may affect all the systems in the plant, identify the system as O (system-independent, per Table 8.1). Similarly, identify the subsystem also as O (per Table 8.3). The component type identifier is EE (to signify that it is an external event, per Table 8.2). We may use the following "external event identifiers" in place of component identifiers (fourth character).

E—Earthquake
L—Lightning
F—Flood
T—Tornado
H—Hurricane

Other external events may also be named in a similar manner.

The failure mode is specified as O (per Table 8.4), because external events, as such, have no failure modes but can produce a variety of failure modes in different components. The seventh and eighth characters are CC, to signify that this external event is a common cause (per Table 8.5). Thus the eight-character name for an earthquake is OEEEOOCC. Using the same procedure, a tornado would be named OEETOOCC.

Once names are assigned to all the terminal events in a fault tree, they are tabulated in alphabetical order. Table 8.6 shows a sample list of names used in a fault tree.

Once names are assigned and tabulated, the names may be shown in the fault tree under each terminal event, for easy identification (see Figure 8.21).

8.7 TRANSFER SYMBOLS

The fault tree of an undesired event in a complex system may not fit on a single sheet of paper. A tree could be so large as to require 10, 20, or more sheets. The proper continuation of the tree from one sheet to another is indicated by *transfer symbols*, which are triangles with the apex

Table 8.6.   Sample Table of Terminal Event Names Used in a Fault Tree

TERMINAL EVENT NAME	SYSTEM	SUBSYSTEM	COMPONENT TYPE	COMPONENT	FAILURE MODE	SPECIAL SITUATION
OEEE00CC	not applicable	not applicable	external event	earthquake	not applicable	common cause
DPIA7R00	steam generation	water inlet	pipe	6-inch pipe	rupture	none
DSV2KCA2	steam generation	steam output	safety valve	4-inch valve (no. 72)	fails closed	mutually exclusive with DSV2KOA7
DSV2KOA7	steam generation	steam output	safety valve	4-inch valve (no. 72)	fails open	mutually exclusive with DSV2KCA2

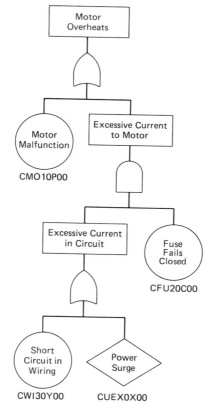

Figure 8.21. Terminal event identification in fault trees.

at the top. A triangle with a line emanating from its top, or just attached to the rectangle of an intermediate event above, is the *transfer-in symbol* (see Figure 8.5). A triangle with a line emanating from the center of one of the inclined sides or from the base is the *transfer-out symbol* (see Figure 8.5). Each transfer symbol in a fault tree is identified by a name (say, A007, D7, AEX, etc.).

The use of transfer symbols is illustrated in Figures 8.22 and 8.23 (extracted from the *Reactor Safety Study* [Nuclear Regulatory Commission (1975)]). Figure 8.22 is the first sheet of a multiple-sheet fault tree. There are six transfer-in symbols appearing in Figure 8.22; they are Q_{7A}, Q_{7B}, C_1 (appears twice), C_2 (appears twice), D_1, and D_2.

Consider the transfer-in symbol D_1. The notation "sht. 4" below it indicates that the portion of the tree below D_1 is given in sheet 4 of the

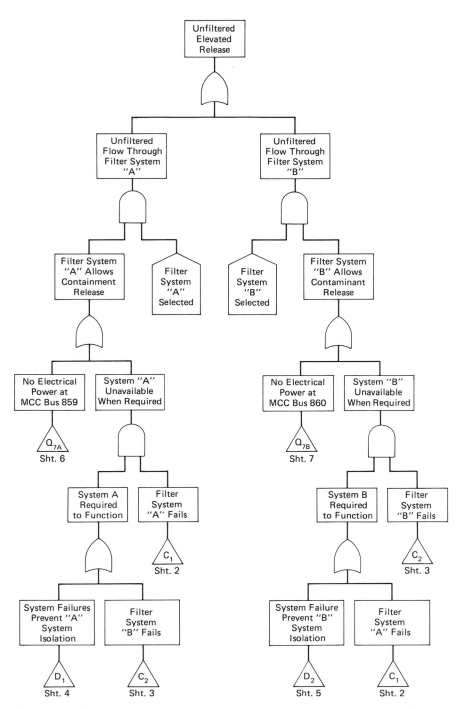

Figure 8.22. Illustration of the use of transfer symbols: first sheet of a fault tree. [*Source:* Nuclear Regulatory Commission (1975).]

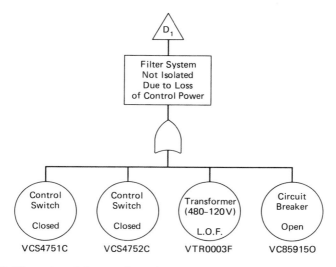

Figure 8.23. Illustration of the use of transfer symbols: fourth sheet of a fault tree. [*Source*: Nuclear Regulatory Commission (1975).]

fault tree. Figure 8.23 shows that fourth sheet. The transfer-out symbol D_1 in Figure 8.22 indicates that the tree below it should go to where the transfer-in symbol D_1 appears in Figure 8.22.

The part of the tree below the transfer-in symbols C_1, C_2, D_2, Q_{7A}, and Q_{7B} are given in sheets 2, 3, 5, 6, and 7 of the fault tree, respectively (as indicated below the symbols in Figure 8.22). These sheets are not shown here in the book.

8.8 COMMON CAUSE FAILURES

Common cause failures are discussed in Section 5.2.9 and Chapter 11 also. Here, we discuss common cause failures in the context of fault tree construction; this is an alternate way of considering common cause failures in system reliability analysis.

An event or mechanism that can cause failures in more than one component simultaneously (or at about the same time, as discussed in Section 11.2) is called a *common cause*, and the failures thus caused are called *common cause failures*. Common causes include the following:

1. *External Events:* earthquakes, tornados, hurricanes, floods, lighting, explosions, etc., that could affect more than one component simultaneously.

2. *Internal Events:* fires, internal explosions, falling objects, etc., that could affect more than one component simultaneously.
3. *Malfunction of Other Systems:* off-site power failure, air-conditioning failure, etc., that could affect more than one component simultaneously.
4. *Abnormal Operating and/or Environmental Conditions:* abnormal temperature, pressure, humidity, vibration, dust, mechanical stress, etc., that could affect more than one component simultaneously.
5. *Gross Errors:* gross errors in design, manufacture, installation, testing, or operation that could affect more than one component simultaneously.

Common cause failures may be treated as secondary failures (see Section 8.6.3.2). Figure 8.24 shows a fault tree in which two components are affected by the same common cause. (The general principles of resolving a component failure into primary, secondary, and command faults is explained in Section 8.6.3.2; no command fault is shown for component Y because this component is not affected by any command fault.) Here, earthquake (magnitude 4 or higher) is the common cause, which appears twice in the same tree as terminal events. If the common cause affects n components in the fault tree, it will appear in the tree at least n times. Common causes could appear in fault trees as terminal events or intermediate events. Common causes that are external events or gross errors are usually terminal events whereas other types of common cause could be terminal events or intermediate events.

Very low probability common causes may be ignored. For example, if the probability of occurrence of a common cause is at least an order of magnitude lower than the expected top-event probability, that common cause may be ignored, provided the sum of all ignored common cause probabilities is small compared to the top-event probability.

8.9 PROPAGATING FAILURES

The failure of one component may lead to the failure of other components or increase the failure probability of other components. Such failures are known as *propagating failures*.

Consider a heavy load being hung from two cables, each cable being subjected to half the load. If one cable fails due to a flaw (primary failure), the other cable will now have to carry the entire load. This will either significantly increase the failure probability of the second cable or cause it to fail. So this is an example of a propagating failure.

Propagating failures can be modelled as secondary failures (see Section 8.6.3.2). Figure 8.25 shows a sample fault tree. (In this figure, the event

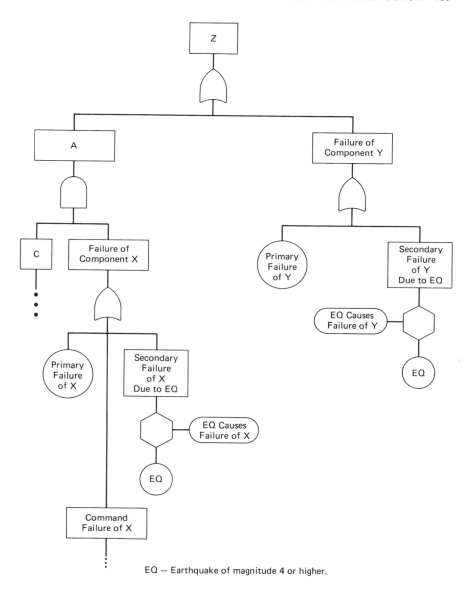

EQ — Earthquake of magnitude 4 or higher.

Figure 8.24. Common cause failure in a fault tree.

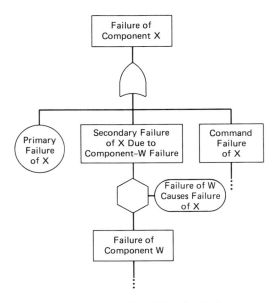

Figure 8.25. Propagating failure in a fault tree.

'failure of component W' could be either an intermediate event or a terminal event.) If the failure of a component (say, component W) causes the failure or increases the failure probability of n components, then component-W failure will appear in at least n places in the tree. When a component failure causes the failure of or increases the failure probability of a number of components, such a propagating failure is a common cause failure and may be treated as a common cause failure (see Section 8.8).

Secondary failures due to propagating failures can be ignored under conditions stated in Section 8.6.3.2 for ignoring secondary failures; if the propagating failure is a common cause, then conditions stated in Section 8.8 are applicable.

8.10 DEPENDENCE BETWEEN TERMINAL EVENTS

If there is statistical dependence between two or more terminal events in a fault tree, it should be considered in the analysis; otherwise, the quantitative results (the probability of the top and/or intermediate events) could be in error significantly. Even in qualitative analyses, it is important to know the statistical dependence between the combination of events that could produce the top event.

Common causes and propagating failures are two of the reasons for statistical dependence between terminal events. Other reasons include

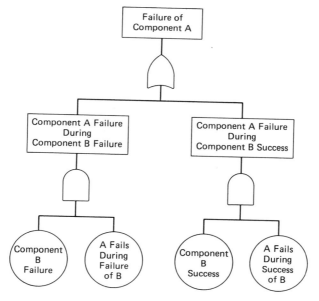

Figure 8.26. Modelling of dependent events.

common design, common manufacture, and common installation. If the reason is a common cause failure or a propagating failure, then the statistical dependence is accounted for automatically if the common cause failure or the propagating failure is modelled as discussed in Sections 8.8 and 8.9. If such modelling is not practical or if the reason for the dependence is neither a common cause nor a propagating failure, then the procedure discussed next may be used.

Consider two components A and B. Let the failure probability of component A be dependent on the failure or success of component B. If such a dependence does not exist, failure of component A may simply be modelled as a terminal event, but because the dependence does exist, failure of component A should be modelled as an intermediate event that is further resolved as shown in Figure 8.26. Note that the terminal event 'component-B success' may have a high probability of occurrence. High-probability terminal events may cause difficulties during quantitative fault tree analysis. Also, note that the terminal events 'component-B success' and 'component-B failure' are complementary events, and they are mutually exclusive.

If the product of the probability of 'component-B failure' and the conditional probability of 'component A fails during component-B failure' (the conditional probability of component-A failure given that component

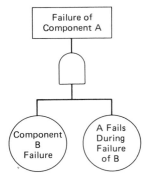

Figure 8.27. Modelling of dependent events: A special case.

B has failed) is much higher (one order of magnitude or more) than the product of the probability of 'component-B success' and the conditional probability of 'component A fails during component-B success' (the conditional probability of component-A failure given that component B has not failed), then the right branch of the tree below the OR gate may be discarded. The resultant tree is shown in Figure 8.27.

Consider another case in which failures of component A and component B are mutually exclusive. Because component-A failure and component-B failure are mutually exclusive, component A cannot fail during component-B failure. So the conditional probability of component-A failure during component-B failure is zero. So the left branch of the tree under the OR gate in Figure 8.26 can be discarded. The resultant fault tree is shown in Figure 8.28.

If two terminal events C and D in a fault tree are mutually exclusive, one of them (say, C) should always be modelled as in Figure 8.28, irrespective of where the other terminal event appears in the fault tree (D could be in one corner of the tree and C could be in another corner).

How does Figure 8.26, representing dependent component failures, fit within a full fault tree?

Consider a simple system as shown in Figure 8.29. Components A and B are in parallel and component C is in series with them. First, consider the case in which all the component failures (failures of components A, B, and C) are independent. The fault tree is shown in Figure 8.30, where the following notation is used:

A, B, C = failure of component A, B, or C, respectively
S = failure of the system
$D1$ = an intermediate event (components A and B in failed state)

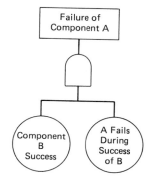

Figure 8.28. Modelling of mutually exclusive failures (component-A and component-B failures are mutually exclusive).

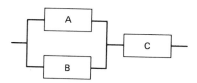

Figure 8.29. Schematic diagram.

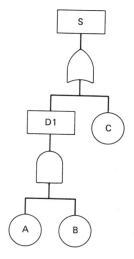

Figure 8.30. Fault tree for system shown in Figure 8.29 (no dependence between failures of components A, B, and C).

Let us now consider the case in which failure of A is dependent on B. The fault tree is shown in Figure 8.31, where the following notation is used:

A, B, C = failure of component A, B, or C, respectively
$\quad S$ = failure of the system
$\quad \overline{B}$ = success of component B
$\quad D1$ = an intermediate event (components A and B in failed state)
$\quad A'$ = failure of component A during component-B failure
$\quad A''$ = failure of component A during component-B success
$(A|B)$ = component A fails during component-B failure (failure of component A given component-B failure)
$(A|\overline{B})$ = component A fails during component-B success (failure of component A given component B success)

Note that the part of the fault tree under the intermediate event A is identical to Figure 8.26.

More complex cases where the failure of component A is dependent on both components B and C may also be modelled using the concepts discussed in the preceding paragraphs. Rumble, Leverenz, and Erdmann (1975) provide examples of fault trees with dependent events.

Two events A and B are said to be fully correlated (statistical correlation = 1) if A occurs if and only if B occurs and B occurs if and only if A occurs; in other words, A occurs whenever B occurs and A does not occur whenever B does not occur. If two such events are present in a fault tree during fault tree quantification, we may simply replace B by A wherever B appears in the fault tree; that will take care of the statistical dependence between A and B during the quantitative analysis.

Can statistical dependence between terminal events be ignored, and can all terminal events be assumed to be statistically independent? Such assumptions could introduce significant errors in the quantitative fault tree analysis. Probability of the top event could be in error by one or more orders of magnitude. Will the top-event probability be underpredicted or overpredicted? It depends on the structure of the fault tree, the number of OR gates and AND gates, the sequence in which these gates appear, and the probabilities of the terminal events.

Consider a very simple fault tree in which the top event is connected to two terminal events by an OR gate. This represents a two-event series system. Let the two terminal events be statistically dependent. In this case, assuming the two terminal events to be statistically independent will provide an upper-bound solution to the top-event probability. Assuming the two terminal events to be fully correlated will yield a lower-bound solution to the top-event probability.

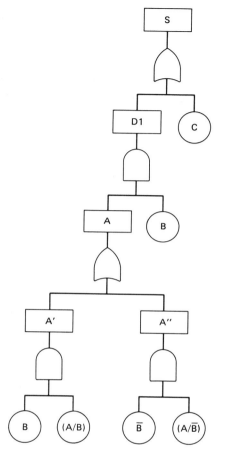

Figure 8.31. Fault tree for system shown in Figure 8.29 (component-A failure is dependent on component-B failure).

Consider another fault tree, in which the top event is connected to two terminal events by an AND gate. This represents a two-event parallel system. In this case, assuming statistical independence between the two terminal events will result in a lower-bound solution to the top-event probability, and assuming full statistical correlation will result in an upper-bound solution. Most fault trees contain a combination of OR gates and AND gates, so it is not possible to foretell whether assuming statistical independence would yield a lower- or upper-bound solution.

8.11 MAINTENANCE OUTAGE

A system or plant could become unavailable not only due to unexpected component failures, but also due to maintenance and testing outages. If the complete plant is shut down for a specified duration and all the components are maintained during that time, then that duration of time divided by the operating time between such maintenance outages is the plant unavailability due to maintenance. However, in many plants, some components are maintained while the plant is operating; for example, if there are two pumps to supply water to the cooling loop and only one pump is sufficient (the second pump is a standby), then routine mainte- nance on these pumps may be carried out one at a time, without having to shut down the plant. However, there is a finite probability that the

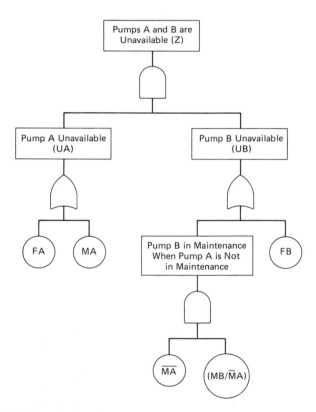

Figure 8.32. Explicit modelling of component maintenance in fault trees (mutually exclusive nature of the maintenance of pumps A and B is included).

operating pump fails while the standby pump is undergoing maintenance and thus both pumps are unavailable; such a situation will result in a plant shut down or an accident.

This scenario of maintaining operating and standby pumps without shutting down the system is modelled in Figure 8.32. The system consists of a number of components, including the two pumps A and B. The following notation is used in Figure 8.32 and in the following discussion.

Z = an intermediate or top event 'pumps A and B are unavailable'

UA = pump A is unavailable (pump A is in a failed state or is undergoing maintenance)

UB = pump B is unavailable (pump B is in a failed state or is undergoing maintenance)

FA = pump A is in a failed state (pump-A failure)

FB = pump B is in a failed state (pump-B failure)

MA = pump A is undergoing maintenance

MB = pump B is undergoing maintenance

$\overline{MA}$ = pump A is not undergoing maintenance (complement of MA)

$(MB|\overline{MA})$ = pump B is undergoing maintenance given that pump A is not undergoing maintenance

$P[FA]$ = the probability that pump A is unavailable due to failure; that is, the unavailability of pump A due to failures (estimation of component unavailabilities is discussed in Chapters 3 and 5)

$P[FB]$ = the probability that pump B is unavailable due to failure; that is, the unavailability of pump B due to failures

$P[MA]$ = the probability that pump A is unavailable due to testing and maintenance; that is, the test–maintenance unavailability of pump A (estimation of component test–maintenance unavailabilities is discussed in Sections 3.4.13 and 5.2.4)

$P[MB]$ = the probability that pump B is unavailable due to testing and maintenance; that is, the test–maintenance unavailability of pump B

The left branch under the higher AND gate (in Figure 8.32) shows that pump A can become unavailable if pump A fails or is undergoing mainte- nance. The corresponding right branch shows that pump B can become unavailable if pump B fails or is undergoing maintenance. The event 'pump B undergoing maintenance' is further resolved through an AND gate (the extreme-right-hand branch of the fault tree in Figure 8.32) to

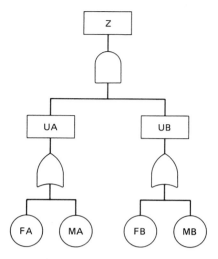

Figure 8.33. A simplified version of the fault tree shown in Figure 8.32 (pump maintenance is included explicitly but the mutually exclusive nature of the maintenance of pumps A and B is ignored).

assure that a scenario in which both pumps A and B are undergoing maintenance is not considered (because plant operating procedures specify that, while the plant is operating, both pumps shall not be undergoing maintenance at the same time; that is, maintenance of pump A and maintenance of pump B are mutually exclusive events).

If the probabilities of maintenance of pumps A and B are substantially less than the probabilities of their failures (that is, $P[FA] \gg P[MA]$ and $P[FB] \gg P[MB]$), then the probability of the event 'pumps A and B undergoing maintenance at the same time' would be much less than the probability of the event 'pumps A and B in failed state at the same time.' In such situations, the fact that MA and MB are mutually exclusive need not be explicitly modelled in the fault tree; that fault tree shown in Figure 8.33 may be used. The probability of Z, as calculated from this fault tree, will be conservative (slightly higher than the actual value).

In addition to the conditions stated in the previous paragraph, if FA, FB, MA, and MB do not occur elsewhere in the fault tree, then FA and MA may be combined and FB and MB may be combined, and the fault tree will be as in Figure 8.34. The event UA represents the unavailability of A due to both failure and maintenance. The probability of UA is equal to $P[FA] + P[MA]$; this is a good approximation if $P[FA]$ and $P[MA]$ are about 0.1 or less. The same argument holds good for UB also.

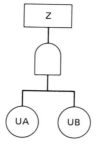

Figure 8.34. A simplified version of the fault tree shown in Figure 8.32 (failure and maintenance of each pump is combined into a single terminal event).

8.12 FAULT TREE REDUCTION (FAULT TREE SIMPLIFICATION)

A fault tree that is constructed using the principles discussed in the preceding sections could be very large if the system has many components. In order to reduce the analysis effort (qualitative and/or quantitative analysis), the fault tree may be simplified (reduced). Some general guidelines for simplifying fault trees are given in the following paragraphs. These guidelines should not be followed blindly; care should be taken to assure that the simplification does not discard any important information or significantly change the probability of the top event or of intermediate events of interest. The rationale for these guidelines may become evident when we discuss quantitative fault tree analysis in Chapter 10.

1. Very high probability events (say, probability greater than 0.99) appearing directly under an AND gate may be discarded. The effect may be to increase slightly the computed probability of the top event; the actual probability will be slightly less than the computed probability. Consider the fault tree shown in Figure 8.35a. Let the probability of the terminal event $E4$ be very high. So we reduce the fault tree to Figure 8.35b.

2. If there are terminal events that by themselves can cause the top event, that is, each of those terminal events can cause the top event (these are called *single-event minimal cut sets* in Chapter 9), then terminal events directly under an AND gate whose combined probability (the product of the probabilities of the terminal events directly under the AND gate) is at least one order of magnitude less than the sum of the probabilities of the single-event cut sets may be discarded, unless there are tens or hundreds of such AND gates. When such terminal events are to be discarded, it is necessary that not only those terminal events but all the terminal events,

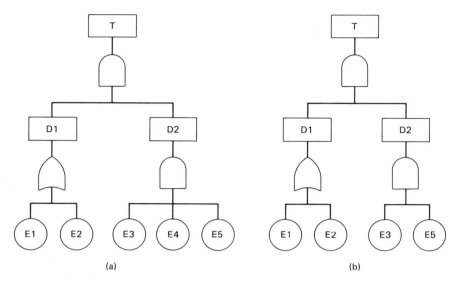

Figure 8.35. Fault tree reduction (case 1): (a) full fault tree (before reduction); (b) reduced fault tree.

intermediate events, and associated gates all the way up to the first-encountered OR gate should be discarded from the tree. [This is similar to reducing a fault tree because of a switched-off house event (see Section 8.6.3.3).] The effect of such a reduction is to reduce the computed top event probability; the computed top event probability will be slightly lower than the actual value.

Consider the fault tree shown in Figure 8.36a. Let the product of the probabilities of terminal events $E5$, $E6$, and $E7$ be very small compared to the sum of the probabilities of single-event minimal cut sets of the tree. We decide to reduce the tree according to the principle described in the preceding paragraph. We discard not only terminal events $E5$, $E6$, and $E7$ from the tree but also discard all the terminal events, intermediate events, and gates up to the first OR gate we encounter as we move up the tree. The reduced fault tree is shown in Figure 8.36b.

3. Any terminal event directly under an OR gate whose probability is at least an order of magnitude less than the sum of the probabilities of the single-event minimal cut sets of the tree may be discarded from the tree unless there are tens or hundreds of such terminal events.

Consider the fault tree shown in Figure 8.37a. The probabilities of terminal events $E3$, $E5$, and $E6$ are very small compared to the sum of the probabilities of the single-event minimal cut sets of the tree. The reduced fault tree is shown in Figure 8.37b.

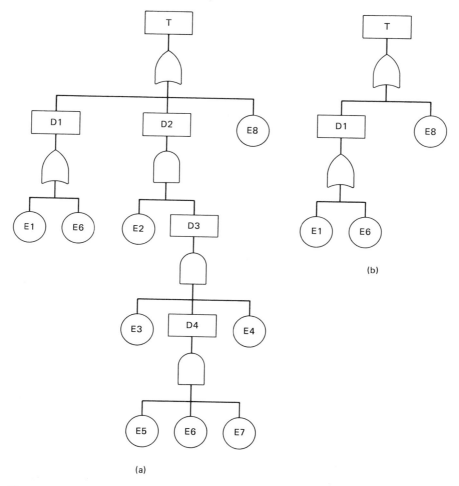

Figure 8.36. Fault tree reduction (case 2): (a) full fault tree (before reduction); (b) reduced fault tree.

4. An approximate, preliminary estimate of the top event probability is sometimes available before the fault tree is analyzed. Let the estimate be P. Any terminal event directly under an OR gate whose probability is at least an order of magnitude less than P may be discarded from the tree unless there are tens or hundreds of such terminal events (this is similar to guideline 3).

If the products of the probabilities of terminal events directly under an AND gate is at least an order of magnitude less than P, these terminal

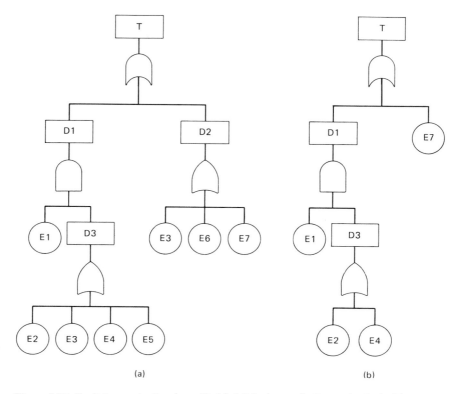

Figure 8.37. Fault tree reduction (case 3): (a) full fault tree (before reduction); (b) reduced fault tree.

events may be discarded from the tree, unless there are tens or hundreds of such AND gates. In addition, all terminal events, intermediate events, and gates up to the first OR gate we encounter as we move up the tree shall be discarded from the tree (this is similar to guideline 2).

If, at a later stage, the preliminary estimate P is found to be in error, all discarded terminal events, intermediate events, and gates should be put back into the tree.

5. If *all* the terminal events directly under an OR gate are to be discarded (per guideline 2, 3, or 4), then we not only discard these terminal events and that OR gate but we should also discard all the terminal events, intermediate events, and gates up to the first OR gate we encounter as we move up the tree.

Consider the fault tree shown in Figure 8.38a. Terminal events $E3$ and $E4$ are of low probability and are to be discarded. We discard not only $E3$

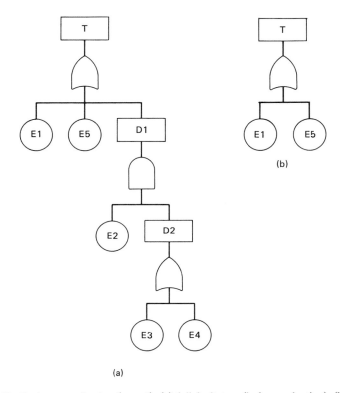

(a)

Figure 8.38. Fault tree reduction (case 4): (a) full fault tree (before reduction); (b) reduced fault tree.

and $E4$ but also all the terminal events, intermediate events, and the associated gates below the first OR gate we encounter as we move up. The reduced fault tree is shown in Figure 8.38b.

6. Terminal events that are statistically dependent on other terminal events should not be discarded unless it can be established that such discarding does not affect the top-event probability significantly. This guideline has precedence over (this guideline overrules) all of the five preceding guidelines.

At the time fault trees are constructed, the reliability analyst may not have probability data for terminal events but may have an approximate idea about these probabilities. That approximate idea is sufficient for fault tree simplification. (For example, some approximate ideas about certain component failure and human error probabilities are given in Section 5.5. Such information may be helpful here.)

The reason for discarding each event should be clearly documented. If these reasons become invalid as more precise data on terminal events become available, one must go over the reduction procedure again to reintroduce any events discarded incorrectly.

8.13 COMPUTERIZED FAULT TREE CONSTRUCTION

So far we have discussed manual fault tree construction. Even very large fault trees can be constructed manually. For example, very large and complex fault trees used in the reliability analysis of nuclear power plants have been constructed manually using the techniques described in the preceding sections.

Fault tree construction algorithms suitable for implementation into computer software are available [Allen and Rao (1980), Chu (1976), Fussell (1973), Kumamoto and Henley (1979), Powers and Tompkins (1974), and Salem, Apostolakis, and Okrent (1977)]. CAT (Computer Automated Tree) is a publicly available fault tree construction computer program (see Appendix I for details).

Do not expect that fault tree construction programs will accept a brief description of the system and produce the required fault tree. The user has to input the effects of each failure mode of each component and subsystem on the functions of the system, subsystems, and components. In order to prepare data for fault tree construction programs, the user must have a good understanding of the system logic; the same level of understanding as that required for manual fault tree construction is needed. Also, data preparation could be almost as time-consuming as manual fault tree construction. One application where computerized methods could be particularly attractive is the construction of fault trees of systems with feedback control loops. Methods developed by Allen and Rao (1980) and Kumamoto and Henley (1979) have the capability to consider control loops.

If a decision is made to computerize the fault tree construction process, select a method or program that has the capability to consider all the features of the fault trees to be constructed. Only some methods and programs have the ability to consider special features such as control loops and statistical dependencies between failures.

8.14 COMPLEMENTARY TREES

Complementary trees are also known as *success trees* and *dual trees*. The concept of a complementary tree is useful in determining the *path sets* of fault trees (discussed in Chapter 9); path sets may be used in both qualitative and quantitative analyses.

The complementary tree of a fault tree is obtained by modifying the fault tree as follows.

1. Change all OR gates to AND gates.
2. Change all AND gates to OR gates.
3. Change all events (terminal, intermediate, and top events) to their complements.

For example, if 'system failure' is the top event of the fault tree, then 'system success' is the top event of the complementary tree. Similarly, if 'auxiliary feedwater system failure' is an intermediate event in the fault tree, then 'auxiliary feedwater system success' is the corresponding intermediate event in the complementary tree. If 'tank nozzle fails in fatigue' is a terminal event in the fault tree, 'tank nozzle does not fail in fatigue' is the corresponding terminal event in the complementary tree. In general, if X is a terminal event in the fault tree, then its complement $\overline{X}$ is the corresponding terminal event in the complementary tree. Note that the probabilities of X and $\overline{X}$ are related by

$$P[\overline{X}] = 1 - P[X]$$

In order to develop a complementary tree using the preceding guidelines, the fault tree should not contain any gates other than OR gates and AND gates; if there are gates of any other type, they should first be transformed to OR gates and AND gates. If statistical dependencies are explicitly modelled in the fault tree (for example, Figure 8.24), extra care should be taken that the complementary tree correctly models the dependence.

8.15 LIMITATIONS

Although fault trees are versatile and are widely used in system reliability analysis, there are some limitations to the fault tree representation of system failures.

1. It is seldom possible to assure completeness of the fault tree. It is quite possible that some failure modes or combination of failure modes are inadvertently missed. If the tree is, in fact, incomplete, the probability of the top event or intermediate events computed from the fault tree could be unconservative. (This drawback is not limited to fault tree analysis; it is true for all system reliability analysis techniques.)

2. Fault tree analysis considers only the binary states of components. That is, we consider only two possible alternatives with respect to each

failure mode: The failure mode exists (occurs) or it does not exist. In some actual situations, however, a continuously increasing state of failure could occur. For example, a pipe may leak at a rate of 1 gallon per minute (gpm), 2 gpm, or 10 gpm, etc.) Fault trees are not best-suited for such nonbinary failure states. This aspect has been discussed in Section 8.6.3.2.

In spite of these limitations, fault tree analysis is still a valuable and effective method of system reliability analysis. As long as we do not lose sight of these limitations, results from fault trees could be very careful in system design and reliability assessment.

8.16 DOCUMENTATION

Documentation is an important part of fault tree construction and it should not be overlooked. Documentation should contain the following information.

1. A list of all documents used in constructing the fault tree (include version number and/or version date of each document, as appropriate). Documents may include the following:

   - System diagrams
   - Piping and instrumentation diagrams
   - Schematic diagrams
   - Block diagrams
   - Logic diagrams
   - Process flow diagrams
   - Installation drawings
   - Parts lists
   - Operating procedures
   - Maintenance and testing procedures
   - Training manuals

2. System description and system functional description (as described in Section 8.3). If assumptions are made, those assumptions and justifications must be stated. Indicate the reference documents, as necessary.
3. Fault tree. State the specific assumptions made during fault tree construction. Justifications for the assumptions must be included.
4. Summary of terminal events (eight-character names and descriptions).
5. Criteria used in fault tree reduction (if the fault tree is reduced).
6. Reduced fault tree (if the fault tree is reduced).

7. A list of discarded terminal events and justifications for discarding those events (if any of the terminal events are discarded).
8. Identification of computer programs (if used); include version number and/or version date, and the source of the program (from whom obtained).
9. Input and output from the computer programs (if computer programs are used).

Some general guidelines on documentation are provided in Section 14.10.

8.17 EXAMPLE PROBLEM[2]

Consider the domestic hot water system described in Section 6.5 and shown in Figure 6.1.

There are a number of possible undesired events, for example, 'hot water tank rupture,' 'gas–air explosion,' and 'gas-leakage poisoning.' We will construct a fault tree for 'hot water tank rupture' in this example. The fault tree is shown in Figure 8.39. Let us go through the fault tree from top to bottom and explain how the fault tree is constructed.

The top event is 'hot water tank rupture' when the hot water faucet is closed. The tank can rupture (i) due to inherent defects in the tank [such as flaws, less-than-expected material strength, etc. (primary failure)] or (ii) due to secondary failures resulting from damage during shipping or installation or (iii) due to internal pressure exceeding 100 psi (this is also a secondary failure but we treat it separately so that the causes for excess pressure can be considered in more detail in the fault tree).

Pressure in the tank can exceed 100 psi if (i) the pressure in the tank rises above 100 psi and (ii) the pressure relief valve (PRV) fails to open even after the pressure exceeds 100 psi. We will expand both of these events further.

The PRV can fail to open even after the pressure exceeds 100 psi (i) due to the PRV failing closed due to inherent defects in the valve (primary failure) or (ii) due to secondary failures due to damage during shipping or installation or (iii) due to the PRV set at a pressure higher than 100 psi (human error in setting the PRV). We treat all three of these events as terminal events.

Pressure in the tank rises to above 100 psi if (i) steam forms within the tank or (ii) the inlet water supply is at an excessive pressure. The latter is due to failures in the water supply system. We decide not to expand this cause and treat it as an undeveloped terminal event.

---

[2]This example is based on Lambert (1973).

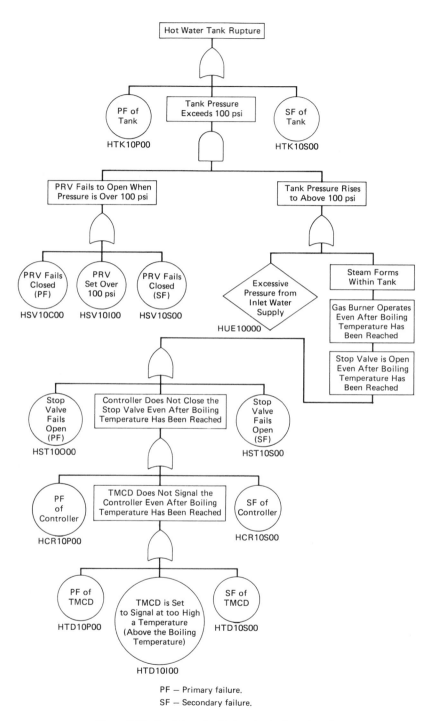

PF — Primary failure.
SF — Secondary failure.

Figure 8.39. Fault tree for 'hot water tank rupture.'

Steam can form inside the tank if the gas burner continues to operate even after the water temperature has reached the boiling point. This can happen if the stop valve is open even after the water temperature has reached the boiling point. Such an event can occur if (i) the stop valve fails open (is stuck open) (primary failure) or (ii) the stop valve fails to close due to secondary causes or (iii) the controller fails to close the stop valve even after the water temperature has reached the boiling point (this is also a secondary failure but we consider it separately so that this event can be expanded further).

The controller can fail to close the stop valve even after the water temperature has reached the boiling point (i) due to primary failure of the controller or (ii) due to secondary failures of the controller or (iii) due to the temperature measuring and comparing device (TMCD) failing to signal the controller to close the stop valve even after the water temperature has reached the boiling point (this is also a secondary failure but we consider it separately so that this event can be expanded further).

The TMCD can fail to signal the controller to close the stop valve even after the water temperature has reached the boiling point (i) due to primary failure of the TMCD or (ii) due to secondary failures of the TMCD or (iii) due to the TMCD being set to signal the controller at too high a temperature (above the boiling temperature) (human error in setting the TMCD). We treat all three of these events as terminal events. This completes the fault tree construction.

Having completed the fault tree construction, we name the terminal events. The eight-character code discussed in Section 8.6.3.4 is used. In addition to the identifiers shown in Tables 8.1–8.5, the following identifiers are also used:

1. System identifier: *H—hot water system*
2. Component-type identifiers:
   CR—Controller
   ST—Stop valve
   SV—Safety valve (pressure relief valve)
   TD—Time measuring and comparing device
3. Failure-mode identifiers:
   I—Improper setting of the device
   P—Primary failure (specific primary failures such as 'fails closed,' 'fails open,' or 'rupture' are identified by specific identifiers such as C, O, or R; P is used when no such specific identifier is used)
   S—Secondary failure

The eight-character identifier is written below each terminal event in the fault tree.

EXERCISE PROBLEMS

8.1. Consider the electrical system shown in Figure 8.1. A fault tree for the top event 'motor overheats' is shown in Figure 8.2. Construct a fault tree for the top event 'motor fails to run.'

8.2. Consider the domestic hot water system described in Section 6.5. A fault tree for the top event 'hot water tank rupture' is shown in Figure 8.39. (i) Construct a fault tree for the top event 'fire damage to properties in the room." (ii) Construct a fault tree for the top event 'gas leakage.' (iii) Construct a fault tree for the top event 'no hot water.'

8.3. Consider the fault tree shown in Figure 8.2. The probability of the terminal event 'power surge' is very small and we decide to discard it. Draw the reduced fault tree.

8.4. Consider the fault tree shown in Figure 8.2. Draw its complementary tree.

8.5. Consider the fault tree shown in Figure 8.20. For the purposes of this problem, assume that E, F, G, J, K, M, N, P, Q, and R are terminal events (instead of intermediate events as indicated in Figure 8.20). (i) The probability of Q is very small and we decide to discard it. Draw the reduced fault tree. (ii) The probability of M is very small and we decide to discard it. Draw the reduced fault tree. (iii) The probabilities of both E and F are very small and we decide to discard them. Draw the reduced fault tree. (iv) The probability of K is 0.9986 and we decide to discard it. Draw the reduced fault tree.

8.6. Consider the fault tree shown in Figure 8.20. For the purposes of this problem, assume that E, F, G, J, K, M, N, P, Q, and R are terminal events (instead of intermediate events as indicated in Figure 8.20). Draw its complementary tree.

8.7. Consider the fault tree shown in Figure 8.39. Draw its complementary tree.

REFERENCES

Allen, D. J. and M. S. M. Rao (1980). New algorithms for the synthesis and analysis of fault trees. *I & EC Fundamentals* **19** 79–85.
Chu, B. B. (1976). *A Computer-Oriented Approach to Fault Tree Construction*. Report No. NP-288. Electric Power Research Institute, Palo Alto, CA.
Fussell, J. B. (1973). *Synthetic Tree Model: A Formal Methodology for Fault Tree Construction*. Report No. ANCR-32. Idaho National Engineering Laboratory, Idaho Falls.
Kumamoto, H. and E. J. Henley (1979). Safety and reliability synthesis of systems with control loops. *AIChE Journal* **25** 108–114.
Lambert, H. E. (1973). *System Safety Analysis and Fault Tree Analysis*. Report No. UCID-16238. Lawrence Livermore National Laboratory, Livermore, CA.
Nuclear Regulatory Commission (1975). *Reactor Safety Study: An Assessment of Accident Risks in U.S. Commercial Nuclear Power Plants* (WASH-1400). Nuclear Regulatory Commission, Washington, DC.

Nuclear Regulatory Commission (1983). *PRA Procedures Guide* (NUREG/CR-2300). Nuclear Regulatory Commission, Washington, DC.

Powers, G. J. and F. C. Tompkins (1974). Fault tree synthesis for chemical processes. *AIChE Journal* **20** 376–387.

Rumble, R. T., F. L. Leverenz, and R. C. Erdmann (1975). *Generalized Fault Tree Analysis for Reactor Safety*. Report No. EPRI 217-2-2. Electric Power Research Institute, Palo Alto, CA.

Salem, S. L., G. E. Apostolakis, and D. Okrent (1977). A new methodology for the computer-aided construction of fault trees. *Annals of Nuclear Energy* **4** 417–433.

Chapter 9
Qualitative Fault Tree Analysis

9.1 INTRODUCTION

Qualitative fault tree analysis consists of (i) determining the minimal cut sets, (ii) determining the minimal path sets, and (iii) qualitatively ranking the terminal events according to their importance. (Minimal cut sets and minimal path sets are defined in Sections 9.2 and 9.3, respectively.) Minimal cut set analysis (determining the minimal cut sets) is the most commonly performed qualitative analysis, but minimal path sets do provide some qualitative information and can also be useful in quantitative fault tree analysis (see Sections 10.3.11 and 10.3.12). Qualitative ranking of terminal events provides helpful information about the importance of the various terminal events in a fault tree. Quantitative ranking, of course, provides more precise and accurate information about the importance of terminal events, but qualitative ranking may be a useful preliminary assessment of terminal events at early stages of a project before quantitative data on terminal event probabilities are available. Quantitative ranking can be performed without a qualitative ranking being performed first.

There are methods of quantitative fault tree analysis (determination of the probabilities of the top and/or intermediate events) that require neither the minimal cut sets nor the minimal path sets; so a quantitative analysis may be performed without first performing a qualitative analysis. There are other methods of quantitative analysis that utilize either the minimal cut sets or the minimal path sets. Whether or not minimal cut sets and minimal path sets are used in the quantitative analysis, they do contain useful information that is not available from a quantitative analysis.

9.2 CUT SETS AND MINIMAL CUT SETS

9.2.1 Cut Sets

A combination of terminal events that is sufficient to cause the occurrence of the top event is called a *cut set*. In other words, if all the terminal events in a cut set occur (or are present), then the top event will also occur. There can be more than one cut set for the top event.

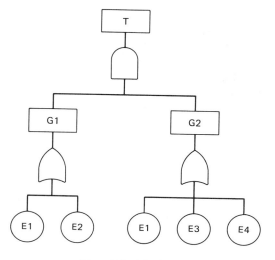

Figure 9.1. A fault tree.

Consider the fault tree shown in Figure 9.1.[1] T is the top event, $G1$ and $G2$ are the intermediate events, and $E1$, $E2$, $E3$, and $E4$ are the terminal events. Cut sets of this simple fault tree may be determined by inspection, without the use of the formal minimal cut set determination procedures discussed in Section 9.6. If any one of the two terminal events under $G1$ and any one of the three terminal events under $G2$ occur, the top event will occur, because $G1$ and $G2$ are connected to the top event by an AND gate and the terminal events under $G1$ and $G2$ are connected through OR gates. There are thus six cut sets, which we may refer as $CS1$, $CS2, \ldots, CS6$.

$$CS1 = (E1, E1) = E1$$
$$CS2 = (E1, E3)$$
$$CS3 = (E1, E4)$$
$$CS4 = (E2, E1)$$
$$CS5 = (E2, E3)$$
$$CS6 = (E2, E4)$$

We are able to enumerate these cut sets easily because it is a rather simple tree. Larger trees may require formal analysis procedures.

---

[1] This example is from Gangadharan, Rao, and Sundararajan (1977).

Consider the second cut set $CS2$. The occurrence of $E1$ causes the intermediate event $G1$ and the occurrence of $E3$ causes the intermediate event $G2$. The occurrence of both $G1$ and $G2$, in turn, causes the occurrence of the top event. So the occurrence of $E1$ and $E3$ results in the top event, and thus the combination of terminal events $(E1, E3)$ is a cut set of the fault tree. (*Note:* The terminal events in the cut set need not necessarily occur simultaneously: They may occur at different times; once all the terminal events in the cut sets have occurred and are present, the top event will occur. For example, in the cut set $CS2$, $E1$ may occur first and then $E3$, or $E3$ may occur first and then $E1$, or $E1$ and $E3$ may occur simultaneously.)

Now, consider the first cut set, $CS1$. The occurrence of $E1$ will cause both $G1$ and $G2$ because it is connected to both of these intermediate events through OR gates. The occurrence of $G1$ and $G2$ will cause the top event because these are connected to the top event through an AND gate. Therefore the occurrence of $E1$ results in the top event. In this cut set, only one terminal event is necessary to cause the top event.

It is worth noting that the fault tree of Figure 9.1 consists of only AND gates and OR gates. If there are gates of other types in a fault tree, they may be transformed to AND gates and OR gates as discussed in Section 8.5. When an inhibit gate is transformed to an AND gate, the inhibit condition will become a terminal event (see Section 8.5.8); therefore inhibit conditions may enter cut sets as if they are terminal events.

One may determine cut sets not only of the top event but also of intermediate events. While determining the cut sets of intermediate events, the part of the tree below the intermediate event is treated as if it is the complete tree and the intermediate event is treated as its top event. Considering the fault tree in Figure 9.1, there are two cut sets for the intermediate event $G1$; they are $E1$ and $E2$, because either terminal event by itself could cause $G1$. Similarly, there are three cut sets for the terminal event $G2$; they are $E1$, $E3$, and $E4$.

9.2.2 Minimal Cut Sets

The smallest subset[2] of a cut set that is sufficient and necessary to cause the occurrence of the top event is called a *minimal cut set*. If all the terminal events in a minimal cut set occur, then the top event will also

---

[2]A set C is said to be a subset of set D if D contains all the elements of C. For example, consider two sets X and Y. Let X contain elements $E1$, $E7$, and $E9$, and let Y contain elements $E1$, $E6$, $E7$, $E9$, $E12$, and $E23$. Then, X is said to be a *subset* of Y, and Y is said to be a *superset* of X.

occur. There can be more than one minimal cut set for the top event, and any one of the minimal cut sets is sufficient to cause the top event. Each minimal cut set represents a possible way the top event can occur.

Consider the fault tree shown in Figure 9.1. The cut sets ($CS1$–$CS6$) are determined in Section 9.2.1. Let us examine each cut set and decide if it is a minimal cut set. (In larger trees with thousands of cut sets, formal computerized procedures are necessary to sort out minimal cut sets from the cut sets.) The first cut set $CS1$ contains only one terminal event, so it cannot contain any other subset. Therefore it is a minimal cut set.

Consider the second cut set $CS2$. The first cut set $CS1$ is a subset of $CS2$. Because $CS1$ is a subset of $CS2$, and $CS1$ is sufficient to cause the top event, $CS2$ is not a minimal cut set. (In general, if a cut set or a minimal cut set is a subset of another cut set, then the latter cut set is not a minimal cut set.) Look at it from another perspective. The cut set $CS2$ contains $E1$ and $E2$. The terminal events $E1$ and $E2$ are sufficient to cause the top event but both of them are not necessary to cause the top event; $E1$ alone is sufficient and $E2$ is unnecessary. So the smallest subset of $CS2$ that is sufficient and necessary to cause the top event is $E1$. Therefore $E1$ is a minimal cut set, which has already been accounted for as $CS1$.

The cut set $CS1$ is a subset of cut set $CS3$. So $CS3$ is not a minimal cut set. The cut set $CS4$ also is not a minimal cut set, for the same reason.

The cut set $CS5$ contains no other cut set or minimal cut set as its subset, so it is a minimal cut set. Similarly, the cut set $CS6$ is also a minimal cut set.

Therefore the minimal cut sets for the top event in Figure 9.1 are

$$MCS1 = (E1)$$

$$MCS2 = (E2, E3)$$

$$MCS3 = (E2, E4)$$

These three minimal cut sets indicate all the possible ways in which the top event can occur.

Minimal cut sets can also be determined for intermediate events. Methods of determining cut sets of intermediate events are described in Section 9.2.1. Once the cut sets for the intermediate events are determined, the determination of minimal cut sets for intermediate events is not different from minimal cut set determination for the top event. Referring to Figure 9.1, there are two minimal cut sets for the intermediate event $G1$; they are $E1$ and $E2$. There are three minimal cut sets for the intermediate event $G2$; they are $E1$, $E3$, and $E4$.

If there are two or more mutually exclusive terminal events in a minimal cut set, such a minimal cut set should be discarded, because those mutually exclusive events cannot exist (cannot be present) simultaneously. In other words, the probability of occurrence of such a minimal cut set is zero and therefore that minimal cut set should be discarded. In the eight-character naming system discussed in Section 8.6.3.4, the seventh and eighth characters identify whether two terminal events are mutually exclusive or not; just by examining the seventh and eighth characters of the terminal events in a minimal cut set, we can determine if there are mutually exclusive events in the minimal cut set.

9.2.3 Order of Minimal Cut Sets

The number of terminal events in a minimal cut set is called the *order of the minimal cut set*. In the minimal cut set example discussed in Section 9.2.2, the order of *MCS*1 is 1, the order of *MCS*2 is 2, and the order of *MCS*3 is also 2. *MCS*1 is called a *first-order minimal cut set*; *MCS*2 and *MCS*3 are *second-order minimal cut sets*.

9.2.4 Number of Minimal Cut Sets

The number of minimal cut sets pertaining to the top event of a fault tree depends on the number of terminal events, the number of AND gates, the number of OR gates, and the configuration of the fault tree. In general, if the gates are mostly OR gates, there will be many small minimal cut sets (by "small" we mean that the order of the minimal cut sets is low); if the gates are mostly AND gates, there will be fewer but larger minimal cut sets (by "larger" we mean that the order of the minimal cut sets is higher). An increase in the number of OR gates increases the number of minimal cut sets; an increase in the number of AND gates increases the order of some or all of the minimal cut sets; an increase in the number of terminal events increases the number of minimal cut sets and/or the order of the minimal cut sets, depending on the configuration of the tree. The number of minimal cut sets of a fairly large tree could be in the thousands or even more.

9.3 PATH SETS AND MINIMAL PATH SETS

9.3.1 Path Sets

Minimal cut sets are more widely used than minimal path sets, but minimal path sets do provide some useful qualitative information that may be of interest to reliability analysts. Top event probability and intermediate event probabilities can be computed using either minimal cut sets or minimal path sets, or without either of them.

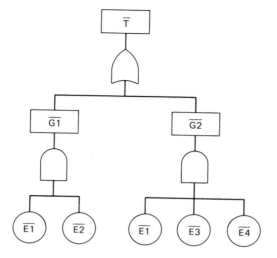

Figure 9.2. Complementary tree of Figure 9.1.

The complement of a terminal event is called the *terminal event comple-ment*. If the terminal event is a component failure, then the terminal event complement is the nonexistence of that particular component failure.

A combination of terminal event complements that is sufficient to assure that the top event does not occur is called a *path set*. If all the terminal event complements of a path set exist, then the top event cannot occur. In other words, if none of the terminal events associated with a path set occur, then the top event does not occur. There can be more than one path set for the top event.

Path sets are determined by examining the complementary tree of the fault tree. Path sets of a fault tree are identical to the cut sets of its complementary tree. The complementary tree of the fault tree shown in Figure 9.1 is given in Figure 9.2. (In this figure, $\overline{T}$, $\overline{G1}$, $\overline{G2}$, $\overline{E1}$, $\overline{E2}$, $\overline{E3}$, and $\overline{E4}$ are the complements of T, $G1$, $G2$, $E1$, $E2$, $E3$, and $E4$, respectively.) Path sets for the top event of the fault tree in Figure 9.1 are obtained by determining the cut sets of this complementary tree. The following cut sets of the complementary tree (which are identical to the path sets of the fault tree) are obtained by examination. (In more complex complementary trees, the cut sets have to be obtained through formal methods of cut set determination.)

$$PS1 = \left(\overline{E1}, \overline{E2}\right)$$

$$PS2 = \left(\overline{E1}, \overline{E3}, \overline{E4}\right)$$

What do these path sets mean? The first path set $PS1$ signifies that if $E1$ and $E2$ do not exist simultaneously, irrespective of whether the other terminal events $E3$ and $E4$ occur or not, the top event cannot occur. Similarly, the second path set $PS2$ signifies that if the terminal events $E1$, $E3$, and $E4$ do not exist simultaneously, irrespective of whether the remaining terminal event $E2$ occurs or not, the top event cannot occur.

Let the top event T be 'system failure,' and let the terminal events $E1$, $E2$, $E3$, and $E4$ represent the failure of components $C1$, $C2$, $C3$, and $C4$, respectively. So the path set $PS1$ signifies that if components $C1$ and $C2$ are operating successfully, then the system will operate successfully irrespective of whether components $C3$ and $C4$ are in success state or in failed state. Similarly, the path set $PS2$ signifies that if components $C1$, $C3$, and $C4$ are operating successfully, the system will operate successfully irrespective of whether component $C2$ is in success state or failed state.

Path set determination becomes difficult if some of the terminal events in the fault tree are success events or inhibit conditions. A blind adherence to the procedures described in Section 8.14 for complementary tree construction could lead to incorrect path sets. Care should be taken that the complementary tree is properly constructed and the path sets are correctly understood and interpreted.

Path sets can be determined for intermediate events also; the same principles outlined for the determination of intermediate event cut sets apply (see Section 9.2.1).

9.3.2 Minimal Path Sets

The smallest subset of a path set that is sufficient and necessary to assure that the top event does not occur is called a *minimal path set*. If all the terminal event complements of a minimal path set exist, then the top event will not occur. In other words, if none of the terminal events associated with a minimal path set occur, then the top event will not occur. There can be more than one minimal path set to the top event of a fault tree, and any one of the minimal path sets is sufficient to assure the nonoccurrence of the top event. Each minimal path set represents a possible way in which the success of the fault tree (nonoccurrence of the top event) can be assured.

Examining the path sets $PS1$ and $PS2$, neither is a subset of the other. So both of these path sets are minimal path sets. So,

$$MPS1 = (\overline{E1}, \overline{E2})$$
$$MPS2 = (\overline{E1}, \overline{E3}, \overline{E4})$$

Formal procedures for sorting out minimal path sets from path sets are available (the same as those for sorting out minimal cut sets from cut sets).

9.3.3 Order of Minimal Path Sets

The number of terminal event complements in a minimal path set is called the *order of the minimal path set*. In the minimal path set example discussed in Section 9.3.2, the order of $MPS1$ is 2 and the order of $MPS2$ is 3. $MPS1$ is called a *second-order minimal path set*, and $MCS2$ is a *third-order minimal path set*.

9.3.4 Number of Minimal Path Sets

The number of minimal path sets pertaining to the top event of a fault tree depends on the number of terminal events, the number of AND gates, the number of OR gates, and the configuration of the fault tree. In general, if the gates are mostly AND gates, there will be many small minimal path sets (by "small" we mean that the order of the minimal path sets is low); if the gates are mostly OR gates, there will be fewer but larger minimal path sets (by "larger" we mean that the order of the minimal path sets is higher). An increase in the number of AND gates increases the number of minimal path sets; an increase in the number of OR gates increases the order of some or all the minimal path sets; an increase in the number of terminal events increases the number of minimal path sets and/or the order of the minimal paths sets, depending on the configuration of the tree.

9.4 QUALITATIVE RANKING OF MINIMAL CUT SETS

An approximate qualitative ranking of minimal cut sets according to their importance in contributing to top event occurrence may be made on the basis of the order of the minimal cut sets. A more accurate quantitative ranking is discussed in Section 12.4. Qualitative ranking is not a prerequisite for quantitative ranking.

Lower-order minimal cut sets are more important than higher-order minimal cut sets if all the terminal events have approximately the same probability (at least in an order-of-magnitude sense) and are also statistically independent.

Consider the fault tree shown in Figure 9.1. Let all the terminal events $E1$, $E2$, $E3$, and $E4$ be statistically independent and also have approximately the same probability of occurrence. The minimal cut sets are determined in Section 9.2.2: They are

$$MCS1 = (E1)$$
$$MCS2 = (E2, E3)$$
$$MCS3 = (E2, E4)$$

The minimal cut set $MCS1$ should be ranked first because it is a first-order minimal cut set and the other two minimal cut sets are of the second order. Because $MCS2$ and $MCS3$ are both second-order minimal cut sets, they tie for the second-place ranking. Thus the qualitative ranking of the minimal cut sets is

$$MCS1, MCS2, MCS3$$

Minimal cut set ranking is of less engineering importance than terminal event ranking. Qualitative ranking of terminal events is discussed in the next section.

9.5 QUALITATIVE RANKING OF TERMINAL EVENTS

An approximate qualitative ranking of terminal events according to their importance in contributing to top event occurrence may be made on the basis of minimal cut sets. The following principles may be used.

1. Terminal events in the lower-order minimal cut sets are usually more important than those in the higher-order minimal cut sets. This principle is valid only if all of the terminal events in the tree have approximately equal probabilities (at least in an order-of-magnitude sense) and are statistically independent.

2. If two terminal events appear only in nth-order minimal cut sets, then the terminal event that appears in more nth-order minimal cut sets than the other is considered more important. This principle is valid only if (i) all the terminal events appearing in the nth-order minimal cut sets have approximately the same probability (at least in an order-of-magnitude sense) and (ii) all of the terminal events in the fault tree are statistically independent.

Consider the fault tree shown in Figure 9.1. Let all the terminal events $E1$, $E2$, $E3$, and $E4$ be statistically independent and also have approximately the same probability of occurrence. The minimal cut sets are determined in Section 9.2.2: They are

$$MCS1 = (E1)$$
$$MCS2 = (E2, E3)$$
$$MCS3 = (E2, E4)$$

The terminal event $E1$ should be ranked first because it appears in a first-order minimal cut set. The remaining three terminal events ($E2$, $E3$, and $E4$) appear in second-order minimal cut sets. Of these three terminal events, $E2$ appears in two second-order minimal cut sets whereas $E3$ and $E4$ appear in one second-order minimal cut set each. So $E2$ should be ranked second, and $E3$ and $E4$ tie for the third rank. Thus the qualitative ranking of terminal events is

$$E1, E2, E3, E4$$

A more accurate ranking of terminal events can be made through quantitative ranking (see Chapter 12). Qualitative ranking may be used in *preliminary studies* before a quantitative analysis is performed. Qualitative ranking is not a prerequisite for quantitative ranking.

9.6 METHODS OF QUALITATIVE ANALYSIS

Even fault trees of moderate complexity require computerized qualitative analysis. A number of excellent qualitative fault tree analysis programs are available (See Section 9.8). We describe here a few methods of qualitative minimal cut set determination. These methods may be used in the manual analysis of simple trees or in the computerized analysis of more complex trees.

Because minimal path sets are the minimal cut sets of the complementary tree, a separate discussion of minimal path set determination is unnecessary.

9.6.1 Boolean Algebra Method

The *method of Boolean algebra* is the most direct approach of minimal cut set determination. However, this approach is practical only in very small trees. Basic concepts of Boolean algebra are presented in Section 2.2. Application of Boolean algebra to minimal cut set determination is discussed here.

An *OR gate* operation between two events (terminal events or intermediate events) is equivalent to a Boolean *union operation*, represented by the symbol $\cup$. (In some of the literature, the symbol $+$ is used instead of $\cup$; there "$+$" does *not* represent arithmetic addition.) An *AND gate* operation between two events (terminal events or intermediate events) is equivalent to a Boolean *intersection operation*, represented by the symbol $\cap$. (In some of the literature, the symbol $\cdot$ is used; there "$\cdot$" does *not* represent arithmetic multiplication.)

Consider the fault tree shown in Figure 9.1. We apply the Boolean algebra operations, starting from the top of the tree:

$$T = G1 \cap G2 \tag{9.1}$$

$$G1 = E1 \cup E2 \tag{9.2}$$

$$G2 = E1 \cup E3 \cup E4 \tag{9.3}$$

Substitution of Equations (9.2) and (9.3) into Equation (9.1) yields

$$
\begin{aligned}
T &= (E1 \cup E2) \cap (E1 \cup E3 \cup E4) \\
&= (E1 \cap E1) \cup (E1 \cap E3) \cup (E1 \cap E4) \cup (E2 \cap E1) \\
&\quad \cup (E2 \cap E3) \cup (E2 \cap E4) \\
&= (E1) \cup (E1 \cap E3) \cup (E1 \cap E4) \cup (E1 \cap E2) \cup (E2 \cap E3) \\
&\quad \cup (E2 \cap E4)
\end{aligned}
\tag{9.4}
$$

Because a Boolean union operation represents an OR gate operation and a Boolean intersection operation represents an AND gate operation, Equation (9.4) may be written as

$$
\begin{aligned}
T &= (E1) \quad \text{or} \\
&= (E1 \text{ and } E3) \quad \text{or} \\
&= (E1 \text{ and } E4) \quad \text{or} \\
&= (E1 \text{ and } E2) \quad \text{or} \\
&= (E2 \text{ and } E3) \quad \text{or} \\
&= (E2 \text{ and } E4)
\end{aligned}
\tag{9.5}
$$

That is, the top event T occurs if any one of the six combinations of events within parentheses occurs. In other words, the six combinations of events within parentheses are the cut sets of T. Therefore the cut sets are

$$
\begin{aligned}
CS1 &= (E1) \\
CS2 &= (E1, E3) \\
CS3 &= (E1, E4) \\
CS4 &= (E1, E2) \\
CS5 &= (E2, E3) \\
CS6 &= (E2, E4)
\end{aligned}
\tag{9.6}
$$

These cut sets are identical to those determined by inspection in Section 9.2.1.

The minimal cut sets are determined by considering the Boolean inter-section operation between the different pairs of cut sets. Consider the ith and jth cut sets, where $i \neq j$. If the intersection is identical to the ith cut set, then the jth cut set is a subset of the ith cut set and so the ith cut set is dropped (it is not a minimal cut set); now the jth cut set has to be tested in the same manner, paired with the other cut sets.

Consider the cut sets $CS1$ and $CS2$:

$$(CS1 \cap CS2) = (E1) \cap (E1 \cap E3) = (E1 \cap E3) \qquad (9.7)$$

Because the intersection of $CS1$ and $CS2$ is identical to $CS2$, the cut set $CS1$ is a subset of $CS2$ and so $CS2$ is not a minimal cut set. $CS1$ is tested similarly, paired with the other cut sets $CS3$, $CS4$, $CS5$, and $CS6$. If we carry out the intersections, we will find that $CS1$ is a minimal cut set and $CS3$ and $CS4$ are not minimal cut sets because $CS1$ is a subset of $CS3$ and $CS4$. We will also find that the other two cut sets $CS5$ and $CS6$ are minimal cut sets. We have to check if either $CS5$ or $CS6$ is a subset of the other:

$$(CS5 \cap CS6) = (E2 \cap E3) \cap (E2 \cap E4)$$
$$= (E2 \cap E3 \cap E4) \qquad (9.8)$$

Because the intersection is not identical to either $CS5$ or $CS6$, neither cut set is a subset of the other. We have already shown that the only remaining minimal cut set $CS1$ is not a subset of $CS5$ or $CS6$. So $CS5$ and $CS6$ are minimal cut sets. Thus we have three minimal cut sets,

$$MCS1 = (E1)$$
$$MCS2 = (E2, E3) \qquad (9.9)$$
$$MCS3 = (E2, E4)$$

These minimal cut sets are identical to the minimal cut sets determined by inspection in Section 9.2.2.

The Boolean algebra method is direct and straightforward, but it is not practical for larger fault trees because such trees could have thousands or even millions of cut sets. For example, a fault tree with 300 terminal events and 300 gates could have over 60 million cut sets. The Boolean algebra approach is more suitable for smaller trees.

9.6.2 Combination Testing Method

First, one by one, each terminal event is tested to see if it could cause the top event. Those that could cause the top event form the first-order minimal cut sets. Next, each possible two-event combination of the terminal events is tested to see if it could cause the top event. (Terminal events that have already been established as first-order minimal cut sets may be omitted in forming the two-event combinations.) Those two-event combinations that could cause the top event are the second-order minimal cut sets. The procedure is repeated with three-event combinations, four-event combinations, and so on, until the n-event combination is finally tested, where n is the number of terminal events.

Consider the fault tree shown in Figure 9.1. First, we test if $E1$ is a minimal cut set, proceeding from the bottom of the tree toward the top. If $E1$ occurs, intermediate events $G1$ and $G2$ occur. Occurrence of $G1$ and $G2$ results in the top event T; therefore $E1$ is a first-order minimal cut set. Next, we test $E2$. If $E2$ occurs, $G1$ occurs but $G2$ does not; the top event T can occur only if both $G1$ and $G2$ occur, so $E2$ by itself cannot cause T; therefore $E2$ is not a minimal cut set. $E3$ and $E4$ are also tested in the same manner and are found not to be minimal cut sets.

Next, two-event combinations are tested; because $E1$ is a minimal cut set, combinations including $E1$ need not be considered. The two-event combinations to be tested are $(E2, E3)$, $(E2, E4)$, and $(E3, E4)$. Consider the combination $(E2, E3)$. If $E2$ occurs, $G1$ occurs; if $E3$ occurs, $G2$ occurs; the occurrence of $G1$ and $G2$ results in the top event T; therefore $(E2, E3)$ is a second-order minimal cut set. Testing in the same manner, $(E2, E4)$ is also found to be a second-order minimal cut set whereas $(E3, E4)$ is found not to be a minimal cut set.

Considering the possible three-event combinations, all such combinations contain in them either the first- or the second-order minimal cut sets already determined. So none of them could be minimal cut sets. Similarly, the four-event combination also contains in it the previously determined minimal cut sets, so it is not a minimal cut set.

In summary, the minimal cut sets determined using the combination testing method are

$$MCS1 = (E1)$$
$$MCS2 = (E2, E3)$$
$$MCS3 = (E2, E4)$$

These minimal cut sets are identical to those determined in Sections 9.2.2 and 9.6.1.

The combination testing method is a straightforward procedure but it cannot be used in manual analyses if the number of terminal events is large. Computerized procedures are available [Vesely and Narum (1970)]. Minimal cut set determination using combination testing could be very time-consuming, even with computers, if the fault tree is large.

9.6.3 Prime Number Method

The prime number method was introduced by Semanderes (1971). We illustrate the method by applying it to the fault tree shown in Figure 9.1.

First, each terminal event is assigned a unique prime number, starting from 2. Thus,

$$E1 = 2 \qquad E2 = 3 \qquad E3 = 5 \qquad E4 = 7 \qquad (9.10)$$

Now, the intermediate event $G1$ is composed of $E1$ and $E2$, connected through an OR gate. Therefore,

$$G1 = (2 \text{ or } 3) \qquad (9.11)$$

Similarly,

$$G2 = (2 \text{ or } 5 \text{ or } 7) \qquad (9.12)$$

The top event T is composed of $G1$ and $G2$, connected through an AND gate. In this approach, the AND gate operation is represented by multiplication. Therefore,

$$T = (G1 \times G2) = (2 \text{ or } 3) \times (2 \text{ or } 5 \text{ or } 7)$$
$$= (4 \text{ or } 10 \text{ or } 14 \text{ or } 6 \text{ or } 15 \text{ or } 21) \qquad (9.13)$$

Each of the numbers in the second line of Equation (9.13) represents a cut set; the cut set numbers are

$$
\begin{aligned}
CS1 &= 4 \\
CS2 &= 10 \\
CS3 &= 14 \\
CS4 &= 6 \\
CS5 &= 15 \\
CS6 &= 21
\end{aligned}
\qquad (9.14)
$$

Some of the cut sets may contain repetitions of events. Such a cut set is represented by a number that is not the product of distinct prime numbers (one or more prime numbers are repeated). Examining the preceding six cut set numbers, $CS1$ is not a product of distinct prime numbers; the prime number 2 is repeated ($CS1 = 4 = 2 \times 2$). The duplication is eliminated by dividing the cut set number by the repeated prime number.

$$CS1 = \tfrac{4}{2} = 2$$

(Suppose a cut set number is 8575. We could expand it as $5 \times 5 \times 7 \times 7 \times 7$, which is not a product of distinct prime numbers. The repetition is eliminated by dividing 8575 by 5 once and by 7 twice. The resulting cut set number is 35, which is a product of distinct prime numbers.)

Examination of the other cut set numbers $CS2$, $CS3$ $CS4$, $CS5$, and $CS6$ shows that they are all products of distinct prime numbers. Therefore the cut set numbers, after eliminating repetitions, are

$$
\begin{aligned}
CS1 &= 2 \\
CS2 &= 10 \\
CS3 &= 14 \\
CS4 &= 6 \\
CS5 &= 15 \\
CS6 &= 21
\end{aligned}
\qquad (9.15)
$$

The supersets among the cut sets are eliminated, and the minimal cut sets are thus sorted out, by discarding those cut set numbers that are divisible by other cut set numbers. (If the quotient is an integer, then the numerator is said to be divisible by the denominator.) Examining the six cut set numbers enumerated in Equations (9.15), $CS1$ is not divisible by any other cut set number, so $CS1$ is a minimal cut set. $CS2$ is divisible by $CS1$ ($[CS2/CS1] = \frac{10}{2} = 5$), so it is not a minimal cut set. Similarly, $CS3$ and $CS4$ are also divisible by $CS1$ and are not minimal cut sets. The cut set numbers $CS5$ and $CS6$ are not divisible by any cut set number, so they are minimal cut sets. Thus the minimal cut set numbers are

$$
\begin{aligned}
MCS1 &= CS1 = 2 \\
MCS2 &= CS5 = 15 \\
MCS3 &= CS6 = 21
\end{aligned}
\qquad (9.16)
$$

The next step is to factor the minimal cut set numbers into their component prime numbers; there is only one way each minimal cut set number can be factored. Once a minimal cut set number is factored into its component prime numbers, the corresponding terminal events in that minimal cut set can be identified:

$$
\begin{aligned}
MCS1 &= 2 = (E1) \\
MCS2 &= 15 = 3 \times 5 = (E2, E3) \\
MCS3 &= 21 = 3 \times 7 = (E2, E4)
\end{aligned}
\qquad (9.17)
$$

These minimal cut sets are the same as those determined in Sections 9.2.2, 9.6.1, and 9.6.2.

The prime number method is an elegant approach, particularly suitable for computerized analysis. The only disadvantage is that when the number of terminal events becomes large, very large prime numbers have to be assigned to the terminal events. The products of such large prime numbers will be very large so that they will become cumbersome even for computerized analysis.

Semanderes (1971) and Wong (1975) use the prime number method in their computer programs for qualitative fault tree analysis.

9.6.4 Binary Bit String Method

The binary bit string method was developed by Gangadharan, Rao, and Sundararajan (1977). The method is illustrated by applying it to the fault tree shown in Figure 9.1.

First, each terminal event is assigned a *binary bit string*. A binary bit string consists of a number of binary bits (a binary bit is either 0 or 1). If there are n terminal events in the fault tree, then the binary bit string should have at least n bits. Binary bit string representation of terminal events is ideal for digital computers because FORTRAN words are represented by binary bit strings. Some computers use 32 binary bits to represent each word, some use 60 bits. If we are using a 32-bit computer and the number of terminal events is 90, we will use three words ($32 \times 3 = 96$) to represent each terminal event. If all the bits of a string are zero, it is called a *null string* (null set).

Let A and B be two events, each n bits long. The AND gate operation between A and B produces another event C of the same length. The ith bit of C is equal to 0 if the ith bits of A and B are both 0; the ith bit of C is equal to 1 if the ith bit of A and/or the ith bit of B is 1, where $i = 1, \ldots, n$. For example, if $A = 0101$ and $B = 1100$, then $C = 1101$. If $A = 0100$ and $B = 1100$, then $C = 1100$. (Note that *logical AND* operations used in some digital computers and the *AND gate* operations used in fault trees are *not* the same.)

The complement of an n-bit-long event A is also n bits long. The ith bit of the complement $\overline{A}$ is 0 if the ith bit of A is 1, and the ith bit of $\overline{A}$ is 1 if the ith bit of A is 0, where $i = 1, \ldots, n$. For example, if $A = 0101$, then $\overline{A} = 1010$. If $A = 0100$, then $\overline{A} = 1011$.

The intersection of two events A and B, each of n-bit length, is an event D of the same length. The ith bit of D is 0 if the ith bit of A and/or the ith bit of B is 0. The ith bit of D is 1 if both the ith bit of A and the ith bit of B are 1, where $i = 1, \ldots, n$. For example, if $A = 0101$ and $B = 1100$, then $D = 0100$. If $A = 0100$ and $B = 1100$, then $D = 0100$. If $A = 0010$ and $B = 1100$, then $D = 0000$.

If the intersection of an event A and the complement of the event B is a null string (null set), then A is a subset of B (that is, B is a superset of A). Consider $A = 0100$ and $B = 1100$. $A \times \overline{B} = (0100) \times (0011) = 0000 = $ null string; therefore B is a superset of A. Now, $B \times \overline{A} = (1100) \times (1011) = 1000$; because it is not a null string, A is not a superset of B.

Consider the fault tree in Figure 9.1. Because there are only four terminal events, we will assign each event a unique 4-bit binary string (if we are using a 32-bit binary string, the 5th through 32 bits will be filled with zeros).

$$
\begin{aligned}
E1 &= 1000 \\
E2 &= 0100 \\
E3 &= 0010 \\
E4 &= 0001
\end{aligned}
\tag{9.18}
$$

We start from the bottom of the fault tree and move toward the top. The intermediate event $G1$ is connected to two terminal events $E1$ and $E2$ by an OR gate. Therefore $G1$ has two cut sets, $G1(1)$ and $G1(2)$:

$$
\begin{aligned}
G1(1) &= E1 = 1000 \\
G1(2) &= E2 = 0100
\end{aligned}
\tag{9.19}
$$

Similarly,

$$
\begin{aligned}
G2(1) &= E1 = 1000 \\
G2(2) &= E3 = 0010 \\
G2(3) &= E4 = 0001
\end{aligned}
\tag{9.20}
$$

The top event T is connected to the intermediate events $G1$ and $G2$ by an AND gate. The cut sets of T are the "AND combinations" of the two $G1$ cut sets and the three $G2$ cut sets.

$$
\begin{aligned}
T(1) &= G1(1) \text{ AND } G2(1) = 1000 \text{ AND } 1000 = 1000 \\
T(2) &= G1(1) \text{ AND } G2(2) = 1000 \text{ AND } 0010 = 1010 \\
T(3) &= G1(1) \text{ AND } G2(3) = 1000 \text{ AND } 0001 = 1001 \\
T(4) &= G1(2) \text{ AND } G2(1) = 0100 \text{ AND } 1000 = 1100 \\
T(5) &= G1(2) \text{ AND } G2(2) = 0100 \text{ AND } 0010 = 0110 \\
T(6) &= G1(2) \text{ AND } G2(3) = 0100 \text{ AND } 0001 = 0101
\end{aligned}
\tag{9.21}
$$

Next, the minimal cut sets are identified by checking which of the preceding cut sets are supersets. Consider $T(1)$ and $T(2)$:

$$
T(1) \times \overline{T(2)} = (1000) \times (0101) = 0000 = \text{null set}
$$

So $T(2)$ is a superset of $T(1)$; therefore $T(2)$ is not a minimal cut set.

Similar operations between the different combinations of the six cut sets show that $T(1)$, $T(5)$, and $T(6)$ are minimal cut sets. These minimal cut sets are resolved into their component terminal events as follows:

$$MCS1 = T(1) = 1000 = (E1)$$

$$MCS2 = T(5) = 0110 = 0100 \text{ AND } 0010 = E2 \text{ AND } E3 = (E2, E3)$$

$$MCS3 = T(6) = 0101 = 0100 \text{ AND } 0001 = E2 \text{ AND } E4 = (E2, E4)$$

$$(9.22)$$

These minimal cut sets are the same as those determined in Sections 9.2.2 and 9.6.1–9.6.3.

The binary bit string method can be used for the manual determination of minimal cut sets of simple fault trees. The method is ideal for the computerized analysis of larger trees. The method has been shown to be very efficient and fast [Gangadharan, Rao, and Sundararajan (1977)].

Another minimal cut set determination procedure using binary bit strings has been presented by Erdmann, Leverenz, and Kirch (1978).

9.6.5 Other Methods

Two other methods, the *Boolean Indicated Cut Sets Method* (BICS method) and the *Monte Carlo Simulation Method* are reviewed by Gangadharan, Rao, and Sundararajan (1977). The former method was developed by Fussell and Vesely (1972) and was implemented in a computer program by Fussell, Henry, and Marshall (1974). The Monte Carlo simulation method has been used in the computer program by Crosetti (1971).

9.7 SELECTIVE DETERMINATION OF MINIMAL CUT SETS

Even when computer programs are used, determining all of the minimal cut sets could be time-consuming and expensive if the tree is very large. In such cases, some reliability analysts choose to determine only lower-order minimal cut sets (for example, up to second-order minimal cut sets, up to fifth-order minimal cut sets, etc.). There is no hard-and-fast rule for choosing the order of minimal cut sets to discard (for example, discard all minimal cut sets above the second order, discard all minimal cut sets above the fifth-order, etc.). An approximate criterion is given in the next paragraph.

Let the highest probability of occurrence of any terminal event in the fault tree be p. (If the actual probabilities are not known at this stage of the analysis, an approximate estimate is sufficient.) Then, an upper-bound estimate of the probability of an nth-order minimal cut set in which all the terminal events are statistically independent is p^n. The number of minimal cut sets above the nth order is seldom known before they are actually determined, but an experienced analyst may provide a rough estimate. Let the number of minimal cut sets above the nth order be m. Then the sum of the probabilities of all the discarded minimal cut sets (above the nth order) is approximately equal to $P' = mp^{n+1}$; this could be a *very conservative* estimate. After a quantitative analysis is performed using the minimal cut sets up to the nth order (see Chapter 10), let the estimate of the probability of the top event be P. If P' is significantly smaller than P (say, one or more orders of magnitude less), then it is acceptable to discard minimal cut sets above the nth order, provided that there is no statistical dependence between terminal events.

This procedure may be suitably modified depending on the specific knowledge of the probabilities of terminal events occurring in the higher-order minimal cut sets. For example, if we know that the highest probability terminal event is not part of minimal cut sets above the nth order, the p value used in computing P' could be lowered.

9.8 COMPUTER PROGRAMS

FTAP, MICSUP, PL-MOD, RAS, SETS, WAMCUT, and WAMCUT-II are some of the commercially or publicly available computer programs for qualitative fault tree analysis (see Appendix I for details). This is only a partial list of available programs. New computer programs also become available. Capabilities and computational efficiencies of the programs differ. Users should compare the available programs and select those that best suit their needs.

9.9 DOCUMENTATION

Documentation of qualitative fault tree analysis should include the following:

1. Minimal cut sets.
2. Minimal path sets (if determined).
3. Justifications for discarding any minimal cut sets or minimal path sets.
4. Identification of computer programs (if used); include version number and/or version date, and the source of the program (from whom obtained).
5. Input and output from the computer programs (if computer programs are used).

Some general guidelines on documentation are provided in Section 14.10.

9.10 EXAMPLE PROBLEM

Consider the domestic hot water system discussed in Section 6.5 and shown in Figure 6.1. A fault tree for the top event 'hot water tank rupture' has already been constructed (Figure 8.34). Here we will determine by inspection the minimal cut sets for the top event. Readers may determine the minimal cut sets using the other methods discussed in Section 6.6 and also by using any computer programs they may have for qualitative fault tree analysis.

First we determine the cut sets for the top event by inspection. There are 26 cut sets.

$$CS1 = (\text{HTK10P00})$$
$$CS2 = (\text{HTK10S00})$$
$$CS3 = (\text{HSV10C00, HUE10000})$$
$$CS4 = (\text{HSV10I00, HUE10000})$$
$$CS5 = (\text{HSV10S00, HUE10000})$$
$$CS6 = (\text{HSV10C00, HST10O00})$$
$$CS7 = (\text{HSV10I00, HST10O00})$$
$$CS8 = (\text{HSV10S00, HST10O00})$$
$$CS9 = (\text{HSV10C00, HST10S00})$$
$$CS10 = (\text{HSV10I00, HST10S00})$$
$$CS11 = (\text{HSV10S00, HST10S00})$$
$$CS12 = (\text{HSV10C00, HCR10P00})$$
$$CS13 = (\text{HSV10I00, HCR10P00})$$
$$CS14 = (\text{HSV10S00, HCR10P00})$$
$$CS15 = (\text{HSV10C00, HCR10S00})$$
$$CS16 = (\text{HSV10I00, HCR10S00})$$
$$CS17 = (\text{HSV10S00, HCR10S00})$$
$$CS18 = (\text{HSV10C00, HTD10P00})$$
$$CS19 = (\text{HSV10I00, HTD10P00})$$
$$CS20 = (\text{HSV10S00, HTD10P00})$$
$$CS21 = (\text{HSV10C00, HTD10I00})$$
$$CS22 = (\text{HSV10I00, HTD10I00})$$
$$CS23 = (\text{HSV10S00, HTD10I00})$$

$$CS24 = (\text{HSV10C00}, \text{HTD10S00})$$
$$CS25 = (\text{HSV10I00}, \text{HTD10S00})$$
$$CS26 = (\text{HSV10S00}, \text{HTD10S00})$$

Examination of these 26 cut sets shows that all of them are minimal cut sets. Thus we have 2 first-order minimal cut sets and 24 second-order minimal cut sets.

EXERCISE PROBLEMS

9.1. Consider the fault tree shown in Figure 8.21. (i) Determine the minimal cut sets. (ii) Determine the minimal path sets.

9.2. Consider the fault tree shown in Figure 8.39. Determine the minimal path sets.

9.3. Consider the fault tree shown in Figure 8.39. (i) Determine the minimal cut sets for the intermediate event 'stop valve is open even after the boiling temperature is reached.' (ii) Determine the minimal path sets for the intermediate event 'stop valve is open even after the boiling temperature is reached.'

9.4. Consider the fault tree shown in Figure 8.20. For the purposes of this problem, assume that E, F, G, J, K, M, N, P, Q and R are terminal events (instead of intermediate events as indicated in Figure 8.20). (i) Determine the minimal cut sets. (ii) Determine the minimal path sets. (iii) Determine the minimal cut sets for the intermediate event B. (iv) Determine the minimal path sets for the intermediate event B.

REFERENCES

Crosetti, P. A. (1971). Fault tree analysis with probability evaluation. *IEEE Transactions on Nuclear Science* **NS-18** 465–471.

Erdmann, R. C., F. L. Leverenz, and H. Kirch (1978). *WAMCUT: A Computer Code for Fault Tree Evaluation*. Report No. NP-803. Electric Power Research Institute, Palo Alto, CA.

Fussell, J. B., E. B. Henry, and N. H. Marshall (1974). *MOCUS: A Computer Program to Obtain Minimal Sets from Fault Trees*. Report No. ANCR-1156. Aerojet Nuclear Company, Idaho Falls.

Fussell, J. B. and W. E. Vesely (1972). A new methodology for obtaining cut sets. *Transactions of the American Nuclear Society* **15** 262–273.

Gangadharan, A. C., M. S. M. Rao, and C. Sundararajan (1977). Computer methods for qualitative fault tree analysis. In *Failure Prevention and Reliability*, S. B. Bennett, A. L. Ross, and P. Z. Zemanick, eds. American Society of Mechanical Engineers, New York, pp. 251–262.

Semanderes, S. N. (1971). ELRAFT: A computer program for the efficient logic reduction analysis of fault trees. *IEEE Transactions on Nuclear Science* **NS-18** 481–487.

Vesely, W. E. and R. E. Narum (1970). *PREP and KITT: Computer Codes for the Automatic Evaluation of a Fault Tree*. Report No. IN-1349. Idaho Nuclear Corporation, Idaho Falls.

Wong, P. Y. (1975). *FAUTRAN: A Fault Tree Analyzer*. Report No. AECL-5182. Atomic Energy of Canada Limited, Chalk River.

Chapter 10
Quantitative Fault Tree Analysis

10.1 INTRODUCTION

A quantitative fault tree analysis may be carried out independently of the qualitative fault tree analysis or by using the minimal cut sets or minimal path sets generated during the qualitative analysis. Both approaches are discussed in this chapter. What quantities are computed during a quantitative fault tree analysis depends on the purpose for which the fault tree is used. If the top event of the fault tree is an accident, we will be interested in the probability that no accident occurred anytime between time zero and time t (that is, we are interested in the reliability of the top event at time t). Here we are interested in reliability, rather than availability, because we are interested in the probability of no accident between time zero and time t. If the top event is 'system failure' ('system shutdown,' 'system inoperational'), we may be interested in the probability that the system is in an operational state (success state) at time t, irrespective of whether the system failed one or more times before time t but was restored to service (repaired) by time t (that is, we are interested in the availability of the top event at time t). In addition to reliability and/or availability, related system parameters such as failure rate, mean time to failure, mean time to repair, and the expected number of failures may also be computed, as necessary.

Reliability, availability, and other related parameters may also be computed for intermediate events of the fault tree. Instead of the complete fault tree, just the portion of the fault tree below the intermediate event of interest is considered in such an analysis. This procedure is similar to the procedure used in determining cut sets of intermediate events; see Section 9.2.1.

Although constant failure rate and constant repair rate are assumed for components in most fault tree analyses, other types of failure rate and repair rate may be considered. It should be noted that even if all of the components have constant failure rates, it does not necessarily mean that the system will also have a constant failure rate.

In this chapter we assume constant failure rates and constant repair rates for components, unless otherwise stated. Various other assumptions may also be made in the process of fault tree quantification, and such assumptions are discussed as we describe the quantification procedures later in this chapter.

Computerized analysis is necessary except for very simple fault trees. Appendix I gives a list of some readily available computer software. However, simple hand calculation techniques will be useful for computing the reliability and availability of small systems or subsystems.

Some simple quantitative analysis procedures used in hand calculations as well as in computerized analyses are described in Sections 10.3 and 10.4. We do not go into the mathematics of more complex analysis procedures but references to the mathematical formulations of these complex techniques are provided in Section 10.6 for the benefit of interested readers.

10.2 COHERENT STRUCTURE

A *coherent set* of minimal cut sets has a monotonically increasing structure and all the terminal events in the minimal cut set are relevant. A terminal event is said to be *relevant* if the occurrence of the terminal event contributes in some way to the occurrence of the top event. A set of minimal cut sets is monotonically increasing if (i) the occurrence of any terminal event of the minimal cut set always increases the probability of occurrence of the top event of the fault tree and (ii) the repair of any failed terminal event of the minimal cut set always decreases the probability of occurrence of the top event. The second condition is, of course, not applicable if the components in the system are nonrepairable.

A fault tree is said to be *coherent* (or is said to have a coherent structure) if all its minimal cut sets are coherent.

Success events (as in the fault tree shown in Figure 8.28b) are not relevant events because success events do not contribute to the occurrence of the top event. The mere fact that there is a success event in a fault tree does not mean that the success event will become part of one or more minimal cut sets of the fault tree.

Consider the fault tree shown in Figure 8.28b. We determine the three minimal cut sets by inspection:

$$MCS1 = (C)$$

$$MCS2 = (B, [A|B], B)$$

$$MCS3 = (\bar{B}, [A|\bar{B}], B)$$

The terminal event B is repeated in the second minimal cut set, so it is rewritten as

$$MCS2 = (B, [A|B])$$

The third minimal cut set contains B and $\bar{B}$, which are mutually exclusive events. This minimal cut set can never occur because mutually exclusive events cannot exist simultaneously. Therefore this minimal cut set should be discarded. Hence the minimal cut sets of the fault tree are

$$MCS1 = (C)$$
$$MCS2 = (B, [A|B])$$

In this example, the success event does not appear in any of the minimal cut sets even though it is present in the fault tree.

In some problems, the success event may be present in a minimal cut set as a terminal event but it can be "removed" by combining it with another terminal event in that minimal cut set. Consider the following minimal cut set:

$$MCS = ([E|\bar{F}], \bar{F}, G)$$

where E and G are terminal events that represent the failure of components E and G, respectively, and $\bar{F}$ is a terminal event that represents the success of component F. This minimal cut set may be transformed to

$$MCS = (E', G)$$

where E' represents a new terminal event, which combines $E|\bar{F}$ and $\bar{F}$. The probability of E' is given by

$$P[E'] = P[E|\bar{F}]P[\bar{F}]$$

Note that the preceding transformation should be done only after minimal cut sets containing mutually exclusive terminal events are discarded. Moreover, such a transformation should be done at the quantitative analysis stage. It is better to leave the minimal cut set in its original form during qualitative analysis because the minimal cut set in its original form provides more qualitative insight into the top event than the transformed form does.

Consider a component failure that is resolved into primary, secondary, and command failures, as in Figure 8.15. In case of a primary failure, the component is restored to its normal state (success state) if the primary failure is repaired. So, if a primary failure is a terminal event in a minimal cut set, repair of the primary failure "repairs" the minimal cut set itself, and that minimal cut set no longer causes the top event.

In case of a command failure, repair of the command failure results in the provision of correct command (signal) to the component at the appropriate time and the component will function correctly. So, if a command failure is a terminal event in a minimal cut set, repair of the command failure repairs the minimal cut set itself, and that minimal cut set no longer causes the top event. (*Note:* a command failure does not damage the component, it simply sends incorrect commands (signals) to the component. Therefore the component starts functioning correctly once the command failure is repaired.)

Now consider a secondary failure. The secondary cause does damage the component. So the repair of a secondary failure (secondary cause) will not restore the component to its normal state (success state); the component itself has to be repaired. Therefore, the repair of a secondary failure (secondary cause) does not repair the minimal cut set; the minimal cut set is repaired only if the secondary failure (secondary cause) as well as the component damage resulting from the secondary cause are repaired. For example, if an air-conditioning failure (a secondary failure) results in the failure of an electronic component due to excessive temperatures, the repair of the air-conditioning system (repair of the secondary failure) does not get the electronic component repaired.

So, if the component failure is resolved into primary, secondary, and command failures (as in Figure 8.15) and a secondary failure is part of a minimal cut set, repair of the secondary failure does not repair the minimal cut set and the top event will continue to be in a failed state. Repair of the secondary failure (without the repair of the component damaged by the secondary failure) will not decrease the top event probability. Therefore the second condition for coherency is not satisfied. Fault trees that contain secondary failures (as in Figure 8.15) do not have a coherent structure if the secondary failure is repairable.

A method of transforming such noncoherent fault trees to coherent trees is discussed in Section 10.4.4. Such a transformation is necessary only if the secondary failure is repairable.

10.3 NONREPAIRABLE SYSTEMS

A system is called a *nonrepairable system* if all the components of the system are nonrepairable.

As discussed in Section 3.3.2, the unreliability at time t, the unavailability at time t, and the failure probability at time t are all identical for nonrepairable components. The same is true for nonrepairable systems also.

As discussed in Section 10.2, the fault tree should have a set of coherent minimal cut sets. Two necessary conditions for coherency are noted in Section 10.2. The second condition is not applicable to nonrepairable systems. A method for satisfying the first condition even when there are success events as terminal events is discussed in Section 10.2.

10.3.1 AND Gate

Consider an AND gate with two terminal events (Figure 10.1). The output event could be the top event or an intermediate event. The input events could be intermediate events or terminal events. We are interested in the probability of the output event Z, in terms of the probabilities of the input events A and B.

The probability of A at time t is the probability that A exists at time t. If the terminal event A represents the failure of component A, then the probability of A at time t is the probability that component A is in a failed state at time t, given that component A was new or as good as new at time zero; that is, the probability of A at time t is the unavailability (equal to the unreliability) of component A at time t. Similar definitions apply to the probability of B and Z also.

The probability of Z at time t is given by

$$P[Z] = P[A \cap B] = P[A]P[B|A] \tag{10.1}$$

where $P[A]$ is the probability of A at time t, and $P[B|A]$ is the conditional probability of B at time t given that A has occurred. If A and B are statistically independent, then $P[B|A] = P[B]$, so Equation (10.1) reduces to

$$P[Z] = P[A]P[B] \tag{10.2}$$

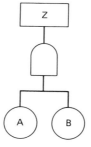

Figure 10.1. AND gate with two terminal events.

The preceding equations can be generalized to AND gates with n input events. Let the input events be $I_1, I_2, \ldots, I_n$. If the input events are statistically independent, we have

$$P[Z] = P[I_1]P[I_2] \cdots P[I_n] \qquad (10.3)$$

If there is statistical dependence between the input events, the corresponding formula is

$$P[Z] = P[I_1]P[I_2|I_1] \cdots P[I_n|I_1, I_2, \ldots, I_{n-1}] \qquad (10.4)$$

where $P[I_n|I_1, I_2, \ldots, I_{n-1}]$ represents the conditional probability of I_n given that $I_1, I_2, \ldots, I_{n-1}$ have occurred. If I_n is statistically independent of $I_1, I_2, \ldots, I_{n-1}$, then

$$P[I_n|I_1, I_2, \ldots, I_{n-1}] = P[I_n]$$

If I_n is statistically independent of I_2 but is dependent on the others, then

$$P[I_n|I_1, I_2, \ldots, I_{n-1}] = P[I_n|I_1, I_3, \ldots, I_{n-1}]$$

10.3.2 OR Gate

Consider an OR gate with two input events (Figure 10.2). The output event Z could be the top event or an intermediate event. The input events A and B could be intermediate events or terminal events. The probability of Z at time t is given by

$$P[Z] = P[A \cup B] = P[A] + P[B] - P[A \cap B] \qquad (10.5)$$

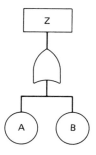

Figure 10.2. OR gate with two terminal events.

This formula is applicable whether A and B are statistically independent or not. It may be expanded by substituting Equation (10.1) or (10.2) for $P[A \cap B]$. If A and B are statistically independent, we have

$$P[Z] = P[A] + P[B] - P[A]P[B] \qquad (10.6)$$

If A and B are statistically dependent, we have

$$P[Z] = P[A] + P[B] - P[A]P[B|A] \qquad (10.7)$$

In Equations (10.5)–(10.7), the third term may be dropped if it is substantially smaller (say, an order of magnitude or more) than the first two terms. In such a case, we have

$$P[Z] = P[A] + P[B] \qquad (10.8)$$

This equation is known as the *small probability approximation* or the *rare event approximation*. If $P[A]$ and $P[B]$ are less than 0.1 and A and B are statistically independent, Equation (10.8) will provide fairly accurate results. Moreover, Equation (10.8) is always conservative; that is, the $P[Z]$ value computed using Equation (10.8) is always higher than that computed using Equation (10.5), (10.6), or (10.7).

Equation (10.5) can be generalized for OR gates with n input events $I_1, I_2, \ldots, I_n$:

$$
\begin{aligned}
P[Z] = {}& P[I_1 \cup I_2 \cup \cdots \cup I_n] \\
= {}& \{P[I_1] + P[I_2] + \cdots + P[I_n]\} \\
& - \{\text{sum of the probabilities of all possible double combinations}\} \\
& + \{\text{sum of the probabilities of all possible triple combinations}\} \\
& - \cdots + \{(-1)^{n-1} P[I_1 \cap I_2 \cap \cdots \cap I_n]\}
\end{aligned}
\qquad (10.9)
$$

There are $\binom{n}{2}$ double combination terms, $\binom{n}{3}$ triple combination terms, $\binom{n}{4}$ quadruple combination terms, and so on, where

$$\binom{n}{r} = \frac{n!}{r!(n-r)!}$$

If $n = 3$, the Equation (10.9) reduces to

$$
\begin{aligned}
P[Z] = {}& P[I_1] + P[I_2] + P[I_3] \\
& - \{P[I_1 \cap I_2] + P[I_2 \cap I_3] + P[I_3 \cap I_1]\} \\
& + P[I_1 \cap I_2 \cap I_3]
\end{aligned}
\qquad (10.10)
$$

If the input events are statistically independent, Equation (10.9) reduces to

$$P[Z] = P[I_1 \cup I_2 \cup \cdots \cup I_n]$$

$$= 1 - \prod_{i=1}^{n} \{1 - P[I_i]\} \tag{10.11}$$

where $\prod_{i=1}^{n}(a_i)$ represents the product of $a_1, a_2, \ldots, a_n$.

If the small probability approximation is applicable, the following equation may be used instead of either Equation (10.9) or (10.11).

$$P[Z] = \sum_{i=1}^{n} P[I_i] \tag{10.12}$$

This will always be a conservative estimate for Equation (10.9) or (10.11) and will usually be quite accurate if all of the $P[I_i]$ are small. However, if there are a large number of higher-order combinations (double combination, triple combination, etc.), their total contribution could become significant and adversely affect the accuracy of Equation (10.12), although it will still be a conservative estimate.

Equation (10.9) can be adapted to compute successively closer upper and lower bounds. If only the first-order terms are used and all the higher-order terms are discarded [that is the same as Equation (10.12)], it will be an upper bound to Equation (10.9) [or Equation (10.11)]. It will be a lower bound if only the first- and second-order terms of Equation (10.9) are used. It will be an upper bound if only the first-, second-, and third-order terms of Equation (10.9) are used, and this upper bound will be closer to the exact solution [Equation (10.9) or (10.11)] than the first-order approximation [Equation (10.12)]. Another lower bound can be obtained by considering only the first-, second-, third-, and fourth-order terms of Equation (10.9), and this lower bound will be closer to the exact solution than that obtained by considering only the first- and second-order terms. In this manner a series of successively closer upper and lower bounds to Equation (10.9) [or Equation (10.11)] can be computed. This principle of computing successively closer upper and lower bounds is known as the *inclusion–exclusion principle*.

Example 10.1

Consider an OR gate with three input events ($n = 3$). Let the probability of each of these three events be 0.2. Compute the exact value and the upper and lower bounds of the output event probability. Assume that the input events are statistically independent.

Solution

Let the three input events be denoted by A, B, and C, and let the output event be denoted by Z.

The upper-bound solution obtained by considering only the first-order terms (small probability approximation) is given by

$$P[Z] = P[A] + P[B] + P[C] = 0.2 + 0.2 + 0.2 = 0.6$$

The lower-bound solution obtained by considering the first- and second-order terms is given by

$$
\begin{aligned}
P[Z] &= \{P[A] + P[B] + P[C]\} \\
&\quad - \{P[A]P[B] + P[B]P[C] + P[C]P[A]\} \\
&= [0.2 + 0.2 + 0.2] - [0.04 + 0.04 + 0.04] \\
&= 0.6 - 0.12 = 0.48
\end{aligned}
$$

The exact solution is

$$
\begin{aligned}
P[Z] &= \{P[A] + P[B] + P[C]\} \\
&\quad - \{P[A]P[B] + P[B]P[C] + P[C]P[A]\} \\
&\quad + \{P[A]P[B]P[C]\} \\
&= [0.2 + 0.2 + 0.2] - [0.04 + 0.04 + 0.04] + [0.008] \\
&= 0.6 - 0.12 + 0.008 \\
&= 0.488
\end{aligned}
$$

Note that the small probability approximation is not very accurate compared to the exact solution:

$$\text{error} = \frac{0.6 - 0.488}{0.488} \times 100 = +22.95\%$$

Such an error is expected because the probabilities of the input events are greater than 0.1. The lower-bound solution obtained by considering the first- and second-order terms is quite good in this problem:

$$\text{error} = \frac{0.48 - 0.488}{0.488} \times 100 = -1.64\%$$

Example 10.2

Solve Example 10.1 with $P[A] = 0.2$, $P[B] = 0.1$, and $P[C] = 0.1$.

Solution

We use the same formulae as in Example 10.1, with $P[A] = 0.1$ and $P[B] = P[C] = 0.1$. The upper-bound solution obtained by considering only the first-order term (small probability approximation) is

$$P[Z] = 0.2 + 0.1 + 0.1 = 0.4$$

The lower-bound solution obtained by considering the first- and second-order terms is

$$P[Z] = [0.2 + 0.1 + 0.1] - [0.02 + 0.01 + 0.02]$$
$$= 0.4 - 0.05 = 0.35$$

The exact solution is given by

$$P[Z] = [0.2 + 0.1 + 0.1] - [0.02 + 0.01 + 0.02] + [0.002]$$
$$= 0.4 - 0.05 + 0.002 = 0.352$$

The error in the preceding upper-bound solution is $+13.64\%$, and the error in the lower-bound solution is -0.57%.

The errors in the solution to Example 10.2 are less than those in Example 10.1 because the probabilities of some of the input events are smaller in Example 10.2.

Example 10.3

Solve Example 10.1 with $P[A] = P[B] = P[C] = 0.1$.

Solution

We use the same procedure used in Example 10.1.

The upper-bound solution obtained by considering only the first-order term (small probability approximation) is 0.3. The lower-bound solution obtained by considering the first- and second-order terms is 0.27. The exact solution is 0.271.

The error in the preceding upper-bound solution is $+10.7\%$, and the error in the lower-bound solution is -0.34%.

Example 10.4

Consider an OR gate with two input events ($n = 2$). Let the probability of each of these two events be 0.1. Compute the exact value and the upper bound for the probability of the output event. Assume that the input events are statistically independent.

Solution

We use the same procedure used in Example 10.1.

The upper-bound solution obtained by considering only the first-order term (small probability approximation) is 0.2. The exact solution is obtained by considering all terms, and it is 0.19.

The error in the preceding upper-bound solution is $+5.26\%$.

Comparing this error with the error in the upper-bound solution in Example 10.3, we see that, even with the same input event probabilities, the error is lower when the number of input events is less.

10.3.3 Inhibit Gate

Transforming an inhibit gate into an AND gate is discussed in Section 8.5.8.

10.3.4 Direct Quantification of Fault Trees

The probabilistic relationships developed between the input and output events of OR gates and AND gates can be utilized in computing the probability of the top event of fault trees. There are two approaches: the direct approach and the minimal cut set approach. The direct approach computes the top event probability directly from the fault tree, without the use of minimal cut sets.

Consider the fault tree shown in Figure 10.3. Let the terminal events $E1, E2,,\ldots, E5$ be statistically independent. We shall first compute the probabilities of the intermediate events $G1$ and $G2$ and then move up to the top event T:

$$P[G1] = P[E1 \cup E2] = P[E1] + P[E2] - P[E1]P[E2]$$
$$P[G2] = P[E3 \cup E4 \cup E5]$$
$$= P[E3] + P[E4] + P[E5] \qquad (10.13)$$
$$- \{P[E3]P[E4] + P[E4]P[E5] + P[E5]P[E3]\}$$
$$+ P[E3]P[E4]P[E5]$$
$$P[T] = P[G1]P[G2] \qquad (10.14)$$

Numerical values of $P[G1]$ and $P[G2]$ are computed using Equation (10.13) and then are substituted into Equation (10.14) to obtain the top-event probability $P[T]$. The small probability approximation may be used in Equations (10.13) if terminal-event probabilities are small (see Section 10.3.2).

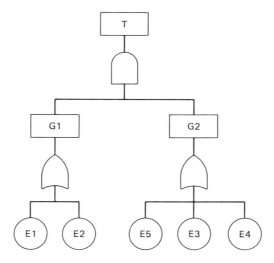

Figure 10.3. A fault tree.

This procedure is direct and straightforward even for very large trees. If, however, some terminal events are repeated within the tree (that is, the same terminal event appears in more than one place in the tree) or if some terminal events are statistically dependent, the procedure is not applicable. Let us replace $E5$ by $E1$ in Figure 10.3 (the fault tree, now, is identical to the tree shown in Figure 9.1). Terminal event $E1$ now appears twice in the tree. This does not change Equations (10.13) except that $E5$ is replaced by $E1$ in those equations. However, Equation (10.14) is no longer valid since the intermediate events $G1$ and $G2$ are now statistically dependent because both of them are a function of the terminal event $E1$. Now the correct version of Equation (10.14) is

$$P[T] = P[G1]P[G2|G1]$$

Although $P[G2|G1]$ can be computed and used in the preceding equation, the procedure is particularly cumbersome if the fault tree is large.

If there are statistical dependencies between terminal events, there will also be statistical dependencies between intermediate events, and thus the same difficulties discussed in the previous paragraph will arise. There are computer programs that could analyze even large fault trees via the direct approach, but the direct approach is not best-suited for hand calculations. The minimal cut set approach is easier in such situations.

10.3.5 Quantification via Minimal Cut Sets

We shall illustrate the minimal cut sets approach by applying it to the fault tree shown in Figure 9.1. Let all the terminal events be statistically independent.

Minimal cut sets of that fault tree are given in Section 9.6. The minimal cut sets are

$$M1 = (E1)$$

$$M2 = (E2, E3)$$

$$M3 = (E2, E4)$$

What do these minimal cut sets mean? The top event occurs if any one of the three minimal cut sets occurs; that is, the top event is connected to each of the minimal cut sets by an OR gate. A minimal cut set occurs if all of the terminal events in that minimal cut set occur; that is, each minimal cut set is connected to its terminal events by an AND gate. Thus the relationship between the top event, the minimal cut sets, and the terminal events may be represented by a fault tree as shown in Figure 10.4. This is a *reorganized form* of the same fault tree shown in Figure 9.1.

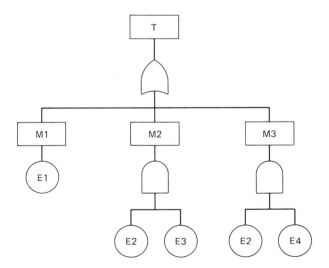

Figure 10.4. Fault tree representation via minimal cut sets.

The probabilities of the intermediate events (minimal cut sets) in Figure 10.4 are given by

$$P[M1] = P[E1]$$
$$P[M2] = P[E2 \cap E3] = P[E2]P[E3] \qquad (10.15)$$
$$P[M3] = P[E2 \cap E4] = P[E2]P[E4]$$

The probability of the top event is given by

$$P[T] = \{P[M1] + P[M2] + P[M3]\}$$
$$- \{P[M1 \cap M2] + P[M2 \cap M3] + P[M3 \cap M1]\}$$
$$+ \{P[M1 \cap M2 \cap M3]\} \qquad (10.16)$$

We may invoke the small probability approximation and discard the double and triple combination terms in Equation (10.16) if it is judged that the discarded terms do not significantly alter the computed top-event probability. Because $M1$ and $M2$ are statistically independent, we have

$$P[M1 \cap M2] = P[M1]P[M2] = \{P[E1]\}\{P[E2]P[E3]\}$$

$M2$ and $M3$ are not statistically independent because $E2$ appears in both $M2$ and $M3$. Therefore,

$$P[M2 \cap M3] = P[E2 \cap E3 \cap E2 \cap E4]$$
$$= P[E2 \cap E3 \cap E4]$$
$$= P[E2]P[E3]P[E4] \qquad (10.17)$$

We can, of course, write down the final step of the preceding equation without going through the intermediate steps. The probability of the combination of two or more minimal cut sets is given by the product of the probabilities of the various terminal events contained in those cut sets, without repeating any of the terminal events (this is applicable only if all of the terminal events are statistically independent). Using this procedure, we have

$$P[M3 \cap M1] = P[E1]P[E2]P[E4]$$
$$P[M1 \cap M2 \cap M3] = P[E1]P[E2]P[E3]P[E4] \qquad (10.18)$$

Consider a general case in which the fault tree has M minimal cut sets. If we discard the double, triple, and higher combination terms and compute $P[T]$ using only the single terms ($P[M1]$, $P[M2]$, $P[M3]$, etc.), we will get an upper bound to the exact value of $P[T]$; this is the small probability approximation, which is a reasonably accurate approximation to the exact solution if the probabilities of the double, triple, and higher combination terms are small compared to the probabilities of the minimal cut sets. We will get a lower bound to the exact value of $P[T]$ if we include the single and double combination terms. We will get an upper bound to $P[T]$ if we include the single, double, and triple combination terms; this will be closer to the exact solution than the small probability approximation. We will get a lower bound to $P[T]$ if we include the single, double, triple, and quadruple combination terms; this lower bound will be closer to the exact solution than the lower bound obtained by including only the single and double terms. We may continue to include successively higher-order terms and get closer and closer upper and lower bounds up to the Mth combination term, which will yield the exact value of $P[T]$. This principle of obtaining successively closer upper and lower bounds is known as the *inclusion–exclusion principle*.

Computation of the top-event probability via minimal cut sets, using the small probability approximation and assuming statistical independence between terminal events, may be generalized as follows.

$$P[T] = \sum_{j=1}^{M} \left\{ \prod_{i \in M_j} P[E_i] \right\} \tag{10.19}$$

where M is the number of minimal cut sets and $\prod_{i \in M_j} P[E_i]$ represents the product of the probabilities of all of the terminal events in the minimal cut set M_j. For example,

$$\prod_{i \in M1} P[E_i] = P[E1]$$

because $E1$ is the only terminal event in the minimal cut set $M1$.

$$\prod_{i \in M2} P[E_i] = P[E2]P[E3]$$

because $E2$ and $E3$ are the terminal events in the minimal cut set $M2$.

Equation (10.19) is an upper bound to the exact value of $P[T]$. A lower bound can be obtained by including the two minimal cut set combinations shown in Equation (10.16).

If system unavailability at time t ($=$ system unreliability at time t) is the top event of the fault tree, Equation (10.19) may be written as

$$Q_s = \sum_{j=1}^{M} \left[\prod_{i \in M_j} (q_i) \right] \tag{10.20}$$

where Q_S is system unavailability at time t, q_i is the unavailability at time t (= unreliability at time t) of the ith terminal event, and M is the number of minimal cut sets.

Note that Equation (10.20) is applicable only if all of the terminal events are nonrepairable and are statistically independent. The equation is based on the small probability approximation and is an upper bound to the exact solution.

If the small probability approximation is not used, the system unavailability at time t (system unreliability at time t) is given by

$$Q_S = \coprod_{j=1}^{M} \left[\prod_{i \in M_j} (q_i) \right] \tag{10.21}$$

where $\coprod_{j=1}^{M}(a_j) = 1 - \prod_{j=1}^{M}(1 - a_j)$. Equation (10.21) may be rewritten as

$$Q_S = \coprod_{j=1}^{M} [z_j] = 1 - \prod_{j=1}^{M} (1 - z_j)$$

$$= 1 - (1 - z_1)(1 - z_2) \cdots (1 - z_M)$$

where $z_j = \prod_{i \in M_j}(q_i)$. Equation (10.21) is "exact" if all of the terminal events are statistically independent and terminal events are not repeated within the fault tree.

According to Esary and Proschan (1963), Equation (10.21) is an upper bound if terminal events are repeated within the fault tree (but the terminal events should be statistically independent). That is,

$$Q_S \le \coprod_{j=1}^{M} \left[\prod_{i \in M_j} (q_i) \right] \tag{10.22}$$

10.3.6 Probabilities of Intermediate Events

The same procedures discussed in Section 10.3.5 may be used to compute the probabilities of intermediate events. Use the minimal cut sets of the intermediate event in place of the minimal cut sets of the top event.

10.3.7 Mutually Exclusive Terminal Events

Mutually exclusive events, by definition, cannot exist simultaneously. Therefore, if two or more terminal events in a minimal cut set are mutually exclusive, then the probability of that minimal cut set is zero.

Similarly, when considering the probability of a combination of minimal cut sets [as in Equation (10.17) or (10.18)], if the combination contains two or more terminal events that are mutually exclusive, the probability of that combination is set to zero.

10.3.8 Dependent Terminal Events

If there is statistical dependence between terminal events, it is best to represent those dependencies explicitly in the fault tree, as shown in Figure 8.28b; then dependencies are automatically accounted for when the fault tree is quantified.

Consider the fault tree in Figure 8.28b. The statistical dependence between A and B is modelled explicitly in that tree. Minimal cut sets for the top event are

$$M1 = (C)$$
$$M2 = (B, [A|B], B) = (B, [A|B])$$
$$M3 = (\bar{B}, [A|\bar{B}], B)$$

$\quad\quad$ = zero-probability minimal cut set (because B and $\bar{B}$ are mutually exclusive and cannot exist simultaneously)

The probabilities of $M1$, $M2$, and $M1 \cap M2$ are

$$P[M1] = P[C]$$
$$P[M2] = P[B]P[A|B]$$
$$P[M1 \cap M2] = P[B]P[A|B]P[C]$$

The top-event probability is equal to

$$P[T] = P[M1] + P[M2] - P[M1 \cap M2]$$
$$= P[C] + P[B]P[A|B] - P[C]P[B]P[A|B]$$

If the small probability approximation is applicable, we may use

$$P[T] = P[M1] + P[M2] = P[C] + P[B]P[A|B]$$

In the opinion of the author, it is best to model statistical dependence between terminal events explicitly in the fault tree. This avoids miscalculations during fault tree quantification. However, explicit modelling of statistical dependencies may make the fault tree very large, so many reliability analysts prefer not to model statistical dependencies in the fault tree but to consider them when quantifying the fault tree.

Suppose we do not explicitly model the statistical dependencies between terminal events A and B explicitly in the fault tree. The fault tree would be as shown in Figure 8.28a. Minimal cut sets of this fault tree are

$$M1 = (C)$$
$$M2 = (A, B)$$

The probabilities of these minimal cut sets are

$$P[M1] = P[C]$$
$$P[M2] = P[A|B]P[B] = P[A]P[B|A]$$

This result is the same as that obtained using Figure 8.28b. A shortcoming of this approach is that the analyst could forget to consider the effect of the statistical dependency between A and B during quantification and use $P[M2] = P[A]P[B]$; this could grossly underpredict the top-event probability. Especially when there are statistical dependencies between a number of terminal events, the chances of the analyst forgetting such dependencies is high.

Note that the fault tree in Figure 8.28b is not coherent, but this does not matter since the minimal cut set $M3$, which contains the success event $\bar{B}$, drops out because B and $\bar{B}$ are mutually exclusive.

10.3.9 Secondary Failures, Propagating Failures, and Common Cause Failures

Secondary failures (see Figure 8.15), propagating failures (see Figure 8.23), and common cause failures (see Figure 8.22) do not pose any difficulties if the components are nonrepairable. The procedures discussed so far are applicable.

Some analysts prefer not to model common cause failures explicitly in the fault tree. The procedures discussed in Chapter 11 are used instead.

10.3.10 Discarding Minimal Cut Sets

A large fault tree could have thousands of minimal cut sets, and computing the top-event probability using all those minimal cut sets could be time-consuming and expensive. Many minimal cut sets may be discarded judiciously in order to decrease the computational effort. Discarding even a single minimal cut set is "unconservative," that is, the top-event probability computed by using only some of the minimal cut sets is less than that computed by using all the minimal cut sets. But a judicious discarding of even many minimal cut sets may not significantly affect the top-event probability. As discussed in Chapter 5, terminal-event data (hardware failure probabilities, software error probabilities, and human error probabilities) used in fault tree quantification are accurate only in an order-of-magnitude sense. Therefore some judicious approximations in fault tree quantification are acceptable.

Lower-order minimal cut sets are usually more important than higher-order minimal cut sets, provided the higher-order minimal cut sets do not contain statistically dependent terminal events. Some analysts include only up to the second-order minimal cut sets; some may prefer to include up to the third- or fourth-order minimal cut sets. A decision has to be made on a case-by-case basis. The aim is to discard very low probability minimal cut sets. Some additional information relevant to the selective use of minimal cut sets during fault tree quantification may be found in Section 9.7.

10.3.11 Quantification via Minimal Path Sets

Let there be N path sets, and let the jth path set be S_j, which contains N_j terminal event complements. Let the terminal events be statistically independent. Note that the terminal event complements in a path set are "success events" as explained in Section 9.3.1. Let the probability of the terminal event complement $\overline{B_i}$ be $a_i = (1 - q_i)$, where q_i is terminal event probability.

If the terminal event is component failure, the corresponding terminal event complement is component success. The terminal-event probability is the component unavailability (= unreliability) at time t and the terminal-event-complement probability is the component availability (= reliability) at time t.

The top event will not occur if all of the terminal event complements in at least one of the minimal path sets exist. If the top event is 'system failure,' the corresponding top event complement is 'system success.' Then the top-event probability is system unavailability (unreliability) at time t and the corresponding top-event-complement probability is system availability (reliability) at time t. If the top event is an accident (for example, 'fire in the control room'), the corresponding top event complement is 'no fire in the control room.' Then the top-event probability is the accident probability (the probability that the accident has occurred between time zero and time t), and the corresponding top-event-complement probability is the probability that no accident has occurred between time zero and time t.

The probability of the jth minimal path set is the probability of simultaneous existence of all the terminal event complements in that minimal path set; this is equal to the product of the probabilities of the terminal event complements in that minimal path set provided the terminal events are statistically independent.

$$P[S_j] = \prod_{i \in S_j} P[B_i] = \prod_{i \in S_j} (a_i) \qquad (10.23)$$

where $\prod_{i \in S_j} (a_i)$ is the product of the availabilities of all of the terminal events contained in the minimal path set S_j.

The top event of the fault tree does not occur (that is, the top event complement occurs) if any one of the minimal path sets exist. This is like an OR gate connecting all the minimal path sets of the top event complement, so the probability of the top event complement (A_S) is equal to the probability of the union of the N minimal path sets:

$$A_S = P[S_1 \cup S_2 \cup \cdots \cup S_N]$$

$$= \{P[S_1] + P[S_2] \cdots + P[S_N]\}$$

$$- \{\text{sum of the probabilities of double combinations of minimal path sets}\}$$

$$+ \{\text{sum of the probabilities of triple combinations of minimal path sets}\}$$

$$\vdots$$

$$+ \{(-1)^{N-1} P[S_1 \cap S_2 \cap \cdots \cap S_N]\} \qquad (10.24)$$

where the $P[S_j]$ are as defined by Equation (10.23).

There will be $\binom{N}{2}$ two-minimal-path-set combinations (double combinations), $\binom{N}{3}$ three-minimal-path-set combinations (triple combinations), and so on, until there is one N-minimal-path-set combination. Note that the small probability approximation is not acceptable in the minimal path set approach because the probabilities of minimal path sets are seldom small.

Consider the case where all the minimal path sets are statistically independent; that is the case if (i) all of the terminal events are statistically independent and (ii) no terminal event complement appears in more than one minimal path set. In such a case, Equation (10.24) can be written as

$$A_S = 1 - \prod_{j=1}^{N} \{1 - P[S_j]\} = \coprod_{j=1}^{N} \{P[S_j]\}$$

Substitution of Equation (10.23) into this equation yields

$$A_S = \coprod_{j=1}^{N} \left[\prod_{i \in S_j} (a_i) \right] \tag{10.25}$$

Therefore,

$$Q_S = 1 - \left\{ \coprod_{j=1}^{N} \left[\prod_{i \in S_j} (1 - q_i) \right] \right\}$$

$$= 1 - \left(1 - \left\{ \prod_{j=1}^{N} \left[1 - \prod_{i \in S_j} (1 - q_i) \right] \right\} \right)$$

$$= \prod_{j=1}^{N} \left[\coprod_{i \in S_j} (q_i) \right] \tag{10.26}$$

Equations (10.25) and (10.26) are exact if the minimal path sets are statistically independent. Equation (10.26) should provide the same results as Equation (10.21), which was derived using minimal cut sets, if (i) the terminal events are statistically independent, (ii) no terminal event appears in more than one minimal cut set, and (iii) no terminal event complement appears in more than one minimal path set.

Esary and Proschan (1963) have shown that Equation (10.25) is an upper bound to system availability (the probability of the top event complement) if the third condition is violated (that is, if some terminal event complements appear in more than one minimal path set). However, the terminal events should be statistically independent. Because the system availability given by Equation (10.25) is an upper bound, the corresponding system unavailability given by Equation (10.26) is a lower bound under such conditions:

$$Q_S \geq \prod_{j=1}^{N} \left[\coprod_{i \in S_j} (q_i) \right] \tag{10.27}$$

Compare this lower bound to the upper bound given by Equation (10.22), which was derived using minimal cut sets.

10.3.12 Lower Bounds and Upper Bounds

Computing the "exact" top-event probabilities (or system availabilities) of large fault trees could be a very tedious process. Some analysts find it useful to compute lower bounds and upper bounds and to use these bounds, instead of the exact value, if the bounds are fairly close. How close should the bounds be? It depends on the required accuracy of the problem and has to be decided on a problem-by-problem basis. A number of lower and upper bounds have been noted in earlier parts of this chapter and they are summarized in the following subsections.

10.3.12.1 Inclusion – exclusion bounds. Computation of successive upper and lower bounds from minimal cut sets using the inclusion–exclusion principle is discussed in Section 10.3.5.

10.3.12.2 Esary – Proschan bounds. The Esary–Proschan upper bound from minimal cut sets is given by Equation (10.22). The corresponding Esary–Proschan lower bound from minimal path sets is given by Equation (10.27).

10.3.12.3 Partial cut set bounds. If there are M minimal cut sets to the top event of the fault tree, a lower bound to the top-event probability can be computed by considering only m minimal cut sets, where $m < M$.

An upper bound to the top-event probability may be obtained from minimal path sets. If there are N minimal path sets to the top event of the fault tree, an upper bound to the top-event probability can be computed by considering only n minimal path sets, where $n < N$.

Usually we start with the lower-order minimal cut sets (say, all first-order sets or all sets up to the third order, etc.). Similarly, we also start with the lower-order minimal path sets. If the lower and upper bounds are not as close as we would like, then more minimal cut sets and minimal path sets are added. The addition of more minimal cut sets and minimal path sets brings the bounds closer and closer to each other and to the exact solution. Thus, by increasing the number of minimal cut sets and minimal path sets, the bounds are made as close as we would like.

Example 10.5

Compute the unavailability of the system represented by the fault tree shown in Figure 10.5a. T is the top event, which represents system failure. Terminal events $E1$, $E2$, $E3$, and $E4$ represent the failures of nonrepairable components 1, 2, 3, and 4, respectively. These terminal events are

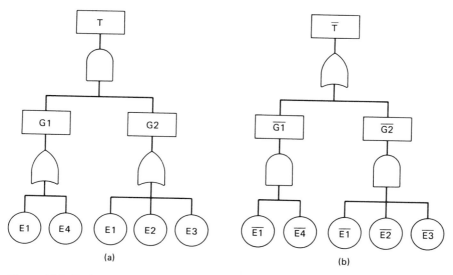

Figure 10.5. Fault tree and complementary tree for Example 10.5: (a) fault tree; (b) complementary tree.

statistically independent. The unavailability of each of the components at time t is q. Compute the various lower bounds, upper bounds, and exact values of system unavailability for $q = 0.01$, 0.1, and 0.2.

Solution

The minimal cut sets of the top event are

$$M1 = (E1)$$
$$M2 = (E2, E4)$$
$$M3 = (E3, E4)$$

The corresponding success tree (the complementary tree) is shown in Figure 10.5b. The minimal path sets are

$$S1 = (\overline{E1}, \overline{E2}, \overline{E3})$$
$$S2 = (\overline{E1}, \overline{E4})$$

where $\overline{E1}$, $\overline{E2}$, $\overline{E3}$, and $\overline{E4}$ are the complements of $E1$, $E2$, $E3$, and $E4$, respectively. That is, $\overline{E1}$, $\overline{E2}$, $\overline{E3}$, and $\overline{E4}$ represent the success (normal state) of components 1, 2, 3, and 4, respectively.

Exact results via minimal cut sets:

$$P[M1] = P[E1] = q$$
$$P[M2] = P[E2]P[E4] = q^2$$
$$P[M3] = P[E3]P[E4] = q^2$$
$$P[M1 \cap M2] = P[E1]P[E2]P[E3] = q^3$$
$$P[M2 \cap M3] = P[E2]P[E3]P[E4] = q^3$$
$$P[M3 \cap M1] = P[E1]P[E3]P[E4] = q^3$$
$$P[M1 \cap M2 \cap M3] = P[E1]P[E2]P[E3]P[E4] = q^4$$
$$P[T] = Q_S = \{P[M1] + P[M2] + P[M3]\}$$
$$- \{P[M1 \cap M2] + P[M2 \cap M3]$$
$$+ P[M3 \cap M1]\} + \{P[M1 \cap M2 \cap M3]\}$$
$$= q + q^2 + q^2 - q^3 - q^3 - q^3 + q^4$$
$$= q(1 + 2q - 3q^2 + q^3)$$

where Q_S is the system unavailability.
Substituting for q, we get the following:

$$\text{If } q = 0.01, \qquad Q_S = 0.01019701$$
$$\text{If } q = 0.1, \qquad Q_S = 0.1171$$
$$\text{If } q = 0.2, \qquad Q_S = 0.2576$$

Exact results via minimal path sets: Let a be the availability of each component. So, $a = 1 - q$. Also, let A_S be the system availability. So, $A_S = 1 - Q_S$.

$$P[S1] = P[E1]P[E2]P[E3] = a^3$$
$$P[S2] = P[E1]P[E4] = a^2$$
$$P[S1 \cap S2] = P[E1]P[E2]P[E3]P[E4] = a^4$$
$$P[T] = A_S = \{P[S1] + P[S2]\} - \{P[S1]P[S2]\}$$
$$= a^3 + a^2 - a^4 = a^2(1 + a - a^2)$$

Substituting for a, where $a = (1 - q)$ and $Q_S = (1 - A_S)$, we get the following:

$$\text{If } q = 0.01, \qquad Q_S = 0.01019701$$
$$\text{If } q = 0.1, \qquad Q_S = 0.1171$$
$$\text{If } q = 0.2, \qquad Q_S = 0.2576$$

Note that the exact results obtained via minimal cut sets are identical to the exact results obtained via minimal path sets.

Inclusion – exclusion bounds via minimal cut sets: First we calculate the upper bound using the small probability approximation:

$$P[T] = Q_S = P[M1] + P[M2] + P[M3] = q + q^2 + q^2$$
$$= q(1 + 2q)$$

Substituting for q, we get the following:

$$\text{If } q = 0.01, \qquad Q_S = 0.0102$$
$$\text{If } q = 0.1, \qquad Q_S = 0.12$$
$$\text{If } q = 0.2, \qquad Q_S = 0.28$$

Now we calculate the lower bound using up to the double combination terms of the minimal cut sets:

$$P[T] = Q_S = \{P[M1] + P[M2] + P[M3]\}$$
$$- \{P[M1 \cap M2] + P[M2 \cap M3] + P[M3 \cap M1]\}$$
$$= q + q^2 + q^2 - q^3 - q^3 - q^3 = q(1 + 2q - 3q^2)$$

Substituting for q, we get the following:

$$\text{If } q = 0.01, \qquad Q_S = 0.010197$$
$$\text{If } q = 0.1, \qquad Q_S = 0.1170$$
$$\text{If } q = 0.2, \qquad Q_S = 0.2560$$

The exact results, which would include the triple combination terms of minimal cut sets, have already been calculated.

Esary – Proschan bounds: First we calculate the Esary–Proschan upper bound from minimal cut sets [using Equation (10.22)]:

$$Q_S \leq \coprod_{j=1}^{M} \left[\prod_{i \in M_j} (q_i) \right]$$
$$= 1 - \left[(1 - q)(1 - q^2)(1 - q^2) \right]$$
$$= q(1 + 2q - 2q^2 - q^3 + q^4)$$

Substituting for q, we get the following:

If $q = 0.01$, $Q_S \leq 0.01020099$

If $q = 0.1$, $Q_S \leq 0.11791$

If $q = 0.2$, $Q_S \leq 0.26272$

Now we calculate the Esary–Proschan lower bound from minimal path sets [using Equation (10.27)].

$$Q_S \geq \prod_{j=1}^{N} \left[\coprod_{i \in S_j} (q_i) \right]$$
$$= 1 - (1 - q)^2 - (1 - q)^3 + (1 - q)^5$$

Substituting for q, we get the following:

If $q = 0.01$, $Q_S \geq 0.00059105$

If $q = 0.1$, $Q_S \geq 0.05149$

If $q = 0.2$, $Q_S \geq 0.17568$

Partial cut set bounds: Let us first compute the lower bounds to system unavailability by considering minimal cut sets.

Considering only the minimal cut set $M1$, we get

$$P[T] = Q_S = P[M1] = q$$

Considering only the minimal cut sets $M1$ and $M2$, we get

$$P[T] = Q_S = P[M1] + P[M2] - P[M1 \cap M2] = q + q^2 - q^3$$

Considering only the minimal cut sets $M1$ and $M3$, we get

$$P[T] = Q_S = P[M1] + P[M3] - P[M1 \cap M3] = q + q^2 - q^3$$

Table 10.1.   Partial Minimal Cut Set Bounds for System Unavailability (Example 10.5)

	SYSTEM UNAVAILABILITY		
q	$M1$ ONLY	$M1$ AND $M2$	EXACT
0.01	0.01	0.010099	0.0101019701
0.1	0.1	0.1090	0.1171
0.2	0.2	0.2320	0.2576

In this problem, it so happens that considering ($M1$ and $M2$) or ($M1$ and $M3$) provides the same top-event probability. This is not always the case.

Considering all the first- and second-order minimal cut sets means considering all the minimal cut sets (because the highest order of the minimal cut sets in this problem is 2), and the resulting top-event probability is the exact value. This value has already been computed. Numerical values of system unavailability considering one minimal cut set ($M1$), two minimal cut sets ($M1$ and $M2$), and all the minimal cut sets (exact value) are tabulated in Table 10.1.

Let us now compute the upper bounds to system unavailability by considering minimal path sets. Considering only the minimal path set $S1$, we get

$$P[T] = A_S = P[E1]P[E2]P[E3] = a^3$$

Considering only the minimal path set $S2$, we get

$$P[T] = A_S = P[E1]P[E4] = a^2$$

where $a = (1 - q)$ and $A_S = (1 - Q_S)$

The exact solution considering all the minimal path sets ($S1$ and $S2$) has already been computed. Numerical results for system unavailability considering $S1$, $S2$, and $S1$ and $S2$ (exact value) are tabulated in Table 10.2.

Accuracy of the various lower and upper bounds varies from problem to problem, so no general conclusions may be made on the basis of the numerical results of this example.

Table 10.2.   Partial Minimal Path Set Bounds for System Unavailability (Example 10.5)

	SYSTEM UNAVAILABILITY		
q	$S1$ ONLY	$S2$ ONLY	EXACT
0.01	0.0199	0.029701	0.0101019701
0.1	0.190	0.2710	0.1171
0.2	0.360	0.4880	0.2576

10.3.13 System Failure Rate

System failure rate can be determined from system reliability ($= 1 -$ system unreliability) using Equations (3.9) and (3.13). However, to compute the system failure rate using Equations (3.9) and (3.13), we need to know the system reliability as a function of time. Computation of system reliability as a function of time is illustrated in Example 10.6.

If system reliability is not known as a function of time but is known only as discrete time points, the differentiation in Equation (3.9) may be carried out numerically and the failure rate may be computed at discrete time points. Even if system reliability is known as a function of time, derivation of the failure rate as a function of time may not necessarily be an easy mathematical task. Again, numerical differentiation may be used, if necessary, and the failure rate may be determined at discrete time points.

It should be remembered that system failure rate is not necessarily constant even if all of the terminal events have constant failure rates.

Example 10.6

Consider the fault tree shown in Figure 10.5a. The top event of the fault tree represents system failure and the system is nonrepairable. Terminal events $E1$, $E2$, $E3$, and $E4$ represent component failures. Let the terminal events $E1$, $E2$, $E3$, and $E4$ have constant failure rates λ_1, λ_2, λ_3, and λ_4, respectively. The terminal events are statistically independent. Determine system reliability.

Solution

Because the system is nonrepairable, all the components are nonrepairable. So terminal event unavailabilities at time t are given by

$$q_i = \lambda_i t \quad \text{where } i = 1, 2, 3, \text{ and } 4$$

Minimal cut sets have already been determined in Example 10.5.

$$M1 = (E1)$$

$$M2 = (E2, E4)$$

$$M3 = (E3, E4)$$

Exact top-event probability ($=$ system unreliability $=$ system unavailability) via minimal cut sets is (see Example 10.5)

$$P[T] = Q_S = \{P[M1] + P[M2] + P[M3]\}$$
$$- \{P[M1 \cap M2] + P[M2 \cap M3] + P[M3 \cap M1]\}$$
$$+ \{P[M1 \cap M2 \cap M3]\}$$
$$= \lambda_1 t + \lambda_2 \lambda_4 t^2 + \lambda_3 \lambda_4 t^2 - \lambda_1 \lambda_2 \lambda_4 t^3$$
$$- \lambda_2 \lambda_3 \lambda_4 t^3 - \lambda_1 \lambda_3 \lambda_4 t^3 + \lambda_1 \lambda_2 \lambda_3 \lambda_4 t^4$$
$$= \lambda_1 t + (\lambda_2 \lambda_4 + \lambda_3 \lambda_4) t^2 - (\lambda_1 \lambda_2 \lambda_4 + \lambda_2 \lambda_3 \lambda_4 + \lambda_1 \lambda_3 \lambda_4) t^3$$
$$+ \lambda_1 \lambda_2 \lambda_3 \lambda_4 t^4$$

$$\text{unreliability } U_S = \text{unavailability } Q_S$$

$$\text{reliability } R_S = (1 - \text{unreliability } U_S)$$

10.3.14 Expected Number of System Failures

The expected number of system failures during the time interval 0 to t is denoted by $E[N_S(t)]$ and is given by

$$E[N_S(t)] = Q_S(t) \tag{10.28}$$

10.4 REPAIRABLE SYSTEMS

A system is classified as a *repairable system* if one or more components in the system are repairable. It is assumed in the following discussions that a component is as good as new after the repair. We also assume that component repairs are independent: The repair of one component does not affect the repair of another component. If one component is in a failed state, one team will repair it; if, say, 10 components are in a failed state at the same time, there will be 10 repair teams available to repair them simultaneously. So, the repair times of components are not dependent on whether one component is in a failed state or more than one component are in a failed state at the same time. This assumption may not introduce significant errors as long as the probability of multiple components in a failed state at the same time is much less than the probability of just one component in a failed state. Finally, in the discussions on repairable systems in this and other chapters, we assume that the failure probabilities of terminal events are statistically independent, unless explicitly stated otherwise.

As stated in Section 10.2, the fault tree should be *coherent*. Two necessary conditions were stated in that section. Whereas only the first condition is relevant to nonrepairable systems, both conditions are relevant to repairable systems. Transformation of noncoherent fault trees to coherent fault trees is discussed in Section 10.4.4. Methods of quantitative analysis presented here (Section 10.4) are applicable to coherent trees only, unless otherwise stated.

Although the procedures and formulae presented in this section refer to "system availability" and "system reliability," these procedures and formulae are equally applicable to the computation of the availability and reliability of any top event whether that top event is a system failure, an accident, or some other undesired event.

10.4.1 System Availability

The same equations given in Section 10.3 for the availability of nonrepairable systems are applicable to repairable systems also, if the terminal events are statistically independent.

Repairable systems may contain both repairable and nonrepairable components. Unavailabilities of these components (terminal events) are required as input to the system unavailability computation.

The unavailability of a repairable component is equal to $\lambda\tau$ (where λ is the failure rate and τ is the mean time to repair) or other applicable relationships (see Section 3.4.15). The unavailability of a nonrepairable component is λt (where λ is the failure rate and t is the time) or other applicable relationships (see Section 3.3.10).

Note that the repairable-component unavailability $\lambda\tau$ is the steady state value, which is applicable for time $t \geq 3\tau$ only; the formula is not applicable for $t < 3\tau$, where the unavailability will be less than $\lambda\tau$ (see Section 3.4.15).

If all of the components are repairable and the steady state terminal event unavailabilities are used in the system unavailability computation, then the system unavailability thus computed is also a steady state value applicable for $t \geq 3\tau_S$, where τ_S is the mean time to repair the system. (System repair time τ_S is discussed in Section 10.4.2.) For $t < 3\tau_S$, the system unavailability will be less than the steady state value.

If the system contains both repairable and nonrepairable components, the unavailabilities of nonrepairable components at time t (where $t =$ the time at which system unavailability is required) and the steady state unavailabilities of repairable components are used in the system unavailability computation. The system unavailability thus computed is valid for $t \geq 3\tau_S$; it will be higher than the actual system unavailability for $t < 3\tau_S$.

10.4.2 System Availability, Reliability, Failure Rate, and Mean Time to Repair

Computation of system reliability is more difficult for repairable systems than for nonrepairable systems. Some approximate formulae are provided by Fussell (1975). More rigorous procedures for the computation of system reliability, failure rate, and repair time are available (see Section 10.6), and some of those methods have been computerized (see Section 10.9).

The method described in this subsection [Fussell (1975)] predicts the system reliability fairly accurately if the terminal event unavailabilities are less than 0.1. Even if some terminal events have higher unavailabilities, the method still provides fairly accurate results if those terminal events occur in third- or higher-order minimal cut sets only. We provide the equations without the derivations. Interested readers may refer to Fussell (1975) for derivations.

First we compute the unavailabilities of minimal cut sets:

$$Q_j(t) = \prod_{i \in M_j} [q_i(t)] \tag{10.29}$$

where $Q_j(t)$ is the unavailability of the jth minimal cut set at time t, M represents the jth minimal cut set, q_i is the unavailability of the ith terminal event, and $i \in M_j$ indicates that all the terminal events in the jth minimal cut set are considered in the product. If all the q_i in the product are independent of time t, then Q_j will also be independent of time.

The probability density of time to first failure of the jth minimal cut set is given by

$$f_j(t) = \left\{ Q_j(t) \sum_{i \in M_j} \frac{\lambda_i}{q_i(t)} \right\} \tag{10.30}$$

where $f_j(t)$ is the probability density of time to first failure of the jth minimal cut set. This formula usually provides an overprediction. In most cases, $f_j(t)$ is either independent of time or is a polynomial of time t.

The unreliability of the jth minimal cut set is given by

$$U_j(t) = \int_0^t f_j(t') \, dt' \tag{10.31}$$

where $U_j(t)$ is the system unreliability at time t, and t' is a dummy variable within the integral. This formula is usually an overprediction and is fairly accurate if $U_j(t) < 0.1$.

The reliability of the jth minimal cut set is

$$R_j(t) = 1 - U_j(t) \tag{10.32}$$

The failure rate of the jth minimal cut set is given by

$$\Lambda_j(t) = \frac{f_j(t)}{R_j(t)} \tag{10.33}$$

System unavailability is given by

$$Q_S(t) = \sum_{j=1}^{M} Q_j(t) \tag{10.34}$$

where M is the number of minimal cut sets. If we substitute Equation (10.29) into Equation (10.34), the resulting equation will be identical to Equation (10.20).

The system unreliability is given by

$$U_S(t) = \left\{ 1 - \prod_{j=1}^{M} [R_j(t)] \right\} \tag{10.35}$$

We may also use

$$U_S(t) = \sum_{j=1}^{M} U_j(t) \tag{10.36}$$

Both Equation (10.35) and (10.36) are usually overpredictions. Equation (10.35) is generally somewhat a better prediction than Equation (10.36).

The system failure rate (or the failure rate of the top event of the fault tree) is given by

$$\Lambda_S(t) = \sum_{j=1}^{M} \Lambda_j(t) \tag{10.37}$$

This equation is not necessarily an overprediction; it could be an underprediction in some cases.

Even when all of the repairable terminal events have constant failure rates and constant repair rates and all of the nonrepairable terminal events have constant failure rates, the failure rate and repair rate of the system could be time-dependent. An approximate estimate of a time-independent mean time to failure (T_S) may be made if all of the terminal events are repairable and have constant failure rates and constant repair rates. Note that the actual mean time to failure is a function of time; what we calculate next is an approximate estimate that is time-independent.

$$T_S = \frac{1}{\Lambda_S(t_m)} \qquad (10.38)$$

where the system failure rate in the denominator of the right-hand side is computed using Equation (10.37) at time $t = t_m$, in which t_m is the maximum system mission time. (For example, $t_m = 5000$ hours if the system will be operating for a total of 5000 hours before it is discarded or completely overhauled.)

Similarly, although the mean time to repair the system (τ_S) will be a function of time, the following time-independent approximate estimate of the mean time to repair may be made if all of the terminal events are repairable and have constant failure rates and constant repair rates.

$$\tau_S = \frac{Q_S}{A_S \Lambda_S(t)} \qquad (10.39)$$

If there are nonrepairable terminal events, Equations (10.38) and (10.39) still may be used if the nonrepairable terminal events appear only in first-order minimal cut sets. The repairable terminal events may appear in the first-order and/or higher-order minimal cut sets. (The nonrepairable terminal events should also have constant failure rates.)

Although the methodology presented in this subsection is approximate, the accuracy of the predicted unreliability are generally acceptable, considering the fact that terminal event failure data (unavailability, failure rate, repair rate, etc.) used in the analysis are themselves approximate. This method may be used in hand calculations and computer programs.

Example 10.7

Compute the unavailability, unreliability, failure rate, mean time to failure, and mean time to repair of the system represented by the fault tree shown in Figure 10.5a; the top event of the tree is 'system failure'. The terminal events $E1$, $E2$, $E3$, and $E4$ are repairable and are statistically independent. The constant failure rates and constant mean time to repair of the terminal events are given in Table 10.3.

Table 10.3.   Terminal-Event Data (Example 10.7)

TERMINAL EVENT	FAILURE RATE (PER 10^6 HOURS)	MEAN TIME TO REPAIR (HOURS)
E1	10	100
E2	100	10
E3	10	10
E4	1	100

Table 10.4.   Terminal Event Unavailabilities and Unreliabilities (Example 10.7)

| | UNAVAILABILITY | | UNRELIABILITY | |
	q	T'	u	T'
E1	10^{-3}	300 to ∞	$10^{-5}t$	0 to 10^4
E2	10^{-3}	30 to ∞	$10^{-4}t$	0 to 10^3
E3	10^{-4}	30 to ∞	$10^{-5}t$	0 to 10^4
E4	10^{-4}	300 to ∞	$10^{-6}t$	0 to 10^5

Solution

The minimal cut sets of the fault tree (determined in Example 10.5) are

$$M1 = (E1)$$
$$M2 = (E2, E4)$$
$$M3 = (E3, E4)$$

First we calculate the terminal event unavailabilities q, unreliabilities u, and the time interval for which these parameters are valid (T' hours) and tabulate them in Table 10.4.

Now we calculate the minimal cut set unavailabilities (Q), probability densities of time to first failure (f), unreliabilities (U), and failure rates (Λ). Equations (10.29)–(10.31) and (10.33) are used. These values are tabulated in Table 10.5.

Finally, we calculate the system unavailability, unreliability, failure rate, mean time to failure, and mean time to repair. Equations (10.34), (10.36), and (10.37) are used.

$$Q_S(t) = 10^{-3} + 10^{-7} + 10^{-8} \approx 10^{-3}$$
$$U_S(t) = (10^{-5} + 1.1 \times 10^{-8} + 1.1 \times 10^{-9})t \approx 10^{-5}t \qquad (10.40)$$
$$\Lambda_S(t) = 10^{-5} + 1.1 \times 10^{-8} + 1.1 \times 10^{-9} \approx 10^{-5} \text{ per hour}$$

Table 10.5.   Minimal Cut Set Parameters (Example 10.7)

	Q	f (PER HOUR)	U	Λ (PER HOUR)
$M1$	10^{-3}	10^{-5}	$10^{-5}t$	10^{-5} (Note 1)
$M2$	10^{-7}	1.1×10^{-8}	$1.1 \times 10^{-8}t$	1.1×10^{-8} (Note 2)
$M3$	10^{-8}	1.1×10^{-9}	$1.1 \times 10^{-9}t$	1.1×10^{-9} (Note 3)

Note 1: Using Equation (10.33), we have $\Lambda_1 = 10^{-5}/(1 - 10^{-5}t)$. We approximate it to 10^{-5}, which introduces an error of 1% for $t = 10^3$ hours and 11.1% for $t = 10^4$ hours.
Note 2: In the same way as in Note 1, error due to the approximation is small for $t \leq 10^6$ hours.
Note 3: In the same way as in Note 1, error due to the approximation is small for $t \leq 10^7$ hours.

Because all of the terminal events are repairable and have constant failure rates and constant repair rates, we may use Equation (10.38) to compute the mean time to failure (T_S) and Equation (10.39) to compute the mean time to repair (τ_S).

$$T_S = \frac{1}{10^{-5}} = 10^5 \text{ hours}$$

Because $\Lambda_S(t)$ is independent of time (it is, of course, an approximation), the value of t_m is immaterial.

$$\tau_S = \frac{10^{-3}}{(1 - 10^{-3})10^{-5}} = 100 \text{ hours}$$

The system parameters calculated so far in this example are valid only for certain time intervals. System unavailability is valid for times greater than or equal to three times the system mean time to repair, so it is valid for $t \geq 300$ hours. System unreliability is valid for values less than 0.1, so it is valid for $t < 10^4$ hours. We know from Note 1 in Table 10.5 that the failure rate of minimal cut set $M1$ is valid for up to 10^4 hours. Because this failure rate is used in computing system failure rate, mean time to failure, and mean time to repair, these three system parameters are valid for $t < 10^4$ hours.

An approximate estimate of system unreliability can be made by assuming that all the terminal events are nonrepairable (that is, assume that even repairable terminal events are nonrepairable) and then calculating the system unreliability using methods discussed in Section 10.3. Let us carry out such an analysis in Example 10.8.

Example 10.8

Compute the unreliability of the system described in Example 10.7, by assuming that all the terminal events are nonrepairable.

Solution

The terminal event unreliabilities at time t are calculated using the formula $u = \lambda t$, where λ is the failure rate:

$$u_1 = 10^{-5}t$$

$$u_2 = 10^{-4}t$$

$$u_3 = 10^{-5}t$$

$$u_4 = 10^{-6}t$$

The minimal cut set unreliabilities are

$$U_1 = u_1 = 10^{-5}t$$

$$U_2 = u_2 u_4 = (10^{-4}t)(10^{-6}t)$$

$$U_3 = u_3 u_4 = (10^{-5}t)(10^{-6}t)$$

The system unreliability is

$$U_S = U_1 + U_2 + U_3 = (10^{-5}t) + (1.1 \times 10^{-10}t^2) \qquad (10.41)$$

Equation (10.41) ignores the effect of the repairability of terminal events. In Example 10.7, we computed system unreliability using Equation (10.40), which includes the effect of the repairability of terminal events.

Results from Equations (10.40) and (10.41) may be tabulated for discrete values of time t (see Table 10.6).

We see that assuming even repairable terminal events to be nonrepairable results in the overprediction of system unreliability. The overprediction increases with time. The overprediction is not significant in this

Table 10.6.   System Unreliability (Example 10.8)

TIME (HOURS)	SYSTEM UNRELIABILITY	
	EQUATION (10.41)	EQUATION (10.40)
1,000	1.011×10^{-2}	1.00121×10^{-2}
10,000	1.11×10^{-1}	1.00121×10^{-1}

example because the unreliability is dominated by the first-order term [a linear function of time, see Equation (10.41)], which is a result of the first-order minimal cut set $M1$. Had there been no dominating first-order minimal cut sets, the unreliability computed using Equation (10.41) would be dominated by higher-order terms in t, and the overprediction could be substantial (even an order of magnitude or more, depending on the problem).

10.4.3 Expected Number of System Failures

In the case of a nonrepairable system, the expected number of system failures at time t is equal to the system unavailability ($=$ system unreliability) at time t [see Equation (10.28)]. That is,

$$E[N_S(t)] = Q_S(t) = U_S(t) \tag{10.42}$$

If the system is repairable, the system unreliability is a lower bound to the expected number of system failures:

$$E[N_S(t)] \geq U_S(t) \tag{10.43}$$

10.4.4 Transformation of Noncoherent Fault Trees

The system unavailability, unreliability, and other related parameter calculations discussed in Sections 10.4.1–10.4.3 are applicable only if the fault tree has a coherent structure.

Necessary conditions for a fault tree to have a coherent structure are discussed in Section 10.2. The difficulty presented by success events and how to eliminate this difficulty during quantitative analysis are also discussed in Section 10.2.

The difficulty presented by resolving a component failure into primary, secondary, and command failures is also noted in Section 10.2. This difficulty does not arise if the system is nonrepairable. If the system is repairable, the fault tree containing secondary failures has to be transformed into a coherent fault tree before a quantitative analysis is carried out. A transformation technique [Fussell et al. (1976)] is discussed here.

Consider the fault tree shown in Figure 10.6a. The primary failure is a terminal event; the secondary failure could be a terminal event or an intermediate event; similarly, the command failure could be a terminal event or an intermediate event. All of these failures have constant failure rates and constant repair rates and are statistically independent. The command failure does not pose any difficulty with reference to coherency (see Section 10.2), and we will not alter the command failure. The fault

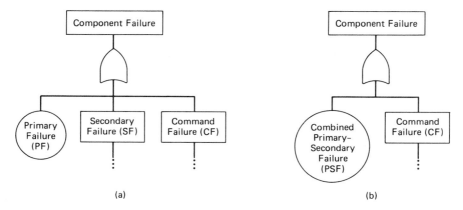

Figure 10.6. Transformation of a noncoherent fault tree to a coherent fault tree: (a) noncoherent fault tree; (b) equivalent coherent fault tree.

tree in Figure 10.6a is transformed to Figure 10.6b in order to make it coherent. The secondary failure (either a terminal event or an intermediate event) is combined with the primary failure as a single terminal event (called the combined primary–secondary failure and denoted by PSF). We need to determine both the failure rate and the mean time to repair for this combined event.

The transformation procedure described in this subsection is valid only if the following conditions are satisfied.

1. The primary failure should have constant failure rate and constant repair rate.
2. If the secondary failure is a terminal event, it should have constant failure rate and constant repair rate. If the secondary failure is an intermediate event, all the terminal events below that intermediate event should have constant failure rates and constant repair rates.
3. The primary failure should be statistically independent of all other terminal events in the fault tree.
4. If the secondary failure is a terminal event, it should be statistically independent of all other terminal events in the fault tree. If the secondary failure is an intermediate event, all the terminal events below that intermediate event should be statistically independent of all other terminal events in the fault tree.
5. The primary failure should not appear elsewhere in the fault tree.
6. If the secondary failure is a terminal event, it should not appear elsewhere in the fault tree. If the secondary failure is an intermediate event, none of the terminal events below that intermediate event should appear elsewhere in the fault tree.

Note: If one or more of conditions 3–6 is not satisfied, there will be statistical dependencies between the combined primary–secondary failure and some of the other terminal events in the fault tree. The procedure described in this subsection can be used even if one or more of conditions 3–6 is not satisfied (i) if it can be shown that such statistical dependencies do not affect the results of the quantitative analysis significantly or (ii) if such statistical dependencies are computed and included in the quantitative fault tree analysis. Computation of such statistical dependencies may not be an easy task. (Many fault trees encountered in practice satisfy all six of the conditions, so the procedure described in this subsection is applicable in a wide variety of situations.)

Let the failure rate and mean time to repair for the combined event be λ_c and τ_c, respectively. Determination of these values involves the following four steps.

Step 1

Compute the mean time to repair the secondary failure (τ_{SF}). If the secondary failure is a terminal event, τ_{SF} is equal to the mean time to repair that terminal event, as given in the data. If it is an intermediate event, the mean time to repair should be calculated as a function of the mean times to repair the terminal events coming under that intermediate event. (Methods of determining the mean time to repair top events and intermediate events are discussed in Section 10.4.2.)

Step 2

Compute the mean time to repair the combined primary–secondary failure (the combined event); let it be denoted by τ_c. This value depends on whether the secondary failure and the primary failure are repaired simultaneously or if the primary failure is repaired only after the repair of the secondary failure is completed.

If the primary and secondary failures are repaired simultaneously,

$$\tau_c = \max(\tau_{PF}, \tau_{SF}) \tag{10.44}$$

where

$\max(a, b) = a$ or b, whichever is higher
τ_{PF} = mean time to repair the primary failure (specified data for the terminal event)
τ_{SF} = mean time to repair the secondary failure (as computed in Step 1)

If the primary failure is repaired after the repair of the secondary failure is completed,

$$\tau_c = \tau_{PF} + \tau_{SF} \tag{10.45}$$

Step 3

Compute the failure rate of the secondary failure (λ_{SF}). If the secondary failure is a terminal event, then λ_{SF} is equal to the failure rate of that terminal event (as given in the failure data). If the secondary failure is an intermediate event, then λ_{SF} is computed as a function of the failure rates of the terminal events under that intermediate event (see Section 10.4.2).

Step 4

Compute the failure rate of the combined primary–secondary failure; let it be denoted by λ_c.

$$\lambda_c = \lambda_{PF} + \lambda_{SF} \qquad (10.46)$$

If two or more secondary failures are connected to the primary failure through an OR gate (as in Figure 10.7), what shall we do? Even in such cases, the transformed coherent tree will be in the same form as shown in Figure 10.6b. The combined primary–secondary failure (the combined event), denoted by PSF, combines the primary failure and all the secondary failures. The following four steps are used to determine the failure rate and the MTTR of the combined event PSF.

Step 1

Let the mean time to repair the ith secondary failure be τ_{SFi}, where $i = 1, 2, \ldots, n$, in which n is the number of secondary failures. Compute each τ_{SFi} by using the same procedure described to compute τ_{SF}.

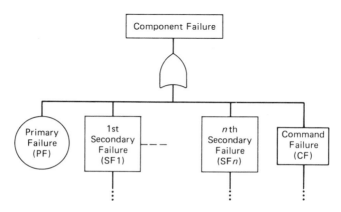

Figure 10.7. Multiple secondary failures.

Step 2

If all the secondary failures and the primary failure are repaired simultaneously, the mean time to repair the combined primary–secondary failure is given by

$$\tau_c = \max(\tau_{PF}, \tau_{SF1}, \tau_{SF2}, \dots, \tau_{SFn}) \qquad (10.47)$$

If all the secondary failures are repaired simultaneously and the primary failure is repaired after the completion of the repair of all the secondary failures, we have

$$\tau_c = \tau_{PF} + \max(\tau_{SF1}, \tau_{SF2}, \dots, \tau_{SFn}) \qquad (10.48)$$

Step 3

Let the failure rate of the ith secondary failure be λ_{SFi}. Each of the λ_{SFi} is calculated using the same procedure discussed for λ_{SF}.

Step 4

The failure rate of the combined primary–secondary event is given by

$$\lambda_c = \lambda_{SF1} + \lambda_{SF2} + \cdots + \lambda_{SFn} \qquad (10.49)$$

A special case of secondary failures is worth noting. This special case must satisfy the following three conditions: (i) the secondary failures are intermediate events and all the gates under those intermediate events are OR gates; (ii) the secondary failures and the primary failure are repaired simultaneously; and (iii) the mean time to repair each of the terminal events under the secondary failures is less than or equal to the mean time to repair the primary failure. In such a special case, the fault tree can be transformed to a coherent tree without having to combine the primary and secondary failures into a combined primary–secondary failure. The fault tree remains the same, but change the mean time to repair the terminal events under the secondary failures to the mean time to repair the primary failure. No change need be made for the failure rates.

A special situation under the special case is when (i) the secondary failures are terminal events, (ii) the secondary failures and the primary failure are repaired simultaneously, and (iii) the mean time to repair each secondary failure is less than or equal to the mean time to repair the primary failure. Under such conditions, no change in the fault tree is needed; simply change the mean times to repair the secondary failures to the mean time to repair the primary failure. No changes are needed for the failure rates.

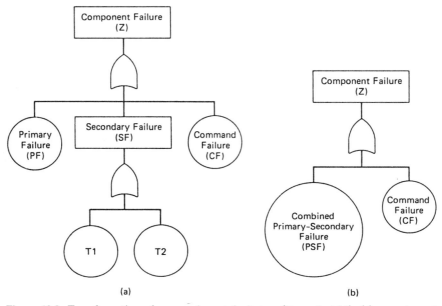

Figure 10.8. Transformation of a noncoherent fault tree (Example 10.9): (a) noncoherent fault tree; (b) equivalent coherent fault tree.

Example 10.9

A part of a fault tree is shown in Figure 10.8a. This fault tree does not have a coherent structure. Transform it to a coherent fault tree. The terminal events are statistically independent, and have constant failure rates and constant repair rates. The primary failure is repaired after the secondary failure is repaired. The failure rates and the mean time to repair the terminal events *PF*, *T*1, *T*2, and *CF* are given in Table 10.7.

Table 10.7.   Failure Rates and Mean Times to Repair (Example 10.9)

TERMINAL EVENT	FAILURE RATE (PER HOUR)	MEAN TIME TO REPAIR (HOURS)
PF	10^{-7}	100
*T*1	10^{-5}	10
*T*2	10^{-4}	10
CF	10^{-6}	10

Solution

First we calculate the failure rate and mean time to repair the secondary failure, which is an intermediate event. Consider the portion of the fault tree below this intermediate event. This intermediate event has two minimal cut sets:

$$M1 = (T1)$$
$$M2 = (T2)$$

We will use the method described in Section 10.4.2 to determine the failure rate (λ_{SF}) and mean time to repair (τ_{SF}). Because the minimal cut sets are of the first order, the failure rate of the minimal cut set is same as the failure rate of the terminal event [we would have used Equation (10.33) in a more complex tree]. Failure rate of the secondary failure (intermediate event) is computed using Equation (10.37):

$$\lambda_{SF} = \lambda_1 + \lambda_2 = 1.1 \times 10^{-4} \text{ per hour}$$

where λ_1 and λ_2 are the failure rates of $T1$ and $T2$, respectively.

The unavailability of the secondary failure (Q_{SF}) is calculated using Equation (10.34).

$$Q_{SF} = \lambda_1\tau_1 + \lambda_2\tau_2 = 1.1 \times 10^{-3}$$

where τ_1 and τ_2 are the mean time to repair $T1$ and $T2$, respectively.

The availability is

$$A_{SF} = 1 - Q_{SF} = 0.9989$$

The mean time to repair the secondary failure is calculated using Equation (10.39).

$$\tau_{SF} = \frac{1.1 \times 10^{-3}}{0.9989 \times 1.1 \times 10^{-4}}$$
$$= 10.011 \text{ hours}$$

[*Note:* The exact value would be 10 hours; as noted earlier, Equation (10.39) is an approximate formula.]

Now we proceed to calculate the mean time to repair the combined primary–secondary failure shown in Figure 10.8b. Using Equation (10.48), we have

$$\tau_c = 100 + 10.011 = 110.011 \text{ hours}$$

The failure rate of the combined primary–secondary failure is given by Equation (10.46):

$$\lambda_c = 10^{-7} + 1.1 \times 10^{-4}$$

$$= 1.101 \times 10^{-4} \text{ per hour}$$

The unavailability of the intermediate event 'component failure' (Z) is computed using the transformed coherent fault tree (Figure 10.8b) and Equations (10.29) and (10.34).

$$Q_Z = (1.101 \times 10^{-4} \times 110.011) + (10^{-6} \times 10)$$

$$= 0.0121222111$$

Finally, we show how the noncoherent fault tree could have provided the wrong results for the unavailability of the intermediate event 'component failure' (Z) if we did not transform it to a coherent tree. We use the noncoherent fault tree (Figure 10.8a) and Equations (10.29) and (10.34):

$$Q_Z = (10^{-7} \times 100) + (10^{-5} \times 10) + (10^{-4} \times 10) + (10^{-6} \times 10)$$

$$= 0.00112$$

We see that the unavailability from the noncoherent fault tree (Figure 10.8a) is grossly erroneous. It is an order of magnitude higher than the unavailability from the transformed, coherent fault tree. Note that the amount of error will depend on the problem.

Example 10.10

Redo Example 10.9 with just one change: The primary failure and the secondary failure are repaired simultaneously.

Solution

This problem satisfies all three conditions for the special case (stated in Section 10.4.4). So we may use the simplified method applicable for the special case.

We need not transform the fault tree in this special case. We use Figure 10.8a itself and change the MTTR of $T1$ and $T2$ to 100 hours. The unavailability of the intermediate event 'component failure' (Z) using Figure 10.8a and the MTTR of $T1$ and $T2$ set at 100 hours is computed using Equations (10.29) and (10.34):

$$Q_Z = (10^{-7} \times 100) + (10^{-5} \times 100) + (10^{-4} \times 100) + (10^{-6} \times 10)$$

$$= 0.01102$$

10.5 PERIODICALLY MAINTAINED SYSTEMS

System reliability and availability are improved if the system is maintained periodically (preventive maintenance). It is a common practice to assume that the system is as good as new immediately after the preventive maintenance. Let T be the periodic maintenance interval. The availability at time t is given by

$$A_S(t) = A'_S(t') \qquad (10.50)$$

where $A_S(t)$ is the system availability that includes the effect of periodic maintenance, $A'_S(t)$ is the system availability if it is not periodically maintained (calculated using the procedures discussed earlier), and the time t is always expressed as

$$t = nT + t' \qquad (10.51)$$

in which n is zero or a positive integer and t' is always kept less than T. For example, if we are interested in system availability at $t = 3800$ hours and the periodic maintenance interval $T = 1200$, we write $3800 = (3 \times 1200) + 200$. This is a unique relationship and there is no other way of expanding $t = 3800$ using Equation (10.51). So, in this example $n = 3$ and $t' = 200$. (Note that system availability is equal to unity immediately after each periodic maintenance because it is assumed that the system is as good as new after the maintenance.) It is assumed in Equation (10.50) that the duration of periodic maintenance (time taken for carrying out the maintenance) is zero; this assumption is valid if the duration of the periodic maintenance is very small compared to the periodic maintenance interval, which is true in most cases.

Note that the availability and reliability of a periodically maintained system are higher than or equal to the availability and reliability, respectively, of the same system under no periodic maintenance.

A plot of $A_S(t)$ is given in Figure 10.9. This is a graphic representation of Equation (10.50). The availability is 1.0 at the time the system is put

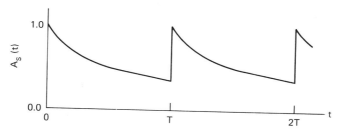

Figure 10.9. Availability of a periodically maintained system.

into service at time $t = 0$, and then it decreases with time. As soon as the periodic maintenance is carried out at time $t = T$, the availability jumps to 1.0. Following the maintenance, the availability decreases with time. Periodic maintenance is carried out again at time $t = 2T$ and availability jumps at 1.0 again. This cycle repeats itself until the system is retired. The availability graph between $t = 0$ and $t = T$ is identical to the graph between $t = T$ and $t = 2T$, and will be identical to the graph between $t = (k - 1)T$ and $t = kT$, where $k \geq 3$, until the system is retired.

System reliability of periodically maintained systems is given by

$$R_S(t) = [R_S'(T)]^n R_S'(t') \tag{10.52}$$

where $R_S(t)$ is the reliability of the periodically maintained system, $R_S'(t)$ is the reliability of the same system without periodic maintenance. Given t and T as data, n and t' are computed using Equation (10.51). System reliability is always less than or equal to system availability for periodically maintained systems.

10.6 OTHER METHODS

We have presented an approximate technique for the quantitative fault tree analysis, which can be used in manual calculations (hand calculations) or computerized analyses. There are other more rigorous methods of quantitative fault tree analysis. We shall briefly identify some of these methods and provide references from which more detailed information may be obtained.

Vesely (1969) developed the kinetic tree theory (KITT) for quantitative fault tree analysis. Details of computer programs employing this technique are described by Vesely and Narum (1970). The method is capable of computing the unavailability, unreliability, expected number of failures, and the failure rate of the top and intermediate events. Constant or time-dependent failure rates may be used. The terminal events are considered statistically independent.

Rumble, Leverenz, and Erdmann (1975) have developed the Boolean arithmetic model (BAM) approach of quantitative fault tree analysis. This method is capable of analyzing fault trees in which statistical dependencies between terminal events are explicitly modelled, as discussed in Section 8.10.

By far, fault tree analysis is the most widely used deductive method for the analysis of large systems. We will mention here just three methods that do not use fault trees. The first is the GO method [Gatelby, Stoddard, and Williams (1968)], which uses GO charts instead of fault trees. The second method is the Markov process method. This method has been used for the analysis of smaller systems; it is especially effective in considering statistical dependencies between terminal events. In recent years, the method has been extended to analyze larger systems also [Papazoglou and Gyftopoulos (1977)]. The third method is Monte Carlo simulation.

Monte Carlo simulation is the most versatile approach of quantitative system reliability analysis. It can be used in conjunction with the fault tree analysis or any other method of system reliability analysis. Monte Carlo simulation can consider complex situations that no other method can handle. The only drawback of Monte Carlo simulation is the cost. Because we usually deal with small probabilities of failure, the number of trials required to obtain fairly accurate results is very large. System failure probabilities of 10^{-4} or less are not uncommon; in such cases at least 10^5 trials, preferably more, are required to get acceptable results. Conducting such a large number of trials is very expensive, so Monte Carlo simulation is recommended only when other methods cannot handle the problem or when the simplifying assumptions to be made to make the problem solvable by other methods are not acceptable. Monte Carlo simulation can consider very complex scenarios of system operations, environmental conditions, and failures. The following are some of the situations that can be included in Monte Carlo simulation:

1. Statistical dependencies between terminal events not only in failure rates but also in repair rates.
2. Explicit accounting of the size of repair crew available in the plant. The size of repair crew does affect system availability and reliability. Suppose there is only one repair person and three components fail simultaneously; the time required to repair all three components is different from that if there were two or more repair personnel available.
3. The effect of spare-parts availability.
4. Delayed effects of failures.
5. Component degradation.
6. Effect of varying degrees of failures (see Section 8.6.3.2)
7. Time-dependent failure rates and repair rates.

We do not imply that none of these situations can be considered using other methods, but all of them cannot be considered explicitly and "exactly" by any other method in use today. Widawsky (1971) and Henley and Polk (1977) provide examples of Monte Carlo simulation applications in quantitative system reliability analysis.

10.7 LIMITATIONS OF QUANTITATIVE FAULT TREE ANALYSIS

Although fault tree analysis is a powerful method of system reliability analysis, results from a fault tree analysis are approximate because of the approximate nature of terminal-event data (terminal-event failure rate, failure-on-demand probability, mean time to repair, etc.) and the approximations involved in fault tree construction and analysis. (Approximations in terminal-event data and in fault tree construction and analysis are discussed in Chapter 5, Chapter 8, and earlier sections of this chapter.) The top-event unreliability and unavailability computed via fault tree analysis may be accurate only in an order-of-magnitude sense. We are dealing with very small probabilities (such as 10^{-4}, 10^{-5}, etc.), and estimates of those probabilities, even within an order-of-magnitude accuracy, are useful in reliability engineering applications.

It is best to know system unreliabilities and unavailabilities at least in an order-of-magnitude sense rather than to be totally ignorant of those values. As will be discussed in Chapters 12 and 13, what may be of interest to us in many applications is a comparison of the contributions of different component failures to top-event unreliability or a comparison of the unreliabilities of different system designs. Estimates of system unreliabilities and unavailabilities in an order-of-magnitude sense are surely acceptable and useful in such applications.

The limitations of fault tree analysis noted in this section are applicable to other methods of system reliability analysis also.

10.8 UNCERTAINTY ANALYSIS

In some very critical applications of reliability engineering, for example, in the computation of nuclear reactor accident probabilities, reliability analysts may be interested in quantitative information about the accuracy of the system reliability analysis results. Such information is obtained through a formal *uncertainty analysis*. We will not discuss the subject of uncertainty analysis in any detail because uncertainty analysis is conducted in only a small fraction of reliability engineering projects. We merely indicate what types of data are required and what types of results may be obtained. We also identify a few selected references from which interested readers may get more information.

In the fault tree analysis procedure discussed so far in this chapter, we use point values for the terminal-event data (for example, terminal-event failure rate = 10^{-3} per year, mean time to repair = 15 hours, etc.) and compute point values for the top-event and/or intermediate-event probabilities (for example, top-event unreliability = 8×10^{-5}).

In an uncertainty analysis, we need both a point value and a confidence interval (say, 90% confidence interval) for each terminal-event datum (Sections 5.2.5.2 and 5.2.6 discuss confidence intervals for component failure data). Final results of an uncertainty analysis are point estimates and confidence intervals for the top-event and intermediate-event probabilities. For example, the point estimate of top-event unavailability is

7×10^{-5} per year, and the 90% confidence interval is 2×10^{-5} to 8×10^{-4}. What does the confidence interval mean? In this example, it means that there is a 90% probability that the actual value of top-event unavailability lies somewhere between 2×10^{-5} and 8×10^{-4}; there is a 5% probability that the actual value of top-event unavailability is less than 2×10^{-5} and there is a 5% probability that the actual value of top-event unavailability is greater than 8×10^{-4}. Although we have given a 90% confidence interval as an example, 80, 95, or 99% confidence intervals may be computed through an uncertainty analysis.

What is the benefit of having a confidence interval in addition to the point estimate? If we have only the point estimate (say, 7×10^{-5}), we do not know how conservative or unconservative this estimate is. What is the probability of this estimate being exceeded? We cannot answer this question if all we have is the point estimate. On the other hand, if we have the 90% confidence interval (say, 2×10^{-5} to 8×10^{-4}), we do know that there is a 5% probability that the actual top-event unavailability is greater than 8×10^{-4}. Although such information is useful, uncertainty analysis is very expensive, so it is conducted only in very critical applications where the consequences of an accident (top event) are very severe.

Methods of uncertainty analysis are described in the *Reactor Safety Study* [Nuclear Regulatory Commission (1975)] and the *PRA Procedures Guide* [Nuclear Regulatory Commission (1983)]. Computer programs for uncertainty analysis include BACFIRE, BOUNDS, MOCARS, SAMPLE, and SPASM (see Appendix I).

10.9 COMPUTER PROGRAMS

BACFIRE, BOUNDS, ICARUS, MOCARS, PL-MOD, RAS, SAMPLE, SETS, SPASM, WAMBAM, WAMCUT, and WAMCUT-II are some of the commercially or publicly available computer programs for quantitative fault tree analysis (see Appendix I for details). This is only a partial list of available programs. New computer programs also become available. Capabilities and computational efficiencies of the programs differ. Users should compare the available programs and select those that best suit their needs.

10.10 DOCUMENTATION

Documentation of quantitative fault tree analysis should include the following information:

1. The fault tree.
2. Minimal cut sets and/or minimal path sets (if they are used).
3. Failure rate and mean time to repair for each terminal event.

4. Identification of the method of analysis, including references to technical papers, books, or reports.
5. Assumptions made in data, fault tree construction, and analysis.
6. Identification of computer programs and the source of the programs (from where obtained); include version number of version date.
7. Listing of computer inputs and outputs.

Some general guidelines on documentation are provided in Section 14.10.

EXERCISE PROBLEMS

10.1. Consider an AND gate with four input events. The probability of each of these events is 0.1. Compute the output event probability. Assume that the input events are statistically independent.

10.2. Consider an AND gate with four input events A, B, C, and D. The probability of each of these events is 0.1. Input events C and D are statistically dependent on each other, and $P[D|C] = 0.8$.

10.3. Consider an OR gate with four input events. The probability of each of these events is 0.1. Compute the exact value and the upper and lower bounds of the output event probability. Assume that the input events are statistically independent.

10.4. A fault tree has two minimal cut sets: $M1 = (E1, E2)$ and $M2 = (E1, E3, E4)$. Its minimal path sets are $S1 = (\overline{E1})$, $S2 = (\overline{E2}, \overline{E3})$, and $S3 = (\overline{E2}, \overline{E4})$. The top event of the fault tree (T) is 'system failure'. The terminal events $E1$, $E2$, $E3$, and $E4$ represent the failures of four nonrepairable components $C1$, $C2$, $C3$, and $C4$, respectively. $\overline{E1}$, $\overline{E2}$, $\overline{E3}$, and $\overline{E4}$ are the complements of $E1$, $E2$, $E3$, and $E4$, respectively. The terminal events are statistically independent. The unavailability of each terminal event at time t is q. Determine the various lower bounds, upper bounds, and the exact value of system unavailability at time t. Compute the system unavailability at time t for $q = 0.01, 0.05$, and 0.1.

10.5. A fault tree has two minimal cut sets: $M1 = (E1)$ and $M2 = (E2, E3, E4)$. The top event of the tree is 'system failure'. The terminal events $E1$, $E2$, $E3$, and $E4$ represent the failures of components $C1$, $C2$, $C3$, and $C4$, respectively. The components are nonrepairable. The terminal events are statistically independent. The constant failure rates of components $C1$, $C2$, $C3$, and $C4$ are λ_1, λ_2, λ_3, and λ_4, respectively. Component unavailabilities q_i at time t are given by $q_i = \lambda_i t$. (i) Determine the system unavailability at time t. (ii) Compute the system unavailability at time $t = 1000$ hours for $\lambda_1 = \lambda_2 = \lambda_3 = \lambda_4 = 10^{-5}$ failures per hour. (iii) Compute the expected number of system failures at time $t = 1000$ hours for $\lambda_1 = \lambda_2 = \lambda_3 = \lambda_4 = 10^{-5}$ failures per hour.

Table 10.8.   Data for Problem 10.7

TERMINAL EVENT	FAILURE RATE (PER HOUR)	MTTR (HOURS)
$E1$	10^{-5}	100
$E2$	10^{-4}	10
$E3$	10^{-5}	10
$E4$	10^{-6}	100

10.6. Redo part (ii) of Problem 10.5, after discarding the minimal cut set $M2$. Compare the results with the results of Problem 10.5(ii).

10.7. A fault tree has two minimal cut sets: $M1 = (E1, E2)$ and $M2 = (E1, E3, E4)$. The top event of the tree is 'system failure'. The terminal events $E1$, $E2$, $E3$, and $E4$ represent failures of components $C1$, $C2$, $C3$, and $C4$, respectively. The components are repairable and have constant failure rates and constant repair rates. The terminal events are statistically independent. Component failure rates and mean time to repair are given in Table 10.8. Compute the unavailability, unreliability, mean time to repair, mean time to failure, and the failure rate of the system.

10.8. Consider the fault tree shown in Figure 8.18. For the purposes of this problem, assume that E, F, G, J, K, M, N, P, Q, and R are terminal events (instead of intermediate events as indicated in Figure 8.18). All of the terminal events (except for the house event HS) have constant failure rates and constant repair rates. Also, the terminal events are statistically independent. Terminal event failure rates and mean time to repair are given in Table 10.9. (i) Compute the unavailability and unreliability of the intermediate event C at time $t = 600$ hours. (ii) Compute the unavailability and unreliability of the top event A at time $t = 600$ hours with the house event HS switched off. (iii) Compute the unavailability and unreliability of the top event A at time $t = 600$ hours with the house event HS switched on.

Table 10.9.   Data for Problem 10.8

TERMINAL EVENT	FAILURE RATE (PER HOUR)	MTTR (HOURS)
E	10^{-6}	100
F	4×10^{-6}	40
G	10^{-5}	100
J	4×10^{-6}	40
K	4×10^{-6}	40
M	8×10^{-5}	30
N	3×10^{-6}	50
Q	10^{-5}	100
R	10^{-5}	100

Table 10.10. Data for Problem 10.9

TERMINAL EVENT	FAILURE RATE (PER HOUR)	MTTR (HOURS)
PF	10^{-6}	10
T1	10^{-5}	100
T2	10^{-5}	100
CF	10^{-6}	10

10.9. Consider the fault tree shown in Figure 10.8a. The terminal events are statistically independent and have constant failure rates and constant repair rates. Failure rates and mean time to repair for the terminal events PF, $T1$, $T2$, and CF are given in Table 10.10. The primary failure is repaired after the secondary failures are repaired. (i) Compute the unavailability of event Z. (ii) What will be the unavailability of Z if the primary failure and the secondary failures are repaired simultaneously? (*Note*: The fault tree shown in Figure 10.8a does not have a coherent structure; it has to be transformed to a coherent fault tree before the unavailability computations.)

10.10. Redo Problem 10.7 under the condition that terminal events $E1$ and $E4$ are mutually exclusive. All other data remain the same as in Problem 10.7.

10.11. Consider the fault tree shown Figure 8.26. Let us use the following notation: $P[FA]$ = the probability of component-A failure; $P[FB]$ = the probability of component-B failure; $P[SB] = (1 - P[FB])$ = the probability of component-B success; $P[FA|FB]$ = the probability of component-A failure during component-B failure (conditional probability of component-A failure given that component B has failed); and $P[FA|SB]$ = the probability of component-A failure during component-B success (conditional probability of component-A failure given that component-B has not failed). The following data are available: $P[FB] = 0.1$, $P[FA|FB] = 0.8$, and $P[FA|SB] = 0.05$. (i) Compute $P[FA]$. (ii) If we compute $P[FA]$ using the fault tree shown in Figure 8.27, what will be the error in that result?

10.12. Redo Problem 10.11 with the following data: $P[FB] = 0.1$, $P[FA|FB] = 0.8$, and $P[FA|SB] = 0.001$.

10.13. Consider a series system with two components (component A and component B). Both of these components are inaccessible for repair during system operation, so the components are considered nonrepairable. Both components have constant failure rates and are statistically independent. The failure rates for component A and component B are 10^{-4} and 3×10^{-5} failures per hour, respectively. The system is maintained periodically; the periodic maintenance interval is 480 hours. The periodic maintenance duration is so small compared to the periodic maintenance interval that it may be assumed to be zero. Compute the system unavailability at times $t = 0, 100, 500, 580, 1000, 1100,$ and 1480 hours.

REFERENCES

Esary, J. D. and F. Proschan (1963). Coherent structures with nonidentical components. *Technometrics* **5** 191.

Fussell, J. B. (1975). How to hand-calculate system reliability and safety characteristics. *IEEE Transactions on Reliability* **R-24**(3) 169–174.

Fussell, J. B., D. M. Rasmuson, J. R. Wilson, G. R. Burdick, and J. C. Zipperer (1976). *A Collection of Methods for Reliability and Safety Engineering.* Idaho National Engineering Laboratory, Idaho Falls.

Gatelby, W., D. Stoddard, and R. L. Williams (1968). *GO: A Computer Program for the Reliability Analysis of Complex Systems.* Kaman Science Corporation.

Henley, E. and R. Polk (1977). A risk/cost assessment of administrative time restrictions on nuclear power plants. In *Nuclear Systems Reliability Engineering and Risk Assessment*, J. J. Fussell and R. D. Burdick, eds. Society for Industrial and Applied Mathematics, Philadelphia, pp. 495–518.

Nuclear Regulatory Commission (1975). *Reactor Safety Study: An Assessment of Accident Risks in U.S. Commercial Nuclear Power Plants* (WASH-1400). Nuclear Regulatory Commission, Washington, DC.

Nuclear Regulatory Commission (1983). *PRA Procedures Guide* (NUREG/CR-2300). Nuclear Regulatory Commission, Washington, DC.

Papazoglou, I. A. and E. P. Gyftopoulos (1977). Markov processes for reliability analyses of large systems. *IEEE Transactions on Reliability* **R-26** 232.

Rumble, E. T., F. L. Leverenz, and R. C. Erdmann (1975). *Generalized Fault Tree Analysis for Reactor Safety.* Report No. EPRI 217-2-2. Electric Power Research Institute, Palo Alto, CA.

Vesely, W. E. (1969). *Analysis of Fault Trees by Kinetic Tree Theory.* Idaho Nuclear Corporation, Idaho Falls.

Vesely, W. E. and R. E. Narum (1970). *PREP and KITT: Computer Codes for the Automatic Evaluation of a Fault Tree.* Report No. IN-1349. Idaho Nuclear Corporation, Idaho Falls.

Widawsky, W. H. (1971). Reliability and maintainability parameters evaluated with simulation. *IEEE Transactions on Reliability* **R-20** 158.

Chapter 11
Common Cause Analysis

11.1 INTRODUCTION

An event or mechanism that can cause more than one failure (terminal event) simultaneously is called a *common cause*, and the failures thus caused are called *common cause failures*. Some common causes may not induce the failures of components but may degrade them so that their joint probability of simultaneous failure is increased.

Because common causes could induce the failures of or degrade a number of components, they have the potential to increase system unreliability, and accident probabilities significantly. So it is important to check whether any common causes affect the system. If common causes are present, either take measures to eliminate the common causes or quantify the effects of common cause failures and assure that system unreliability, unavailability, and accident probabilities are within design requirements.

One way of considering the effect of common causes is to include them explicitly in the fault tree. This is discussed in Sections 8.8 and 10.3.9. If there are a number of common causes and they affect a large number of components, the inclusion of these common causes could make the size of the fault tree virtually unmanageable. Therefore some reliability analysts prefer to perform a common cause analysis separately after the fault tree analysis. This approach is discussed in this chapter.

11.2 DEFINITIONS

An event or mechanism that can cause two or more failures simultaneously[1] is called a *common cause* or a *common cause event*. The failures

---

[1]Some common causes induce the failures of a number of components at the very same instant; for example, a severe earthquake could induce a number of pump failures at the very same instant of the earthquake. Some common causes may induce a number of component failures not at the same instant but within a fairly short duration of time so that all the components are in a failed state (due to the common cause) at some point of time. For example, abnormally high temperatures due to an air-conditioning system failure may not induce a number of electronic component failures at the very same instant but may do so
(footnote continued on next page)

(or terminal events) caused by common causes are called *common cause failures*, *common mode failures*, or *common mode events*. Note that common cause failures are secondary failures of components, not primary failures.

A terminal event or failure is called a *neutral event* with respect to a common cause if it is independent of that common cause; that is, a terminal event or failure that is not a common mode event with respect to a common cause is called a neutral event with respect to that common cause.

A common cause that induces the simultaneous occurrence of all the terminal events in at least one minimal cut set is called a *significant common cause*. If all the terminal events in a minimal cut set can occur simultaneously because of a common cause, such a minimal cut set is called a *prime common cause candidate*.

Consider an air-conditioning system failure. If the resulting rise in temperature induces the failures of a number of electronic components, then 'high temperature' or 'air-conditioning system failure' is a common cause. Another example of common cause is maintenance error. If the same maintenance person services a number of components and commits the same error in all of the components, then the maintenance error becomes a common cause. The following is a partial list of common causes:

1. Mechanical common causes

   - Abnormally high or low temperature (beyond design limits)
   - Abnormally high or low pressure (beyond design limits)
   - Stress (above design limits)
   - Impact (above design limits)
   - Vibration (above design limits)

---

over a relatively short duration of time before any of those components could be repaired. Some common causes just degrade a number of components; although these components may not fail at the very same instant, the joint probability of those component failures existing simultaneously increases. In the context of common cause failures, the phrase "simultaneous failure" or "simultaneous occurrences of failures" is used rather loosely to refer to all three of these situations; we are actually referring to the "simultaneous existence of failures" as a result of the failure of a number of components over a relatively short duration of time.

2. Electrical common causes

  - Abnormally high voltage (beyond design limits)
  - Abnormally high current (beyond design limits)
  - Electromagnetic interference

3. Chemical common causes

  - Corrosion
  - Chemical reaction

4. External common causes

  - Earthquake
  - Tornado
  - Flood
  - Lightning
  - Fire

5. Miscellaneous common causes

  - Radiation
  - Moisture
  - Dust
  - Design or production defect
  - Test–maintenance–operation error

In some instances, a common cause may not induce multiple component failures simultaneously (may not cause the occurrence of more than one terminal event simultaneously) but may degrade more than one component, thus increasing the joint probability of simultaneous failures (joint probability of the simultaneous occurrence of more than one terminal event). Such common causes increase not only the joint probability of simultaneous failure but also the individual failure probabilities.

A component is said to be *susceptible* to a common cause if the component would fail or its failure probability would increase if it were exposed to the common cause. A component is said to have been *affected* by a common cause if it is susceptible to the common cause and is exposed to it. For example, a valve may be susceptible to corrosion (common cause). The valve is exposed to corrosion if it is part of a flow loop that carries a corrosive chemical.

For a minimal cut set to be a prime common cause candidate, two conditions need to be satisfied: (i) All the terminal events in the minimal cut set are *susceptible* to the common cause. (ii) All the terminal events in the minimal cut set are *exposed* to the common cause.

Let us say that there are four components (terminal events) in a minimal cut set that are all susceptible to 'dust' (that is, these components will fail if exposed to dust). Let one of the components be located in one room and the other three components be located in another room. If dust is produced only in the second room and there is no way the dust can get into the first room, then only three of the components are exposed to dust (common cause) and fail due to the common cause 'dust.' So the minimal cut set is not a prime common cause candidate. Such minimal cut sets in which all the terminal events (components) are susceptible to a common cause but not all of them are exposed to the common cause are called *common cause candidates* (as opposed to prime common cause candidates). (*Note:* Some authors refer to prime common cause candidates simply as common cause candidates.)

Whether a component is exposed to a common cause or not depends on whether it is located in the *domain of the common cause* or not. In the example of 'dust' as the common cause, if dust is produced in a room and there is no way it can spread to other rooms, then that room is its domain. If it can spread to other rooms through doors and openings, all the rooms that can be exposed to dust become its domain. Physical proximity is not a necessary condition for an area to be a domain of a common cause. For example, a chemical flow loop could be the domain of the common cause 'corrosion.'

Consider three components A, B, and C. Components A and B are located just 1 foot away from each other but A is within a flow loop carrying a corrosive chemical and B is outside that flow loop. Although A and B are in close proximity, only A is in the domain of the common cause 'corrosion.' Component C is located 800 feet away from A and B but is located within the flow loop. Although C is separated from A by several hundred feet, A and C are within the domain of the common cause 'corrosion.'

A *common link* is a condition that closely links a number of components (terminal events). If a number of components are produced in the same manufacturing facility, the 'common production facility' is the common link between these components. If a number of components are maintained by the same maintenance team, then the 'common maintenance team' is the common link.

A common link is not a common cause. A common link may introduce statistical dependencies between the components (terminal events) and increase their joint probability of simultaneous failure. If a number of components are produced in the same manufacturing facility, it does not necessarily increase their joint failure probability. However, if one of the components is found to have a manufacturing defect, it is likely that other components manufactured in the same facility may also be defective; so the joint failure probability of these components increases.

Common links are not as serious a problem as common causes, but reliability analysts should identify the common links and, *if necessary and possible*, avoid common link situations that affect all or most of the terminal events in a minimal cut set. Common links indicate a possibility of common cause failures.

The following is a partial list of common links:

1. Common power supply
2. Common water supply
3. Common fuel supply
4. Common calibration
5. Common manufacturing facility
6. Common design or production team
7. Common test or maintenance crew
8. Common operating crew
9. Common testing or maintenance procedures
10. Common operating procedures

11.3 PRELIMINARY COMMON CAUSE ANALYSIS

A preliminary common cause analysis is performed at the time a preliminary hazard analysis or a failure modes and effects analysis is performed. Preliminary common cause analysis is not performed in some projects, and only a more detailed qualitative and/or quantitative common cause analysis is carried out.

Table 11.1. Preliminary Common Cause Analysis Sheet

NUMBER	COMMON CAUSE	REMARKS
1	earthquake	all the pumps and motors will fail if an earthquake of magnitude 5 or more occurs
2	machinery	vibrations at frequencies below 2 hertz may induce failures in fragile components located in the vicinity of the source of vibration
3	corrosion	the chemical used during the X7-4 process in flow loop FL-2 could damage all steel components exposed to it

A preliminary common cause analysis consists of identifying all possible common causes to which the system is exposed and the potential effects these common causes may have on system failure. Table 11.1 gives a sample preliminary common cause analysis sheet. Probabilities of the common causes may or may not be estimated at this stage.

The preliminary common cause analysis alerts design engineers to potential problems at an early stage of design. Changes in design may be made to avoid common cause failures. For example, noncorrosive coatings may be included in selected components exposed to a corrosive chemical.

11.4 QUALITATIVE COMMON CAUSE ANALYSIS

A qualitative common cause analysis consists of the following four steps:

(i) Identification of common causes and their domains.
(ii) Identification of terminal events susceptible to the common causes and their exposure to the common causes.
(iii) Identification of common cause candidates, prime common cause candidates, and significant common causes.
(iv) Identification of partially affected minimal cut sets (this step is omitted in some common cause analyses).

(i) *Identification of Common Causes and their Domains:* Prepare a list of common causes to which the system is exposed. The list of common causes presented in Section 11.2 may be a useful aid in this step. Note that the list in Section 11.2 is not all-inclusive, and there could be other common causes that are pertinent to the systems under investigation.

Once a list of common causes is prepared, identify the domain of each common cause. A plan of the plant area may be used as an aid in identifying common cause domains; it may be necessary to use floor plans at different elevations. A sample plan at the ground level is given in Figure 11.1.

Let us consider a hypothetical industrial facility (Figure 11.1) in which four common causes have been identified: earthquake, corrosion, machinery vibration (machine 1), and machinery vibration (machine 2). The whole plant is the domain for 'earthquake' because the earthquake can affect all the susceptible components in the plant. The flow loop Z is the domain for the common cause 'corrosion.' Machine 1 is located in room $A2$, and all the components in all the rooms in that building (rooms $A1$, $A2$, $A3$, and $A4$) feel the vibration. Machine 2 is located in room $B3$, and the vibration is felt by all the components in rooms $B2$ and $B3$ (but not in room $B1$). This information is tabulated in Table 11.2.

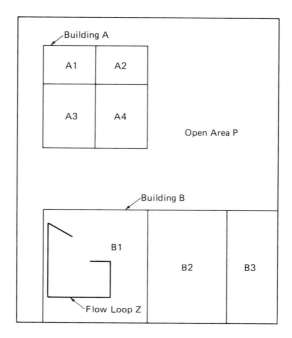

Figure 11.1. Diagram of plant area.

(ii) *Identification of Terminal Events Susceptible to the Common Causes and their Exposure to the Common Causes:* Prepare a table of all terminal events that are susceptible to the identified common causes, and their locations. Identify if they are exposed to (that is, in the domain of) the common causes or not. As an example, let the terminal events $E7$, $E11$, $E12$, $E16$, $E22$, and $E23$ be susceptible to some of the common causes listed in Table 11.2. The terminal event $E7$ represents the failure of a

Table 11.2. Common Cause Analysis Sheet
(common causes and their domains)

NUMBER	COMMON CAUSE	IDENTIFICATION CODE	DOMAIN
1	earthquake	E	A, B, P
2	corrosion	C	Z
3	vibration (machine 1)	V1	A
4	vibration (machine 2)	V2	$B2, B3$

Table 11.3.   Common Cause Analysis Sheet (terminal event
susceptibility and exposure to common causes)

NUMBER	TERMINAL EVENT	SUSCEPTIBILITY	LOCATION	EXPOSURE
1	$E7$	C	Z	yes
2	$E11$	V1	B1	no
		V2	B1	no
		E	B1	yes
3	$E12$	E	P	yes
4	$E16$	V1	A2	yes
		V2	A2	no
		E	A2	yes
5	$E22$	C	Z	yes
6	$E23$	C	Z	yes

component that is susceptible to corrosion (C) and is located in the flow loop Z. The terminal event $E11$ represents the failure of a component that is susceptible to machinery vibrations and earthquakes, and is located in room $B1$. The terminal event $E12$ represents the failure of a component susceptible to earthquakes and is located in the open area P. The terminal event $E16$ represents the failure of a component that is susceptible to machinery vibrations and earthquakes and is located in room $A2$. The terminal events $E22$ and $E23$ represent two failure modes of a component that is susceptible to corrosion and is located in the flow loop Z [both of the terminal events (failure modes) $E22$ and $E23$ are susceptible to corrosion]. This information is concisely tabulated in Table 11.3. Note that a terminal event may be susceptible to more than one common cause. Also, a component may have more than one failure mode, and each failure mode will be represented by a different terminal event. One failure mode may be susceptible to a common cause and another failure mode of the same component may or may not be susceptible to that common cause.

(iii) *Identification of Common Cause Candidates, Prime Common Cause Candidates, and Significant Common Causes:* The procedure of identifying common cause candidates, prime common cause candidates and significant common causes is straightforward. However, the procedure becomes tedious if the number of minimal cut sets and the number of terminal events are large. Some compute programs capable of performing qualitative common cause analysis are noted in Section 11.9.

Each minimal cut set is examined, one by one, to see which, if any, of the minimal cut sets are common cause candidates. The procedure is best illustrated by an example. Let the minimal cut sets for the top event of a fault tree (T) be as follows:

$$M1 = (E23)$$

$$M2 = (E7, E11)$$

$$M3 = (E11, E16)$$

$$M4 = (E7, E12, E22)$$

Let us examine each minimal cut set with the aid of Table 11.3. Minimal cut set $M1$ has only one terminal event, and that terminal event is susceptible to the common cause 'C' and is also in its domain. So $M1$ is a prime common cause candidate with respect to C, and C is a significant common cause.

Both the terminal events in $M2$ are susceptible to V1, but neither of them is in its domain. So $M2$ is a common cause candidate with respect to V1 but it is not a prime common cause candidate. Similarly, $M2$ is a common cause candidate, but not a prime common cause candidate, with respect to V2. Neither V1 nor V2 is a significant common cause.

Both the terminal events in minimal cut set $M3$ are susceptible to and in the domain of E. So $M3$ is a prime common cause candidate with respect to E, and E is a significant common cause.

The minimal cut set $M4$ is neither a common cause candidate nor a prime common cause candidate.

We are able to examine the minimal cut sets manually in this example. Computerized analysis will be needed if there are hundreds of minimal cut sets and terminal events.

(iv) *Identification of Partially Affected Minimal Cut Sets:* Identification of partially affected minimal cut sets may not be important if there are one or more prime common cause candidates. If two or more, but not all, of the terminal events in a minimal cut set are susceptible to and in the domain of a common cause, that minimal cut set is said to be a *partially affected minimal cut set*.

Let us examine the four minimal cut sets $M1$, $M2$, $M3$, and $M4$ and see if any of them are partially affected. Minimal cut sets $M1$, $M2$, and $M3$ are not partially affected. The minimal cut set $M4$ is partially affected by the common cause C, because terminal events $E7$ and $E22$ are both susceptible to and in the domain of C.

Some qualitative common cause analysis computer programs have the capability to identify partially affected minimal cut sets. For example, the computer program WAMCOM [Putney (1981)] identifies not only prime common cause candidates but also (i) minimal cut sets triggered by two common causes and (ii) minimal cut sets triggered by one common cause and one terminal event.

What do we mean by a "minimal cut set triggered by two common causes"? If there are n terminal events in a minimal cut set and m of these terminal events are susceptible to and in the domain of a common cause X and the remaining $(n - m)$ terminal events are susceptible to and in the domain of another common cause Y, such a minimal cut set is called a minimal cut set triggered by two common causes.

What do we mean by a "minimal cut set triggered by one common cause and one terminal event"? If there are n terminal events in a minimal cut set and $(n - 1)$ terminal events are susceptible to and in the domain of a common cause and the remaining one terminal event is not susceptible to and in the domain of any common causes, such a minimal cut set is called a minimal cut set triggered by one common cause and one terminal event.

Are minimal cut sets triggered by two common causes important? Consider the following minimal cut set:

$$M = (E1, E2, E3, E4, E5)$$

Let the terminal events $E1$ and $E2$ be triggered by (susceptible to and in the domain of) a common cause X, and the terminal events $E3$, $E4$, and $E5$ be triggered by another common cause Y. Let the contributions of these two common causes to the probability of the minimal cut set M be denoted by $P[M']$, where

$$P[M'] = P[X]P[Y]$$

If the probabilities of $P[X]$ and $P[Y]$ are comparable or higher than the probabilities of the terminal events in the fault tree and there are no or few first-order minimal cut sets and/or prime common cause candidates, the contribution $P[M']$ due to the two common causes X and Y could be important. Neglecting that contribution could result in the underprediction of the top-event probability. So the identification of minimal cut sets triggered by two common causes could be important.

The same arguments hold for minimal cut sets triggered by one common cause and one terminal event. So identification of such minimal cut sets could also be important.

A decision on whether to identify partially affected minimal cut sets should be made on a problem-by-problem basis.

The qualitative common cause analysis procedure discussed in this section requires a knowledge of all the minimal cut sets of the fault tree. Determination of very high order minimal cut sets is a rather expensive task if the fault tree is large. An alternate method of qualitative common cause analysis that circumvents the need for a minimal cut set analysis of large fault trees is discussed in Section 11.7.

11.5 QUALITATIVE COMMON LINK ANALYSIS

Qualitative common link analysis procedure is very similar to the qualitative common cause analysis procedure described in Section 11.4. Here we consider the common links instead of the common causes.

Common links are not as serious as common causes but the enumeration of minimal cut sets affected by common links will help identify potential statistical dependencies between terminal events in each minimal cut set (see the discussion of common links in Section 11.2). Statistical dependencies between terminal events in a minimal cut set increase the probability of the minimal cut set, and this increases the probability of the top event. Whenever possible, it is better to avoid system designs in which all the terminal events of a minimal cut set are affected by a common link.

11.6 QUANTITATIVE COMMON CAUSE ANALYSIS

11.6.1 A Rigorous Method

Consider a minimal cut set M:

$$M = (A, B, C) \tag{11.1}$$

Let all three terminal events A, B, and C be affected by a common cause X, and let the conditional joint probability of the simultaneous existence of A, B, and C due to X, given that X has occurred, be g.

We introduce an additional minimal cut set M' in order to account for the effect of the common cause X:

$$M' = (ABC \text{ due to } X) \tag{11.2}$$

The probability of M' is given by

$$P[M'] = P[X]P[ABC|X] = P[X]g \tag{11.3}$$

where $P[ABC|X]$ is the conditional joint probability of A, B, and C occurring simultaneously due to X, given that X has occurred.

Consider the following case. Events A, B, and C represent the failures of three identical components, and there are no statistical dependencies between them except for the common cause X. Let the conditional probability of A occurring due to X, given that X has occurred, be

$$P[A|X] = p$$

Because A, B, and C are identical, we have

$$P[A|X] = P[B|X] = P[C|X] = p$$

The conditional probability of A, B, and C occurring simultaneously due to X, given that X has occurred, is

$$g = P[ABC|X] = P[A|X]P[B|X]P[C|X] = p^3 \qquad (11.4)$$

The probability p may be determined from historical data or through other means. The value of g is then computed using Equation (11.4). Alternatively, g may be determined directly from historical data, if available, or through other means.

Example 11.1

The top event (T) of a fault tree has two minimal cut sets:

$$M1 = (A) \quad \text{and} \quad M2 = (C, D)$$

where A, C, and D are terminal events. C and D are susceptible to and in the domain of a common cause X. The probabilities of the primary failures of A, C, and D are $P[a]$, $P[c]$, and $P[d]$. The primary failures are statistically independent. The following numerical data are available from past experience:

$$P[a] = P[c] = P[d] = 0.02$$
$$P[X] = 0.01$$
$$P[CD|X] = 1.0$$

Compute the top-event probability $P[T]$.

Solution

Because the minimal cut set $M2$ is a prime common cause candidate with respect to X, we add a new minimal cut set $M2'$. (Note that only prime common cause candidates should be considered in common cause quantification; common cause candidates should not be considered.) So the minimal cut sets are

$$M1 = (A)$$

$$M2 = (C, D)$$

$$M2' = (CD \text{ due to } X) = (X) \quad \text{because } P[CD|X] = 1.0$$

The top-event probability can be computed using these three minimal cut sets and the method described in Section 10.3.5.

$$P[M1] = P[a] = 0.02$$

$$P[M2] = P[c]P[d] = 0.02 \times 0.02 = 4 \times 10^{-4}$$

$$P[M2'] = P[X] = 0.01$$

$$P[M1 \cap M2] = P[a]P[c]P[d] = 8 \times 10^{-6}$$

$$P[M2 \cap M2'] = P[c]P[d]P[X] = 4 \times 10^{-6}$$

$$P[M2' \cap M1] = P[X]P[a] = 2 \times 10^{-4}$$

$$P[M1 \cap M2 \cap M2'] = P[a]P[c]P[d]P[X] = 8 \times 10^{-8}$$

$$P[T] = \{P[M1] + P[M2] + P[M2']\}$$

$$-\{P[M1 \cap M2] + P[M2 \cap M2'] + P[M2' \cap M1]\}$$

$$+\{P[M1 \cap M2 \cap M2']\}$$

$$= 0.03018808 \tag{11.5}$$

Because the probabilities of double and triple combination terms are very small compared to the minimal cut set probabilities, we could have dropped those terms. In that case, we have the small probability approximation as follows:

$$P[T] = P[M1] + P[M2] + P[M2'] = 0.0304 \tag{11.6}$$

This is only slightly higher than the exact result of Equation (11.5).

Let us now examine the contribution of the second minimal cut set. Because it is a prime common cause candidate, its contribution consists of those due to primary failures ($M2$) and the common cause failure ($M2'$).

The primary failure contribution is two orders of magnitude less than the common cause contribution, so it can be dropped. Dropping $M2$, we get

$$P[T] = P[M1] + P[M2'] = 0.03 \qquad (11.7)$$

This is a good approximation to the top-event probability.

Now, can we drop the contribution of the first minimal cut set ($M1$), which is not affected by any common cause? The answer, in this problem, is "no," because its contribution is comparable to that of the common cause. This is so because $M1$ is a first-order minimal cut set and the terminal event probabilities and the common cause probability are of the same order of magnitude. In general, first-order minimal cut set contributions are comparable to prime common cause candidate contributions if the terminal events and common causes have approximately the same or the same-order probabilities.

Example 11.2

The top event (T) of a fault tree has three minimal cut sets:

$$M1 = (A) \qquad M2 = (C, D) \quad \text{and} \quad M3 = (E, F)$$

where A, C, D, E, and F are terminal events. C and D are susceptible to and in the domain of a common cause X. The probabilities of the primary failures of A, C, D, E, and F are $P[a]$, $P[c]$, $P[d]$, $P[e]$, and $P[f]$. The primary failures are statistically independent.

$$P[a] = P[c] = P[d] = 0.02$$
$$P[e] = P[f] = 0.015$$
$$P[X] = 0.01$$
$$P[CD|X] = 1.0$$

Compute the top-event probability.

Solution

We add a new minimal cut set $M2'$ as in Example 11.1. The probabilities $P[M1]$, $P[M2]$, and $P[M2']$ are the same as in Example 11.1.

$$P[M3] = P[e]P[f] = 0.015 \times 0.015 = 2.25 \times 10^{-4}$$

We calculate the top-event probability using the small probability approximation:

$$P[T] = P[M1] + P[M2] + P[M3] + P[M2']$$

$$= 0.030625 \qquad\qquad (11.8)$$

We concluded in Example 11.1 that the contribution of $M2$ may be dropped without significantly affecting the computed top-event probability. Can we drop the contribution of the other second-order minimal cut set, $M3$? The minimal cut set $M3$ is not a first-order minimal cut set and its terminal events have approximately the same probability as the common cause. So its contribution is substantially smaller than that of the common cause and we may drop it.

Example 11.3

Redo Example 11.1 with $P[X] = 0.9$. All other data remain the same.

Solution

Because $P[X]$ is changed, we recompute all quantities involving $P[M2']$, which is equal to $P[X]$:

$$P[M2'] = P[X] = 0.9$$

$$P[M2 \cap M2'] = 3.6 \times 10^{-4}$$

$$P[M2' \cap M1] = 0.018$$

$$P[M1 \cap M2 \cap M2'] = 7.2 \times 10^{-6}$$

We see that the probabilities of some of the double and triple combination terms have increased substantially over Example 11.1. In fact, $P[M2' \cap M1]$ is greater than $P[M2]$ and is almost equal to $P[M1]$. But the contributions of the double and triple combination terms are much smaller than $P[M2']$, so they may be dropped in the top-event probability computation.

$$P[T] = P[M1] + P[M2] + P[M2'] = 0.02 + 0.0004 + 0.9$$

$$= 0.9204$$

In this example, the common cause dominates every other failure because it has a very high probability compared to the terminal-event probabilities. In fact, we may discard all other minimal cut sets ($M1$ and

$M2$) and consider only the common cause contribution ($M2'$), and the top-event probability thus computed will not be significantly different from the exact result.

Example 11.4

Redo Example 11.1 with $P[X] = 0.0001$. All other data remain unchanged.

Solution

Because $P[X]$ is changed, we recompute $P[M2']$:

$$P[M2'] = P[X] = 0.0001$$

The discussion in Example 11.1 that the double and triple combination terms may be dropped is applicable in this example also.

$$P[T] = P[M1] + P[M2] + P[M2'] = 0.02 + 0.0004 + 0.0001$$
$$= 0.0205 \tag{11.9}$$

In this example, the probability of the common cause is so small compared to the terminal-event probabilities that we may ignore the common cause without significantly affecting the top-event probability.

When can we ignore a common cause? if the probability of a common cause is less than (say, by an order of magnitude) the probability of one or more of the minimal cut sets (including those involving other common causes), such a common cause may be ignored. A common cause analysis (either qualitative or quantitative) is unnecessary for such a common cause.

Example 11.5

Redo Example 11.1 with $P[CD|X] = 0.7$. All other data remain unchanged.

Solution

The probabilities of minimal cut sets $M1$ and $M2$ remain the same as in Example 11.1; only the probability of $M2'$ changes.

$$M2' = (CD \text{ due to } X)$$
$$P[M2'] = P[X]P[CD|X] = 0.01 \times 0.7 = 0.007 \tag{11.10}$$

As discussed in Example 11.1, double and triple combination terms may be dropped in the computation of $P[T]$.

$$P[T] = P[M1] + P[M2] + P[M2'] = 0.02 + 0.0004 + 0.007$$
$$= 0.0274 \qquad (11.11)$$

Example 11.6

The top event (T) of a fault tree has two minimal cut sets:

$$M1 = (C, D) \quad \text{and} \quad M2 = (E, F, G)$$

where C, D, E, F, and G are terminal events. All these terminal events are affected by a common cause X. The probabilities of the primary failures of C, D, E, F, and G are $P[c]$, $P[d]$, $P[e]$, $P[f]$, and $P[g]$, respectively. The primary failures are statistically independent.

$$P[c] = P[d] = P[e] = P[f] = P[g] = 0.02$$
$$P[X] = 0.01$$
$$P[CD|X] = 1.0$$
$$P[EFG|X] = 1.0$$

Compute the top-event probability.

Solution

The minimal cut sets due to primary failures are

$$M1 = (C, D)$$
$$M2 = (E, F, G)$$

The minimal cut sets $M1$ and $M2$ are prime common cause candidates with respect to X. So two additional minimal cut sets are added:

$$M1' = (X)$$
$$M2' = (X)$$

At the very outset, we may say that $M1'$ and $M2'$ are identical, so one of them can be dropped. Instead, let us proceed without dropping one of them and see how it does automatically get dropped as we proceed with the analysis.

First the probabilities of minimal cut sets and combinations of minimal cut sets are computed:

$$P[M1] = 4 \times 10^{-4}$$

$$P[M2] = 8 \times 10^{-6}$$

$$P[M1'] = 0.01$$

$$P[M2'] = 0.01$$

$$P[M1 \cap M2] = P[C \cap D \cap E \cap F \cap G] = 3.2 \times 10^{-9}$$

$$P[M1 \cap M1'] = P[C \cap D \cap X] = 4 \times 10^{-6}$$

$$P[M1 \cap M2'] = P[C \cap D \cap X] = 4 \times 10^{-6}$$

$$P[M2 \cap M1'] = P[E \cap F \cap G \cap X] = 8 \times 10^{-8}$$

$$P[M2 \cap M2'] = P[E \cap F \cap G \cap X] = 8 \times 10^{-8}$$

$$P[M1' \cap M2'] = P[X \cap X] = P[X] = 0.01$$

$$P[M1 \cap M2 \cap M1'] = P[C \cap D \cap E \cap F \cap G \cap X]$$

$$= 3.2 \times 10^{-11}$$

$$P[M1 \cap M2 \cap M2'] = P[C \cap D \cap E \cap F \cap G \cap X]$$

$$= 3.2 \times 10^{-11}$$

$$P[M1 \cap M1' \cap M2'] = P[C \cap D \cap X \cap X] = P[C \cap D \cap X]$$

$$= 4 \times 10^{-6}$$

$$P[M2 \cap M1' \cap M2'] = P[E \cap F \cap G \cap X \cap X]$$

$$= P[E \cap F \cap G \cap X] = 8 \times 10^{-8}$$

$$P[M1 \cap M2 \cap M1' \cap M2'] = P[C \cap D \cap E \cap F \cap G \cap X \cap X]$$

$$= P[C \cap D \cap E \cap F \cap G \cap X]$$

$$= 3.2 \times 10^{-11}$$

Now we compute the top-event probability as follows:

$$P[T] = [\text{single terms}] - [\text{double combinations}]$$
$$+ [\text{triple combinations}] - [\text{quadruple combination}]$$
$$= \left[(4 \times 10^{-4}) + (8 \times 10^{-6}) + 0.01 + 0.01\right]$$
$$- \left[(3.2 \times 10^{-9}) + (4 \times 10^{-6}) + (4 \times 10^{-6})\right.$$
$$+ (8 \times 10^{-8}) + (8 \times 10^{-8}) + 0.01\right]$$
$$+ \left[(3.2 \times 10^{-11}) + (3.2 \times 10^{-11}) + (8 \times 10^{-8}) + (4 \times 10^{-6})\right]$$
$$- \left[(3.2 \times 10^{-11})\right] \tag{11.12}$$

Consolidating this equation by cancelling identical terms with opposite signs, we have

$$P[T] = \left[(4 \times 10^{-4}) + (8 \times 10^{-6}) + 0.01\right]$$
$$- \left[(3.2 \times 10^{-9}) + (4 \times 10^{-6}) + (4 \times 10^{-6})\right.$$
$$+ (8 \times 10^{-8}) + (8 \times 10^{-8})\right]$$
$$+ \left[(3.2 \times 10^{-11}) + (4 \times 10^{-6}) + (8 \times 10^{-8})\right] \tag{11.13}$$

We indicated in the beginning of the solution of this example that $M1'$ and $M2'$ are identical and that the effect of one of them (say, $M2'$) will be cancelled out in due course of the analysis. We see that $P[M2']$ is cancelled out by $P[M1' \cap M2']$. Similarly, $P[M1 \cap M2']$ is cancelled out by $P[M1 \cap M1' \cap M2']$, and $P[M2 \cap M2']$ is cancelled out by $P[M2 \cap M1' \cap M2']$. Also, $P[M1 \cap M2 \cap M2']$ is cancelled out by $P[M1 \cap M2 \cap M1' \cap M2']$. In summary, all terms involving $M2'$ are cancelled out as we consolidate Equation (11.12) into Equation (11.13). This indicates the importance of including the double and higher combination terms when the minimal cut sets are *not statistically independent* ($M1'$ and $M2'$ are statistically dependent in this example). Instead of including all the double and higher-order terms, we could have included only those double and higher-order terms involving statistically dependent minimal cut sets, provided all the minimal cut sets are of small probability.

We could have dropped the minimal cut set $M2'$ from the very beginning. In that case, we need not have considered the double and higher-order combinations, and the top-event probability computation would have been much quicker. (Note that the double and higher combination terms need to be considered even in the absence of identical minimal cut sets if there are other statistical dependencies between minimal cut sets or the minimal cut set probabilities are not small.)

If there are m identical minimal cut sets $M1' = (X)$, $M2' = (X), \ldots, Mm' = (X)$, we can drop $(m - 1)$ of those minimal cut sets and keep only $M1' = (X)$. This will greatly reduce the effort of computing the top-event probability.

Example 11.7

The top event (T) of a fault tree has two minimal cut sets:

$$M1 = (A) \quad \text{and} \quad M2 = (C, D)$$

where A, C, and D are terminal events. A, C, and D are susceptible to and in the domain of a common cause X. The probabilities of the primary failures of A, C, and D are $P[a]$, $P[c]$, and $P[d]$. The primary failures are statistically independent.

$$P[a] = P[c] = P[d] = 10^{-6}$$
$$P[X] = 0.01$$
$$P[A|X] = 0.9$$
$$P[CD|X] = 0.8$$

Compute the top-event probability.

Solution

Because the minimal cut sets $M1$ and $M2$ are prime common cause candidates with respect to X, add two new minimal cut sets $M1'$ and $M2'$. So the minimal cut sets are

$$M1 = (A)$$
$$M2 = (C, D)$$
$$M1' = (A \text{ due to } X)$$
$$M2' = (CD \text{ due to } X)$$

The top-event probability can be computed using these four minimal cut sets and and the method described in Section 10.3.5.

$$P[M1] = P[a] = 10^{-6}$$
$$P[M2] = P[c]P[d] = 10^{-12}$$
$$P[M1'] = P[X \cap (A|X)] = P[X]P[A|X]$$
$$= 0.01 \times 0.9 = 0.009$$
$$P[M2'] = P[X \cap (CD|X)] = P[X]P[CD|X]$$
$$= 0.01 \times 0.8 = 0.008$$

$P[M1]$ and $P[M2]$ are many orders of magnitude smaller than $P[M1']$ and $P[M2']$, so we may discard them at this stage. We are left with $M1'$ and $M2'$. The presence of X in both these minimal cut sets introduces statistical dependencies between them, so the double combination term of $M1'$ and $M2'$ could be important.

$$P[M1' \cap M2'] = P[X \cap (A|X) \cap X \cap (CD|X)]$$
$$= P[X \cap (A|X) \cap (CD|X)]$$
$$= 0.01 \times 0.9 \times 0.8$$
$$= 0.0072 \tag{11.14}$$

The top-event probability is given by

$$P[T] = P[M1'] + P[M2'] - P[M1' \cap M2']$$
$$= 0.009 + 0.008 - 0.0072$$
$$= 0.0098 \tag{11.15}$$

The double combination term $P[M1' \cap M2']$ is not insignificant in this problem. Whether the double combination terms are significant depends on, among other things, the conditional probabilities. This becomes evident in Example 11.8.

Example 11.8

Redo Example 11.7 with $P[A|X] = 0.1$ and $P[CD|X] = 0.05$. All other data remain the same.

Solution

$P[M1]$ and $P[M2]$ are the same as in Example 11.7, only $P[M1']$ and $P[M2']$ change:

$$P[M1'] = P[X \cap (A|X)] = P[X]P[A|X]$$
$$= 0.01 \times 0.1 = 0.001$$
$$P[M2'] = P[X \cap (CD|X)] = P[X]P[CD|X]$$
$$= 0.01 \times 0.05$$
$$= 0.0005$$

$P[M1]$ and $P[M2]$ may be dropped as in Example 11.7. We are left with $M1'$ and $M2'$. The probability of the double combination of $M1'$ and $M2'$ is given by

$$P[M1' \cap M2'] = P[X \cap (A|X) \cap X \cap (CD|X)]$$
$$= P[X \cap (A|X) \cap (CD|X)]$$
$$= 0.01 \times 0.1 \times 0.05$$
$$= 0.00005$$

This probability is at least an order of magnitude smaller than $P[M1']$ and $P[M2']$, so it may be discarded.

The top-event probability is given by

$$P[T] = P[M1'] + P[M2'] = 0.001 + 0.0005 = 0.0015$$

Example 11.9

The top event (T) of a fault tree has two minimal cut sets:

$$M1 = (A, B, C) \quad \text{and} \quad M2 = (D, E, F, G)$$

Terminal events B, C, D, and E are affected by a common cause X, and terminal events F and G are affected by another common cause Y. The primary failure probabilities are $P[a] = 0.15$ and $P[b] = P[c] = P[d] = P[e] = P[f] = P[g] = 0.01$. The common cause probabilities are $P[X] = 0.2$ and $P[Y] = 0.1$. The conditional probabilities $P[BC|X] = P[DE|X] = P[FG|Y] = 1$. Compute the top-event probability.

Solution

The minimal cut sets are

$$M1 = (A, B, C)$$
$$M2 = (D, E, F, G)$$

Because some or all of the terminal events in each of these two minimal cut sets are affected by common causes, we introduce two new minimal cut sets.

$$M1' = (A, X)$$
$$M2' = (X, Y)$$

The probabilities of these minimal cut sets are

$$P[M1] = P[a]P[b]P[c]$$
$$= 0.15 \times 0.01 \times 0.01$$
$$= 1.5 \times 10^{-5}$$
$$P[M2] = P[d]P[e]P[f]P[g]$$
$$= 0.15 \times 0.01 \times 0.01 \times 0.01$$
$$= 1.5 \times 10^{-7}$$
$$P[M1'] = P[a]P[X] = 0.15 \times 0.2 = 0.03$$
$$P[M2'] = P[X]P[Y] = 0.2 \times 0.1 = 0.02$$

Because $P[M1]$ and $P[M2]$ are very small compared to $P[M1']$ and $P[M2']$, we discard $M1$ and $M2$. The presence of X in both $M1'$ and $M2'$ introduces statistical dependencies between them, so the double combination term between $M1'$ and $M2'$ could be significant.

$$P[M1' \cap M2'] = P(a \cap X \cap Y \cap X) = P(a \cap X \cap Y)$$
$$= P[a]P[X]P[Y] = 0.15 \times 0.2 \times 0.1$$
$$= 0.003$$

This is about an order of magnitude less than $P[M1']$ and $P[M2']$, so it may be discarded in the top-event probability computation.

$$P[T] = P[M1'] + P[M2'] = 0.03 + 0.02 = 0.05$$

11.6.2 Approximate Methods

In the previous section, we discussed a rigorous approach for the quantification of common cause failures when the probabilities of common causes and the conditional joint probabilities of the simultaneous existence of a number of terminal events due to common causes are known. There are situations where these probabilities are not known. For example, consider a number of components exposed to a corrosive environment.

The corrosive environment is definitely present (probability $= 1$), but what is the joint probability of 2, 3, or n components simultaneously failing due to corrosion? Such simultaneous failure data are seldom available although the probability of one component failing due to corrosion may be available. This scenario is particularly true for common links. Consider four pumps from the same manufacturer being used in a system. The common link (common manufacturer) is definitely present, but what is the joint probability that 2, 3, or 4 pumps fail simultaneously? Such joint probabilities are seldom known although the probability of one pump failing may be readily available. These situations are handled through the approximate techniques discussed in this section. Three methods are available [Nuclear Regulatory Commission (1975)]:

1. Lower bound method
2. Upper bound method
3. Square root method

11.6.2.1 Lower bound method. Let there be n terminal events $E_1, E_2, \ldots, E_n$ in a minimal cut set M, and let these terminal events be affected by a common cause (or let these terminal events have a common link). There is some statistical dependence between the terminal events because of the common cause (or common link) but the joint probability of simultaneous existence of the n terminal events is not known. However, the probability of occurrence of each terminal event (due to primary causes plus the common cause or common link) is known. Let these probabilities be $P[E_1], P[E_2], \ldots, P[E_n]$.

The lower bound method assumes that there is no statistical dependence between the terminal events. The minimal cut set probability is given by

$$P[M]_{\text{LB}} = P[E_1]P[E_2] \cdots P[E_n] \qquad (11.16)$$

The subscript "LB" is used to indicate that it is a lower bound to the actual minimal cut set probability [the joint probability of simultaneous existence of the n terminal events due to primary causes and/or the common cause (or common link)]. This approach is used when it is judged that the statistical dependencies between terminal events due to the common cause (or common link) are very weak.

11.6.2.2 Upper bound method. An upper bound to the probability of the minimal cut set is given by

$$P[M]_{\text{UB}} = \min\{P[E_1], P[E_2], \ldots, P[E_n]\} \qquad (11.17)$$

the subscript "UB" is used to indicate that it is an upper bound. This formula is used if the statistical dependencies between terminal events due to the common cause (or common link) are strong.

11.6.2.3 Square root method. When the effect of the common cause or common link is neither strong nor very weak, but is somewhat in-between, the square root of the lower bound and upper bound solutions may be used as an approximation.

$$P[M]_{SR} = \{P[M]_{LB}P[M]_{UB}\}^{1/2} \qquad (11.18)$$

where $P[M]_{LB}$ and $P[M]_{UB}$ are as defined by Equations (11.16) and (11.17), respectively. The subscript "SR" is used in Equation (11.18) to indicate that it is the square root solution. Equation (11.18) may underpredict the minimal cut set probability in some cases and overpredict the minimal cut set probability in some other cases. This *square-root formula*, Equation (11.18), is also known as the *geometric-mean formula*.

Example 11.10

The top event (T) of a fault tree has just one minimal cut set:

$$M = (A, B)$$

where A and B are terminal events. Primary failures of A and B are represented by a and b, respectively. $P[a] = P[b] = 0.05$. These primary failures are statistically independent. Both of the terminal events are affected by a common cause Z. $P[Z] = 0.4$, $P[A|Z] = 0.5$, $P[B|Z] = 0.5$, and $P[AB|Z] = 0.5$. Compute the top-event probability using the lower bound method, the upper bound method, the square root method, and the rigorous method.

Solution

In order to apply the lower bound method, the upper bound method, and the square root method, we need the total probability of each terminal event (the total probability is equal to the probability due to primary failure and the probability due to the common cause failure). Let the total probability of A and B be $P[A]$ and $P[B]$, respectively.

$$P[A] = P[a] + P[Z]P[A|Z] = 0.05 + (0.4 \times 0.5) = 0.25$$

$$P[B] = P[b] + P[Z]P[B|Z] = 0.05 + (0.4 \times 0.5) = 0.25$$

The top-event probability is equal to the minimal cut set probability because there is only one minimal cut set.

$$P[T]_{LB} = P[M]_{LB} = P[A]P[B] = 0.25 \times 0.25$$
$$= 0.0625$$
$$P[T]_{UB} = P[M]_{UB} = \min\{P[A], P[B]\} = \min(0.25, 0.25)$$
$$= 0.25$$
$$P[T]_{SR} = \{P[M]_{LB}P[M]_{UB}\}^{1/2} = (0.0625 \times 0.25)^{1/2}$$
$$= 0.125$$

Now we compute the top-event probability using the rigorous method described in Section 11.6.1.

The minimal cut set is

$$M = (A, B)$$

Because M is a primary common cause candidate, we introduce a new minimal cut set M'.

$$M' = (AB \text{ due to } Z)$$
$$P[M] = P[a]P[b] = 0.05 \times 0.05 = 0.0025$$
$$P[M'] = P[Z]P[AB|Z] = 0.4 \times 0.5 = 0.2$$

Discarding the double combination terms, we get the top-event probability as

$$P[T] = P[M] + P[M'] = 0.0025 + 0.2 = 0.2025$$

Comparing the different results obtained through the upper bound method, the lower bound method, the square root method, and the rigorous method, we see that the upper bound method result is the closest to the rigorous method result in this example. The reason for the closeness lies in the fact that the common cause failures are dominant over the primary failures (it is evident from the fact that $P[M']$ is about two orders of magnitude higher than $P[M]$). Of course, if we know only the total probability of each terminal event (instead of the primary and common cause failure probabilities separately) we will not have $P[M]$ and $P[M']$ separately. In such a situation, the decision on whether common cause failures are dominant over primary failures has to be made on the basis of engineering experience and judgement. Then a decision has to be made whether to apply the upper bound method, the lower bound method, or the square root method.

Example 11.11

Redo Example 11.10 with the following data: $P[Z] = 0.01$, $P[AB|Z] = P[A|Z] = P[B|Z] = 0.02$. All other data remain the same.

Solution

We repeat the steps of Example 11.10 using the revised data.

$$P[A] = P[a] + P[Z]P[A|Z] = 0.05 + (0.01 \times 0.02)$$
$$= 0.0502$$
$$P[B] = P[b] + P[Z]P[B|Z] = 0.05 + (0.01 \times 0.02)$$
$$= 0.0502$$

The top-event probability is equal to the minimal cut set probability because there is only one minimal cut set.

$$P[T]_{LB} = P[M]_{LB} = P[A]P[B] = 0.0502 \times 0.0502$$
$$= 0.0025$$
$$P[T]_{UB} = P[M]_{UB} = \min\{P[A], P[B]\} = \min(0.0502, 0.0502)$$
$$= 0.0502$$
$$P[T]_{SR} = \{P[M]_{LB}P[M]_{UB}\}^{1/2} = (0.0025 \times 0.0502)^{1/2}$$
$$= 0.011$$

Now we compute the top-event probability through the rigorous method described in Section 11.6.1.

$$P[M] = P[a]P[b] = 0.05 \times 0.05 = 0.0025$$
$$P[M'] = P[Z]P[AB|Z] = 0.01 \times 0.02 = 0.0002$$

Discarding the double combination terms, we get the top-event probability as

$$P[T] = P[M] + P[M'] = 0.0025 + 0.0002 = 0.0027$$

Comparing the results obtained through the upper bound method, the lower bound method, the square root method, and the rigorous method, we see that the lower bound method result is the closest to the rigorous method result in this example. The reason for the closeness lies in the fact that the statistically independent primary failures are dominant over the common cause failures (it is evident from the fact that $P[M]$ is over an order of magnitude higher than $P[M']$). Of course, if we know only the total probability of each terminal event (instead of the primary and common cause probabilities separately) we would not have $P[M]$ and $P[M']$ separately. An assessment of whether primary failures are dominant over common cause failures has to be made on the basis of engineering experience and judgement. Then a decision has to be made on whether to apply the upper bound method, the lower bound method, or the square root method.

Example 11.12

A minimal cut set consists of three terminal events A, B, and C. The total probability of these terminal events is $P[A] = 0.02$, $P[B] = 0.01$, and $P[C] = 0.01$. These probabilities include both primary failures, common cause failures, and common-link-related failures. Compute the minimal cut set probability using the lower bound method, the upper bound method, and the square root method.

Solution

$$P[M]_{LB} = P[A]P[B]P[C] = 0.02 \times 0.01 \times 0.01$$
$$= 2 \times 10^{-6}$$
$$P[M]_{UB} = \min\{P[A], P[B], P[C]\} = \min(0.02, 0.01, 0.01)$$
$$= 1 \times 10^{-2}$$
$$P[M]_{SR} = \{P[M]_{LB}P[M]_{UB}\}^{1/2} = (2 \times 10^{-6} \times 1 \times 10^{-2})^{1/2}$$
$$= 1.4 \times 10^{-4}$$

The three results are significantly different. We should use the lower bound result only if the common causes and/or the common links are expected to have very little effect on the terminal-event probabilities. The upper bound result is used if the effect of common causes and/or common links is dominant. The square root result may be used if the effect of common causes and/or common links is neither very weak nor dominant.

The *Reactor Safety Study* [Nuclear Regulatory Commission (1975)] uses the square root method in many common cause and common link situations.

Some analysts apply the upper bound formula for all the minimal cut sets involving common causes or common links and use them in the top-event probability computation. If the top-event probability thus computed is acceptable (that is, meets systems reliability requirements), no further calculations are carried out. If the top-event probability thus computed is not acceptable, each minimal cut set involving common causes or common links are examined to see if either the lower bound formula or the square root formula may be appropriate.

Example 11.13

A minimal cut set contains three terminal events A, B, and C. The terminal events A and B are affected by a common link. $P[A] = 0.01$, $P[B] = 0.03$, and $P[C] = 0.02$. $P[A]$ and $P[B]$ include the contributions of both the primary failure and the common link. $P[C]$ is not affected by the common link and is exclusively due to primary failure. Compute the minimal cut set probability using the lower bound method, the upper bound method, and the square root method.

Solution

The minimal cut set probability $P[M]$ is given by

$$P[M] = P[ABC] = P[AB]P[C] \qquad (11.19)$$

We may use this relationship because A and B are statistically dependent on each other due to the common link, and C is statistically independent of A and B. The joint probability $P[AB]$ may be computed using the lower bound method, the upper bound method, and the square root method.

$$P[AB]_{LB} = P[A]P[B] = 0.01 \times 0.03$$
$$= 0.0003$$
$$P[AB]_{UB} = \min\{P[A], P[B]\} = \min(0.01, 0.03)$$
$$= 0.01$$
$$P[AB]_{SR} = \{P[AB]_{LB}P[AB]_{UB}\}^{1/2} = (0.0003 \times 0.01)^{1/2}$$
$$= 0.0017$$

Substituting these values into Equation (11.19), we get

$$P[M]_{LB} = 0.0003 \times 0.02 = 0.000006$$
$$P[M]_{UB} = 0.01 \times 0.02 = 0.0002$$
$$P[M]_{SR} = 0.0017 \times 0.02 = 0.000034$$

11.6.3 Other Methods

Other methods of common cause quantification include the Marshall–Olkin method [Fleming and Raabe (1978)], the beta-factor method [Fleming and Raabe (1978)] and the Chu–Gaver method [Chu and Gaver (1977)]. These methods are based on Markov process models.

11.7 TREES WITH LARGE MINIMAL CUT SETS

In the quantitative fault tree analysis using minimal cut sets, we may usually consider only the lower-order minimal cut sets and discard the higher-order sets (see Chapter 10). Discarding the higher-order minimal cut sets does not significantly affect the results of fault tree quantification if there are no statistical dependencies between terminal events. On the contrary, we need to consider all the minimal cut sets, irrespective of their order, during both qualitative and quantitative common cause analyses.

Consider the top event (T) of a fault tree that has two minimal cut sets.

$$M1 = (E1, E2)$$
$$M2 = (E3, E4, \ldots, E22)$$

Here $M1$ is a minimal cut set of order 2 and $M2$ is a minimal cut set of order 20. $E1, E2, \ldots, E22$ are terminal events.

Let the probability of each terminal event due to primary failures be 0.01, and let the primary failures be statistically independent. Let a common cause Z, of probability $P[Z] = 0.02$, affect $E3, E4, \ldots, E22$. This makes $M2$ a prime common cause candidate, so we introduce a new minimal cut set $M2'$:

$$M2' = (\text{`}E3, E4, \ldots, \text{ and } E22\text{'} \text{ due to } Z) = (Z)$$

Let us consider the probabilities of the three minimal cut sets.

$$P[M1] = 0.0001$$
$$P[M2] = 10^{-40} (\text{very small; discard it})$$
$$P[M2'] = 0.02$$

The top-event probability is computed using the small probability approximation.

$$P[T] = P[M1] + P[M2'] \quad (\text{we have discarded } P[M2])$$
$$= 0.0201$$

Although we may discard the contribution of the second minimal cut set due to the statistically independent primary failures ($P[M2]$), we may not discard the contribution of the second minimal cut set due to the common cause ($P[M2']$) because it contributes significantly to the top-event probability. This illustration shows the importance of considering all the minimal cut sets in the common cause analysis, however high the order may be. Had we not determined the 20th-order minimal cut set $M2$, we would never have known that it is a prime common cause candidate and that the common cause Z is a significant common cause. Thus we would have significantly underpredicted the top-event probability. The lesson learned is that minimal cut sets of all orders must be considered in qualitative and quantitative common cause analyses. This introduces some difficulties to the reliability analyst because the determination of very high order minimal cut sets is not an easy task. In fault trees with a large number of terminal events and a number of AND gates, the order of some of the minimal cut sets could be very high, and the determination of these minimal cut sets could be very expensive and time-consuming. If there are common causes or common links that affect some of the terminal events, and if there is a possibility that there may be one or more prime common cause candidates among the higher-order minimal cut sets, what can we do? Henley and Kumamoto (1981) have introduced an efficient method. Their method is applicable if there are no statistical dependencies between terminal events other than those dependencies introduced by the common causes.

The Henley–Kumamoto method consists of three steps. All the common causes are identified first, and then the three steps are repeated for each common cause.

Step 1

Identify all the common mode terminal events; the remaining terminal events are neutral terminal events.

Step 2

Because the neutral terminal events are not affected by the common cause, their failure probabilities due to the common cause are zero.

Because the neutral terminal events have zero probability within the context of the common cause under consideration, they may be discarded from the fault tree. When such terminal events are to be discarded, it is necessary that not only those terminal events but all the terminal events, intermediate events, and associated gates all the way up to the first-encountered OR gate should be discarded from the tree. (This is similar to reducing a fault tree because of a switched-off house event; see Section 8.6.3.3.)

The resulting reduced fault tree would usually be very small compared to the original tree.

Step 3

Determine the minimal cut sets of the reduced fault tree. All the minimal cut sets thus obtained are prime common cause candidates with respect to the common cause under consideration because all the terminal events in the reduced tree, and thus all the terminal events in its minimal cut sets are common mode events. The minimal cut set determination should be relatively easy because the reduced fault tree will usually be very small compared to the original fault tree.

These three steps are repeated for each common cause and the prime common cause candidates with respect to each common cause are thus determined.

Example 11.14

Consider the top event of the fault tree shown in Figure 11.2. Two common causes (X and Y) are present. Terminal events $E3$, $E4$, $E5$, and $E7$ are common mode events for X, and terminal events $E1$ and $E8$ are common mode events for Y. Determine the prime common cause candidate for X and Y, using the full fault tree (Figure 11.2), as discussed in Section 11.4, and also by using the reduced fault tree, as discussed in Section 11.7.

Solution

(i) *Full Tree Approach (per Section 11.4):* Minimal cut sets of the full fault tree (Figure 11.2) are

$$M1 = (E1, E2)$$
$$M2 = (E3, E4, E5)$$
$$M3 = (E3, E4, E6)$$
$$M4 = (E3, E4, E7)$$
$$M5 = (E1, E7, E8, E9)$$

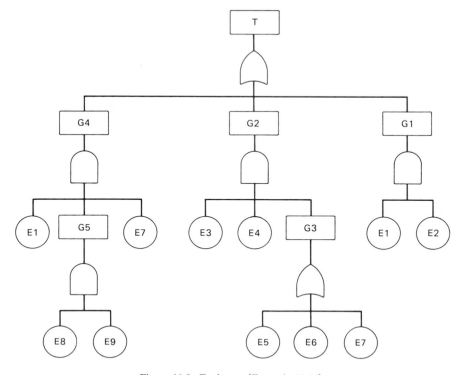

Figure 11.2. Fault tree (Example 11.14).

We determine the prime common cause candidates by inspection because only a few minimal cut sets and terminal events are involved. Prime common cause candidates with respect to X are $M2$ and $M4$. Minimal cut set $M2$ is a prime common cause candidate with respect to X because the terminal events $E3$, $E4$, and $E5$ are common mode events for X. Minimal cut set $M4$ is a prime common cause candidate with respect to X because the terminal events $E3$, $E4$, and $E7$ are common mode events for X.

By examining the minimal cut sets, we determine that there are no prime common cause candidates with respect to Y.

(ii) *Reduced Tree Approach (per Section 11.7):* The reduced fault tree with respect to common cause X is shown in Figure 11.3. Minimal cut sets of this tree are

$$M1 = (E3, E4, E5)$$
$$M2 = (E3, E4, E7)$$

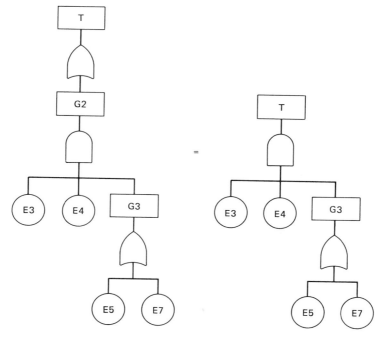

Figure 11.3. Reduced fault tree with respect to common cause *A* (Example 11.14).

Both minimal cut sets are prime common cause candidates with respect to *X*. This result is the same as that obtained from the full fault tree.

The reduced fault tree with respect to common cause *Y* is shown in Figure 11.4. There are no terminal events in that tree; it is a *null tree*. It has no minimal cut sets and the top-event probability is zero. Because there are no minimal cut sets, it means that there are no prime common cause candidates with respect to *Y*. Again, this result is the same as that obtained from the full fault tree.

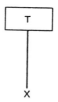

Figure 11.4. Reduced fault tree with respect to common cause *B* (Example 11.14).

In Example 11.14, we are able to determine the minimal cut sets of the full fault tree easily because the tree is small. In very large trees, with hundreds of terminal events and gates, this may not be easy. The fault tree reduction approach will considerably reduce the size and complexity of the tree and make the identification of prime common cause candidates much easier. The prime common cause candidates determined through the reduced tree approach can be added to the lower-order minimal cut sets (not related to common causes) determined from the full tree. The top-event probability can then be computed using all those minimal cut sets.

Example 11.15

A fault tree was analyzed for the first- and second-order minimal cut sets of the top event T. Minimal cut sets thus obtained are

$$M1 = (E1)$$
$$M2 = (E2, E4)$$
$$M3 = (E3, E6)$$

In addition, a reduced fault tree was constructed with respect to common cause Z. Minimal cut sets of this reduced tree are

$$M4 = (E7, E8, \ldots, E16)$$
$$M5 = (E7, E8, E9, E21, E22, \ldots, E30)$$

The terminal events are statistically independent, except for the effect of common cause Z. The probability of each terminal event due to primary causes is 0.01. The probability of Z is 0.05. The conditional probability of the occurrence of $M4$ given that Z has occurred is 1.0. The conditional probability of the occurrence of $M5$ given that Z has occurred is 1.0.

Compute the top-event probability.

Solution

Because the minimal cut sets $M4$ and $M5$ are determined from the reduced fault tree with respect to common cause Z, these minimal cut sets are prime common cause candidates with respect to Z. So we introduce the following two new minimal cut sets:

$$M4' = (Z)$$
$$M5' = (Z)$$

Because both of these minimal cut sets are identical $(= Z)$, we may drop one of them (see Example 11.6). We may also discard the minimal cut sets $M4$ and $M5$ because their probabilities are very small $(P[M4] = 10^{-20}$ and $P[M5] = 10^{-26})$. So the minimal cut sets to be considered in the top-event probability computation are

$$M1 = (E1)$$
$$M2 = (E2, E4)$$
$$M3 = (E3, E6)$$
$$M4' = (Z)$$

Substituting the terminal-event and common cause probabilities, we get

$$P[M1] = 0.01$$
$$P[M2] = 0.0001$$
$$P[M3] = 0.0001$$
$$P[M4'] = 0.05$$

Because these probabilities are small and the minimal cut sets are statistically independent, we may use the small probability approximation.

$$P[T] = P[M1] + P[M2] + P[M3] + P[M4']$$
$$= 0.01 + 0.0001 + 0.0001 + 0.05$$
$$= 0.0602$$

The fault tree reduction approach of common cause analysis is a powerful tool, but it has a drawback. Although it is capable of determining prime common cause candidates, it cannot provide partially affected minimal cut sets.

11.8 CORRECTIVE ACTIONS AGAINST COMMON CAUSE FAILURES

If a common cause increases system failure probabilities beyond acceptable limits, we may reduce the effect of the common cause by one of the following actions.

1. Build "barriers" between susceptible components.
2. Strengthen the components against the common cause.
3. Eliminate the common cause.
4. Reduce the common cause probability.

11.8.1 Build Barriers between Susceptible Components

A common cause affects a system most seriously if it produces prime common cause candidates. If a common cause produces a prime common cause candidate, the contribution of the common cause to top-event probability is equal to the product of the "probability of the common cause" and the "conditional probability of the simultaneous existence of all the terminal events in the prime common cause candidate, given that the common cause has occurred" [see Equation (11.3)]. This contribution can be significantly reduced if barriers are provided between the terminal events of the prime common cause candidate so that not all the terminal events in the minimal cut set (the prime common cause candidate) are in the domain of the common cause; at least one terminal event of the minimal cut set should be outside the domain of the common cause. (When we refer to barriers between terminal events, we are actually referring to barriers between components associated with the terminal events.) Terminal events outside the domain of the common cause are still susceptible to the common cause but are not exposed to it. So the minimal cut set is no longer a prime common cause candidate, but it is now either partially affected by the common cause or not affected at all.

How do we provide such a "barrier?" Consider the top event of a fault tree that has the following minimal cut set:

$$M1 = (E7, E8, E12, E14)$$

Let the terminal events $E7$, $E8$, $E12$, and $E14$ be susceptible to 'dust' (that is, the components related to these terminal events will fail if they are exposed to dust). Let all these components be located in a room in which a piece of equipment that generates dust during its operation is present. So $M1$ is a prime common cause candidate. How can we build a barrier between the components? We may simply build a dust-proof enclosure and place some of the components (at least one) in it. The enclosure acts as a barrier between the components located outside the enclosure and those located inside the enclosure. The minimal cut set $M1$ is still a common cause candidate, but it is not a prime common cause candidate. Of course, we can build a big enclosure and place all the components inside it to protect them from dust, but all that is required is an enclosure big enough to hold one of the components.

A barrier need not be a physical enclosure or wall. Consider the following minimal cut set:

$$M2 = (E1, E2, E3, E4)$$

All of the components relating to terminal events $E1$, $E2$, $E3$, and $E4$ are maintained by the same maintenance team. If one or some of the components are maintained by a different team, a barrier is said to have been built between the components with respect to the common link 'common maintenance team.'

Consider another example. A minimal cut set consists of three terminal events,

$$M3 = (E11, E19, E42)$$

The terminal events $E11$, $E19$, and $E42$ relate to the miscalibration of three identical instruments located at different locations in the plant. These terminal events are dependent on each other because photocopies of the same typed "instructions for calibration" are posted next to each instrument. If there is an error in the typed instructions, it will be found in all three photocopies, so all three instruments will be miscalibrated even though they are located far apart and are calibrated by different personnel. A barrier may be built by typing two sets of instructions (preferably by two different typists) and placing photocopies of the first set of typed instructions near two of the instruments and a photocopy of the second set of typed instructions near the third instrument. Thus the possibility of all three instruments being miscalibrated due to the same wrong instructions is eliminated or reduced considerably.

11.8.2 Strengthen the Components

The conditional joint probability of the simultaneous existence of component failures due to a common cause may be reduced by strengthening one or more components against the common cause. For example, if a number of components are susceptible to failure during an earthquake of intensity 4 or higher, their conditional joint probability of simultaneous failure may be decreased by strengthening a few of the components to withstand forces due to earthquakes of intensity 4.

11.8.3 Eliminate the Common Cause

Some common cause events and mechanisms may be eliminated by design changes. For example, consider a flow loop. A number of components in the flow loop are affected by chemical corrosion and thus produce a prime common cause candidate. This common cause may be eliminated by using a different, noncorrosive chemical (if possible). Elimination of common cause events or mechanisms may not always be possible.

11.8.4 Reduce the Common Cause Probability

The effect of a common cause on the top-event probability is a function of its probability of occurrence. As this probability is decreased, its contribution to the top-event probability is also decreased.

Consider the example of common cause failures due to excessive temperature caused by an air-conditioning system failure. Here, 'excessive temperature' is the common cause mechanism. The probability of this common cause may be reduced by increasing the reliability of the air-conditioning system.

Reduction of the probability of common cause events or mechanisms may not always be possible.

Example 11.16

The top event (T) of a fault tree has the following minimal cut set:

$$M1 = (A, B, C, D, E)$$

where A, B, C, D, and E are terminal events. This minimal cut set is a prime common cause candidate with respect to a common cause Z. The primary failures associated with A, B, C, D, and E are denoted by a, b, c, d, and e, respectively. These primary failures are statistically independent of each other and are also statistically independent of Z. The probabilities are

$$P[a] = P[b] = P[c] = P[d] = P[e] = 0.5 \qquad P[Z] = 0.1$$
$$P[ABCDE|Z] = 0.9 \quad \text{and} \quad P[ABCD|Z] = 0.9$$

Four alternatives are considered for reducing the effect of the common cause Z:

1. Build a barrier to protect the terminal event E from exposure to Z.
2. Strengthen the components against the common cause so that $P[ABCDE|Z] = 0.4$.
3. Eliminate the common cause Z.
4. Reduce the probability of the common cause, $P[Z]$, to 0.05.

Determine the contribution of the common cause Z to the top-event probability under each of those four alternatives.

Solution

Let G be the contribution of the common cause Z to the top-event probability. The original contribution (with none of the four alternatives implemented) is denoted by G'. Contributions under alternatives 1, 2, 3, and 4 are denoted by $G1$, $G2$, $G3$, and $G4$, respectively.

Contribution of the common cause Z to the top-event probability is given by

$$G = P[Z]P[ABCDE|Z]$$
$$G' = P[Z]P[ABCDE|Z] = 0.1 \times 0.9 = 0.09 \qquad (11.20)$$

Under the first alternative, the terminal event E is no longer exposed to the common cause, so it is statistically independent of A, B, C, and D. Therefore, $P[ABCDE|Z] = P[ABCD|Z]P[E]$. Substituting this relationship into Equation (11.20), we get

$$G1 = P[Z]P[ABCDE|Z] = P[Z]P[ABCD|Z]P[E]$$
$$= 0.1 \times 0.9 \times 0.05$$
$$= 0.0045$$

Under the second alternative, $P[ABCDE|Z]$ is reduced to 0.4. So,

$$G2 = P[Z]P[ABCDE|Z] = 0.1 \times 0.4 = 0.04$$

Under the third alternative, the common cause Z is eliminated, that is, $P[Z] = 0.0$. So,

$$G3 = P[Z]P[ABCDE|Z] = 0.0 \times 0.9 = 0.0$$

Under the fourth alternative, the common cause probability is reduced to 0.05. So,

$$G4 = P[Z]P[ABCDE|Z] = 0.05 \times 0.9 = 0.045$$

In this example we find that eliminating the common cause (third alternative) is the best option. The second-best option is to build a barrier between the affected components belonging to the minimal cut set (first alternative). Not only in this example, but also in most problems, these two alternatives rank as the two best options. Among these two alternatives, elimination of the common cause may not always be possible. Building a barrier between the components may be a more practical option in many problems.

The cost of implementing each of the alternatives may enter as a factor in deciding which alternative to choose or to choose 'not to do anything' at all.

11.9 COMPUTER PROGRAMS

BACFIRE, COMCAN-II, SETS, and WAMCOM are some of the commercially or publicly available computer programs for common cause analysis (see Appendix I for details). This is only a partial list of available programs.

New computer programs also become available. Capabilities and computational efficiencies of the programs differ. Users should compare the available programs and select those that best suit their needs.

11.10 DOCUMENTATION

Documentation of a common cause analysis should include the following, as applicable:

1. A list of common causes and common links affecting the system.
2. Common causes and common links that are not considered in the analysis, and reasons for omitting them.
3. Domain of each common cause and common link.
4. A list of terminal events that are susceptible to each common cause or common link.
5. Locations of the susceptible terminal events (components).
6. Minimal cut sets of the full fault tree.
7. Common cause candidates and prime common cause candidates.
8. Partially affected minimal cut sets.
9. Reduced fault tree with respect to each common cause or common link.
10. Prime common cause candidates from the reduced fault trees.
11. Contribution of each common cause to the top-event probability (if determined).
12. Total top-event probability (due to primary failures and common cause failures).
13. Recommendations for reducing the effects of common causes.
14. Identification of computer programs and the source of the programs (from whom obtained); include version number or version date, and the source of the program (from whom obtained).
15. Listings of computer inputs and outputs.

Some general guidelines on documentation are provided in Section 14.10.

EXERCISE PROBLEMS

11.1. Consider the floor plan of the plant area shown in Figure 11.1. A piping system, consisting of pipes, pipe supports, valves, pumps, and a pressure vessel, runs from room $A1$, to room $A3$, to room $A4$, to room $B2$, and finally to room $B3$. This piping system is not shown in Figure 11.1. The main components of this piping system are identified as $C1$, $C2$, ..., $C11$. These components have just one failure mode each. Failures of components $C1$, $C2$, ..., $C11$ are identified as $E1$, $E2$, ..., $E11$.

Locations of the components are as follows:

$C1$—room $A1$	$C2$—room $A1$
$C3$—room $A3$	$C4$—room $A4$
$C5$—open area P	$C6$—open area P
$C7$—open area P	$C8$—room $B2$
$C9$—room $B2$	$C10$—room $B3$
$C11$—room $B3$	

Five common causes are present. The common cause and their identifiers are as follows:

C = corrosion due to chemicals flowing through the piping system
D = dust produced by a machine
E = earthquake
F = flooding
V = vibration due to a machine when the machine is unbalanced

The domain of each common cause is as follows:

C—piping system (components $C3$, $C4$, $C8$, and $C10$ are exposed to corrosion)
D—room $B2$
E—all areas of the plant
F—all areas of the plant
V—building A (all rooms in that building)

The susceptibility of each component to one or more of the common causes are as follows:

$C1$—D, E, F
$C2$—D, E, F
$C3$—C, D, E, V
$C4$—C, D, E, V
$C5$—E
$C6$—E
$C7$—E
$C8$—C, D, E, V
$C9$—E
$C10$—C, D, E, V
$C11$—E, F

Prepare the common cause analysis sheet (similar to Table 11.3).

11.2. Consider the system described in Problem 11.1. A fault tree for this system has two minimal cut sets: $M1 = (E1, E2)$, $M2 = (E4, E11)$. Identify the common cause candidates, the prime common cause candidates, the significant common causes, and the partially affected minimal cut sets, if any.

11.3. Consider the system described in Problem 11.1. A fault tree for this system has four minimal cut sets: $M1 = (E1, E2)$, $M2 = (E3, E8, E11)$, $M3 = (E4, E10, E11)$, and $M4 = (E5, E7, E9)$. Identify the common cause candidates, the prime common cause candidates, the significant common causes, and the partially affected minimal cut sets, if any.

11.4. Consider the fault tree described in Problem 11.2. The probabilities of the primary failures of $E1$, $E2$, $E4$, and $E11$ are denoted by $P[e1]$, $P[e2]$, $P[e4]$, and $P[e11]$, respectively. These primary failures are statistically independent. $P[e1] = P[e2] = 0.01$, $P[e4] = 0.03$, and $P[e11] = 0.0008$. The probabilities of occurrences of common causes C, D, E, F, and V are denoted by $P[C]$, $P[D]$, $P[E]$, $P[F]$, and $P[V]$, respectively. $P[C] = 1.0$, $P[D] = 0.01$, $P[E] = 0.001$, $P[F] = 0.01$, and $P[V] = 0.01$. The conditional probability of simultaneous failure of all the components susceptible to and exposed to C given that C has occurred is 0.008; this probability may be used in this problem as an approximation to the probability of simultaneous failures of any combination of such components due to C. (*Note:* Such an approximation could be unconservative in some problems.) The conditional probability of simultaneous failure of all the components susceptible to and exposed to D given that D has occurred is 0.1; this probability may be used in this problem as an approximation to the probability of simultaneous failures of any combination of such components due to D. (*Note:* Such an approximation could be unconservative in some problems.) The conditional probability of simultaneous failure of all the components susceptible to and exposed to E given that E has occurred is 1.0. The conditional probability of simultaneous failure of all the components susceptible to and exposed to F given that F has occurred is 1.0. The conditional probability of simultaneous failure of all the components susceptible to and exposed to V given that V has occurred is 0.2; this probability may be used in this problem as an approximation to the probability of simultaneous failures of any combination of such components due to V. (*Note:* Such an approximation could be unconservative in some problems.) Compute the top-event probability. [*Note:* If a minimal cut set (M) is a prime common cause candidate with respect to n common causes $(n > 0)$, then n new minimal cut sets $(M', M'', M''', \dots)$ should be added. For example, in Example 11.1, the minimal cut set $M2$ is a prime common cause candidate with respect to one common cause and so we introduced one new minimal cut set $M2'$.]

11.5. Redo Problem 11.4 with $P[D] = 0.1$. All other data remain the same as in Problem 11.4.

11.6. A minimal cut set consists of four terminal events $E1$, $E2$, $E3$, and $E4$. The probabilities of these terminal events are $P[E1] = P[E2] = P[E3] = P[E4] = 0.1$. These probabilities include the effects of primary failures, common cause failures, and common-link-related failures. Compute the minimal cut set probability using the lower bound method, the upper bound method, and the square root method.

11.7. A minimal cut set consists of four terminal events $E1$, $E2$, $E3$, and $E4$. Terminal events $E1$ and $E2$ are affected by a common cause X. Terminal events $E3$ and $E4$ are affected by another common cause Y. The probabilities of the terminal events are $P[E1] = 0.1$, $P[E2] = 0.08$, $P[E3] = 0.14$, and $P[E4] = 0.06$. These probabilities include the effects of both primary failures and common cause failures. Compute the minimal cut set probability using the lower bound method, the upper bound method, and the square root method.

11.8. Consider the fault tree shown in Figure 8.20. For the purposes of this problem, assume that E, F, G, J, K, M, N, P, Q, and R are terminal events (instead of intermediate events as indicated in Figure 8.20). Terminal events E, G, J, and K are common mode events for a common cause Z. The probability of each of the nine terminal events due to causes other than the common cause is 0.02. The terminal events are statistically independent but for the effect of the common cause. The probability of occurrence of the common cause Z is 0.1. The conditional probability of the simultaneous occurrence of the common mode events (E, G, J, and K) given that Z has occurred is 1.0. (i) Determine the prime common cause candidates using the full fault tree (as discussed in Section 11.4) and also using the reduced fault tree (as discussed in Section 11.7). (ii) Compute the top-event probability.

11.9. Consider the fault tree shown in Figure 8.20. For the purposes of this problem, assume that E, F, G, J, K, M, N, P, Q, and R are terminal events (instead of intermediate events as indicated in Figure 8.20). Terminal events E, F, G, J, M, N, P, and Q are common mode events for a common cause Y. Determine the prime common cause candidates using the full fault tree (as discussed in Section 11.4) and also using the reduced fault tree (as discussed in Section 11.7).

11.10. Consider the fault tree shown in Figure 8.39. (i) The terminal events 'PRV fails closed (SF),' 'SF of controller,' and 'SF of TMCD' are common mode events for a common cause A. Determine the prime common cause candidates with respect to common cause A, using the full fault tree (as discussed in Section 11.4) and also using the reduced fault tree (as discussed in Section 11.7). (ii) The terminal events 'PRV set over 100 psi' and 'TMCD is set to signal at too high a temperature (above the boiling temperature)' are common mode events for a common cause B. Determine the prime common cause candidates with respect to common cause B, using the full fault tree (as discussed in Section 11.4) and also using the reduced fault tree (as discussed in Section 11.7).

11.11. A system can fail if either components $C1$, $C2$, $C3$, and $C4$ fail or component $C5$ fails; that is, the minimal cut sets for 'system failure' are $M1 = (E1, E2, E3, E4)$ and $M2 = (E5)$, where the terminal events $E1$, $E2$, $E3$, $E4$, and $E5$ denote the failures of components $C1$, $C2$, $C3$, $C4$, and $C5$, respectively. Terminal events $E1$, $E2$, $E3$, and $E4$ are common mode events with respect to a common cause Y. The probabilities of $E1$, $E2$, and $E3$ due to causes other than the common cause are 0.04, each. The probability of $E4$ due to causes other than the common cause is 0.1. The probability of $E5$ is 0.05. All component failures due to causes other than the common cause are statistically independent. The probability of occurrence of the common cause $P[Y] = 0.1$. The following conditional probabilities are also known. $P[E1 \cap E2 \cap E3 \cap E4|Y] = 0.8$ and $P[E1 \cap E2 \cap E3|Y] = 0.9$. Three alternatives are considered for reducing the system failure probability: (1) Build a barrier to protect the terminal event $E4$ from the common cause. (2) Strengthen components $C1$, $C2$, $C3$, and $C4$ against the common cause so that $P[E1 \cap E2 \cap E3 \cap E4|Y] = 0.4$. This also reduces the failure probabilities of components $C1$, $C2$, $C3$, and $C4$ due to causes other than the common cause; the reduced probability for $E1$, $E2$, and $E3$ are 0.2, each, and the reduced probability for $E4$ is 0.05. (3) Reduce the probability of occurrence of the common cause to 0.05.

Compute the system failure probability when no changes are made in the system (none of the three alternatives is implemented) and when each of the three alternatives is implemented.

REFERENCES

Chu, B. B. and D. P. Gaver (1977). Stochastic models for repairable redundant systems susceptible to common mode failures. In *Nuclear Systems Reliability Engineering and Risk Assessment*, J. B. Fussell and G. R. Burdick, eds. Society for Industrial and Applied Mathematics, Philadelphia.

Fleming, K. B. and P. H. Raabe (1978). A comparison of three methods for the quantitative analysis of common cause failures. In *Probabilistic Analysis of Nuclear Reactor Safety* (vol. 3). American Nuclear Society, LaGrange Park, IL.

Henley, E. J. and H. Kumamoto (1981). *Reliability Engineering and Risk Assessment*. Prentice-Hall, Inc., Englewood Cliffs, NJ.

Nuclear Regulatory Commission (1975). *Reactor Safety Study: An Assessment of Accident Risks in U.S. Commercial Nuclear Power Plants* (WASH-1400). Nuclear Regulatory Commission, Washington, DC.

Putney, B. F. (1981). *WAMCOM: Common Cause Methodologies using Large Fault Trees*. Report No. EPRI-NP-1851. Electric Power Research Institute, Palo Alto, CA.

Chapter 12
Importance Analysis

12.1 INTRODUCTION

Qualitative fault tree analysis provides the reliability analyst with information on how a system failure can happen or how system success can be assured [what combinations of component failures (terminal events) can initiate a system failure (top event), or what combinations of component successes can assure system success]. Quantitative fault tree analysis provides the probability of system failure (top-event probability), which can be used to decide if system performance (reliability, availability, and/or safety) is acceptable or if changes need to be made. Importance analysis, the subject of this chapter, deals with assessing the importance of the various terminal events with reference to their contribution to system unreliability, unavailability, or accident probabilities. A high-probability terminal event does not necessarily contribute more to system failure than a lower-probability terminal event. It depends on both terminal-event probabilities and system configuration.

An important analysis is not always carried out as part of a reliability analysis project; whether or not an important analysis is performed depends on the scope and purpose of the reliability analysis project. An importance analysis is unnecessary if the purpose of the reliability project is just to compute accident probabilities, system unreliability, or system unavailability. An importance analysis will be useful if the purpose of the project is to design systems to meet specified reliabilities, to develop reliability improvement strategies, to develop reliability-based maintenance strategies, etc. Some applications of importance analysis are included in Chapter 13.

12.2 QUALITATIVE RANKING

Qualitative ranking of minimal cut sets according to their importance in contributing to top-event occurrence is discussed in Section 9.4, and qualitative ranking of terminal events is discussed in Section 9.5. Qualitative ranking does not require a knowledge of terminal-event probabilities. Qualitative ranking is a good measure of the importance of minimal cut sets and terminal events if terminal-event probabilities are about the same, at least in an order-of-magnitude sense. Quantitative ranking is more precise than qualitative ranking because it includes the effect of terminal event probabilities.

12.3 QUANTITATIVE MEASURES OF IMPORTANCE

Terminal-event importance or minimal cut set importance may be defined with respect to top-event reliability or top-event availability at specified times. A number of importance measures have been defined by different researchers. Importance measures for terminal events include the following:

1. Birnbaum's measure
2. Criticality measure
3. Upgrading function measure
4. Vesely–Fussell measure
5. Barlow–Proschan measure
6. Sequential contributory measure

Statistical independence between terminal events is assumed in the computation of these importance measures.

The different importance measures may not necessarily rank the terminal events in the same order. A terminal event that is the most important with respect to one measure may not necessarily be the most important terminal event with respect to another measure. Also, depending on the time dependencies of terminal-event probabilities, it is possible that a terminal event that is the most important at time t_1 may not necessarily be the most important terminal event at time t_2, even with respect to the same measure of importance.

Minimal cut sets may also be ranked according to their importance. Measures of minimal cut set importance include the following:

1. Vesely–Fussell measure
2. Barlow–Proschan measure

Importance ranking with respect to one measure may not necessarily be identical to the ranking with respect to another measure. Also, importance ranking may change with time.

The Vesely–Fussell measure is one of the better importance measures, and we shall discuss it, in some detail, in the next three sections.

Readers interested in the other importance measures are referred to Lambert (1975) and Lambert and Davis (1980).

12.4 VESELY–FUSSELL MEASURE OF IMPORTANCE FOR MINIMAL CUT SETS

The Vesely–Fussell measure of importance of the jth minimal cut set at time t is defined as the ratio

$$\frac{\text{unreliability (or unavailability) of the } j\text{th minimal cut set at time } t}{\text{top-event unreliability (or unavailability) at time } t}$$

Let us use the following symbols:

$I_{A,j}(t)$ = Vesely–Fussell measure of importance of the jth minimal cut set with respect to top-event availability at time t

$I_{R,j}(t)$ = Vesely–Fussell measure of importance of the jth minimal cut set with respect to top-event reliability at time t

If the top event is 'system failure,' then these importance measures are with respect to system availability and system reliability, respectively.

These importance measures are computed using the following equations:

$$I_{A,j}(t) = \frac{Q_j(t)}{Q_S(t)} \tag{12.1}$$

$$I_{R,j}(t) = \frac{U_j(t)}{U_S(t)} \tag{12.2}$$

where

$Q_j(t)$ = unavailability of the jth minimal cut set at time t
$U_j(t)$ = unreliability of the jth minimal cut set at time t
$Q_S(t)$ = top-event unavailability at time t
$U_S(t)$ = top-event unreliability at time t

Equations (12.1) and (12.2) are based on the assumption that terminal events are statistically independent. It is also assumed that the probability of more than one minimal cut set in a failed state at time t is very small compared to the probability of only one minimal cut set in a failed state at time t; it is a reasonable assumption for most systems.

12.5 VESELY–FUSSELL MEASURE OF IMPORTANCE FOR TERMINAL EVENTS

The Vesely–Fussell measure of importance of the kth terminal event at time t is defined as the ratio of the sum of the unreliabilities (or unavailabilities) of the minimal cut sets containing the kth terminal event at time t divided by the top-event unreliability (or unavailability) at time t. Let us use the following symbols:

$i_{A,k}(t)$ = Vesely–Fussell measure of importance of the kth terminal event with respect to top-event availability at time t

$i_{R,k}(t)$ = Vesely–Fussell measure of importance of the kth terminal event with respect to top-event reliability at time t

If the top event is 'system failure,' then these importance measures are with respect to system availability and system reliability, respectively. These importance measures are computed using the following equations:

$$i_{A,k}(t) = \frac{\Sigma_j Q_j(t)}{Q_S(t)} = \sum_j I_{A,j}(t) \qquad (12.3)$$

$$i_{R,k}(t) = \frac{\Sigma_j U_j(t)}{U_S(t)} = \sum_j I_{R,j}(t) \qquad (12.4)$$

where all the notation is as defined before. Summations in these equations are over all the minimal cut sets containing the kth terminal event.

Equations (12.3) and (12.4) are based on the assumption that the terminal events are statistically independent. It is also assumed that the probability of more than one minimal cut set in a failed state at time t is very small compared to the probability of only one minimal cut set in a failed state at time t; it is a reasonable assumption for most systems.

Once the importance measures are computed, terminal events may be ranked according to their importance in descending order. A terminal event with higher importance is more important in contributing to top-event unavailability (or unreliability) than a terminal event with lower importance. What does it mean in practical terms?

Consider two terminal events A and B. Let A be ranked seventh and let B be ranked eighth with respect to top-event availability at time $t = 3000$ hours. Let us decrease the unavailability of A at $t = 3000$ hours by 60% and keep the unavailability of B unchanged. Let this change produce a 28% decrease in the top-event unavailability at $t = 3000$ hours. If we decrease the unavailability of B at $t = 3000$ hours by 60% and keep the unavailability of A unchanged, the resulting decrease in the top-event unavailability at $t = 3000$ hours will be less than 28%.

Consider two other terminal events C and D. Let C be ranked fourth and let D be ranked third with respect to top-event reliability at time $t = 2500$ hours. Let us decrease the unreliability of C at $t = 2500$ hours by 50% and keep the unreliability of D unchanged. Let this change produce a 35% decrease in the top-event unreliability at $t = 2500$ hours. If we decrease the unreliability of D at $t = 2500$ hours by 50% and keep the unreliability of C unchanged, the resulting decrease in the top-event unreliability at $t = 2500$ hours will be greater than 35%.

So, if we are interested in decreasing the top-event unavailability (or unreliability), it is prudent to decrease the unavailabilities (or unreliabilities) of the higher-ranked terminal events rather than those of the lower-ranked terminal events, provided cost is not of consideration or the costs of upgrading the terminal events are about the same. Terminal event ranking is useful even when the cost of upgrading is to be considered. [Some examples in the next chapter (Examples 13.2–13.4) illustrate the use of importance ranking in such situations.]

Example 12.1

Consider a fault tree with the top event 'system failure.' The minimal cut sets are

$$M1 = (E1, E2)$$
$$M2 = (E1, E3)$$
$$M3 = (E3, E4)$$

The terminal events are statistically independent. The unreliabilities of the minimal cut sets at time $t = 4000$ hours and system unreliability at $t = 4000$ hours have already been computed using the methods discussed in Section 10.4.2: $U_1 = 10^{-2}$, $U_2 = 7 \times 10^{-3}$, $U_3 = 2 \times 10^{-2}$, $U_S = 3.7 \times 10^{-2}$. Compute the Vesely–Fussell measure of importance with respect to system reliability at time $t = 4000$ hours.

Solution

Minimal cut set importance with respect to system reliability at $t = 4000$ hours is as follows:

$$I_{R,1} = \frac{U_1}{U_S} = 0.27$$

$$I_{R,2} = \frac{U_2}{U_S} = 0.19$$

$$I_{R,3} = \frac{U_3}{U_S} = 0.54$$

We rank the minimal cut sets on the basis of the preceding importance values:

$$M3, M1, M2$$

Terminal-event importance with respect to system reliability at $t = 4000$ hours is as follows:

$$i_{R,1} = \frac{U_1 + U_2}{U_S} = 0.46$$

$$i_{R,2} = \frac{U_1}{U_S} = 0.27$$

$$i_{R,3} = \frac{U_2 + U_3}{U_S} = 0.73$$

$$i_{R,4} = \frac{U_3}{U_S} = 0.54$$

We rank the terminal events on the basis of the preceding importance values:

$$E3, E4, E1, E2$$

Example 12.2

The top event of a fault tree is 'system failure.' The minimal cut sets of the fault tree are

$$M1 = (E1)$$

$$M2 = (E2)$$

$$M3 = (E2, E3)$$

$$M4 = (E3, E4)$$

The unavailabilities of the terminal events at $t = 2400$ hours are

$$q_1 = 0.04 \qquad q_2 = 0.02 \qquad q_3 = 0.05 \quad \text{and} \quad q_4 = 0.08$$

The terminal events are statistically independent. Compute the Vesely–Fussell measures of importance with respect to system availability at $t = 2400$ hours.

Solution

Because we are not provided with minimal cut set and system unavailabilities, we calculate these quantities first, using the equations given in Section 10.4.2.

The minimal cut set unavailabilities at $t = 2400$ hours are

$$Q_1 = q_1 = 0.04$$

$$Q_2 = q_2 = 0.02$$

$$Q_3 = q_2 q_3 = 0.02 \times 0.05 = 0.001$$

$$Q_4 = q_3 q_4 = 0.05 \times 0.08 = 0.004$$

The system unavailability at $t = 2400$ hours is

$$Q_S = Q_1 + Q_2 + Q_3 + Q_4 = 0.04 + 0.02 + 0.001 + 0.004$$

$$= 0.065$$

Now we compute the Vesely–Fussell measures of minimal cut set importance with respect to system availability at $t = 2400$ hours.

$$I_{A,1} = \frac{Q_1}{Q_S} = 0.62$$

$$I_{A,2} = \frac{Q_2}{Q_S} = 0.31$$

$$I_{A,3} = \frac{Q_3}{Q_S} = 0.02$$

$$I_{A,4} = \frac{Q_4}{Q_S} = 0.06$$

We rank the minimal cut sets on the basis of the preceding importance values:

$$M1, M2, M4, M3$$

The Vesely–Fussell measures of terminal event importance with respect to system availability at $t = 2400$ hours are

$$i_{A,1} = \frac{Q_1}{Q_S} = 0.62$$

$$i_{A,2} = \frac{Q_2 + Q_3}{Q_S} = 0.32$$

$$i_{A,3} = \frac{Q_3 + Q_4}{Q_S} = 0.077$$

$$i_{A,4} = \frac{Q_4}{Q_S} = 0.06$$

We rank the terminal events on the basis of the preceding importance values:

$$E1, E2, E3, E4$$

Example 12.3

For the system described in Example 12.2, decrease the unavailability (at $t = 2400$ hours) of each terminal event by 50% and determine the corresponding percentage decrease in system unavailability at $t = 2400$ hours.

Solution

Let us first derive expressions for minimal cut set unavailabilities (Q_1, Q_2, Q_3, and Q_4) and system unavailability (Q_S) in terms of terminal-event unavailabilities (q_1, q_2, q_3 and q_4).

$$Q_1 = q_1$$
$$Q_2 = q_2$$
$$Q_3 = q_2 q_3$$
$$Q_4 = q_3 q_4$$
$$Q_S = Q_1 + Q_2 + Q_3 + Q_4$$
$$= q_1 + q_2 + q_2 q_3 + q_3 q_4 \qquad (12.5)$$

(i) Decrease q_1 by 50%:

$$q_1 = 0.02$$

Other terminal-event probabilities remain the same as before:

$$q_2 = 0.02 \qquad q_3 = 0.05 \quad \text{and} \quad q_4 = 0.08$$

Substituting these values into Equation (12.5), we get

$$Q_S = 0.045$$

percentage decrease in system unavailability

$$= \left[\frac{0.065 - 0.045}{0.065} \right] \times 100 = 30.8\%$$

where 0.065 is the original system unavailability, as computed in Example 12.2.

(ii) Decrease q_2 by 50%:

$$q_2 = 0.01$$
$$q_1 = 0.04 \qquad q_3 = 0.05 \quad \text{and} \quad q_4 = 0.08$$

Substituting into Equation (12.5), we get

$$Q_S = 0.0545$$

percentage decrease in system unavailability

$$= \left[\frac{0.065 - 0.0545}{0.065} \right] \times 100 = 16.2\%$$

(iii) Decrease q_3 by 50%:

$$q_3 = 0.025$$

$$q_1 = 0.04 \qquad q_2 = 0.02 \quad \text{and} \quad q_4 = 0.08$$

Substituting into Equation (12.5), we get

$$Q_S = 0.0625$$

percentage decrease in system unavailability

$$= \left[\frac{0.065 - 0.0625}{0.065} \right] \times 100 = 3.8\%$$

(iv) Decrease q_4 by 50%:

$$q_4 = 0.04$$

$$q_1 = 0.04 \qquad q_2 = 0.02 \quad \text{and} \quad q_3 = 0.05$$

Substituting into Equation (12.5), we get

$$Q_S = 0.063$$

percentage decrease in system unavailability

$$= \left[\frac{0.065 - 0.063}{0.065} \right] \times 100 = 3.1\%$$

We may rank the terminal events according to the effect a 50% decrease in their unavailabilities has on system unavailability. The ranking is as follows:

$$E1, E2, E3, E4$$

This ranking is the same as the ranking obtained on the basis of the Vesely–Fussell measure of importance with respect to system availability at $t = 2400$ hours (see Example 12.2).

Example 12.4

Consider the minimal cut sets given in Example 12.2. Qualitatively rank the minimal cut sets and terminal events according to their importance. The terminal events are statistically independent. Assume that we do not know the terminal-event probabilities at the time of qualitative ranking.

Solution

We shall use the guidelines provided in Section 9.4 for the qualitative ranking of minimal cut sets.

Because $M1$ and $M2$ are first-order minimal cut sets, they are more important than $M3$ and $M4$, which are second-order minimal cut sets. So the qualitative ranking is

$$M1, M2, M3, M4$$

We do not know whether $M2$ should be ranked before $M1$ because both are first-order minimal cut sets. We have arbitrarily ranked $M1$ before $M2$. Similarly, we do not know if $M4$ should be ranked before $M3$ or not; we have arbitrarily ranked $M3$ before $M4$.

We shall use the guidelines provided in Section 9.5 for the qualitative ranking of terminal events.

Because $E1$ and $E2$ appear in first-order minimal cut sets, they are more important than $E3$ and $E4$, which appear in second-order minimal cut sets. Between $E1$ and $E2$, which one is more important? Because $E1$ appears in only one minimal cut set (a first-order minimal cut set) and $E2$ appears not only in a first-order minimal cut set but also in one more minimal cut set (a second-order minimal cut set), $E2$ is ranked before $E1$.

Between $E3$ and $E4$, which is more important? $E3$ appears in two second-order minimal cut sets whereas $E4$ appears in only one second-order minimal cut set. So $E3$ is ranked before $E4$. Summarizing, the qualitative ranking of terminal events is

$$E2, E1, E3, E4$$

Now we compare this qualitative ranking with the quantitative ranking given in Example 12.2. Qualitative minimal cut set ranking is identical to the quantitative ranking. Of course, the ranking could be different if the quantitative data in Example 12.2 (the terminal-event unavailabilities) are different.

We find that qualitative and quantitative rankings of terminal events are different. (Again, they could have been identical if the quantitative data in Example 12.2 were different.) It is worth noting that the qualitative ranking has correctly identified $E3$ and $E4$ as the least-important terminal events. So qualitative ranking does have some use in identifying the important terminal events and minimal cut sets, at an early stage when quantitative data are not yet available.

12.6 VESELY–FUSSELL MEASURE OF IMPORTANCE FOR COMPONENTS

The concept of terminal-event importance can be extended to component importance.

If each component has only one terminal event (failure mode) associated with it, then the Vesely–Fussell measure of importance of the component will be identical to the Vesely–Fussell measure of importance of the corresponding terminal event. However, there are components that have two or more terminal events (failure modes) associated with them. What we need is a formula that relates terminal-event importances to component importances irrespective of whether one or more terminal events are associated with each component. The Vesely–Fussell measures of component importance with respect to top-event availability and reliability at time t are defined next.

$$i^*_{A,j}(t) = \sum_k i_{A,k}(t) \tag{12.6}$$

$$i^*_{R,j}(t) = \sum_k i_{R,k}(t) \tag{12.7}$$

where $i^*_{A,j}(t)$ is the Vesely–Fussell measure of importance of the jth component with respect to top-event availability at time t, $i^*_{R,j}(t)$ is the Vesely–Fussell measure of importance of the jth component with respect to top-event reliability at time t, and $i_{A,k}(t)$ and $i_{R,k}(t)$ are terminal-event importances, as defined by Equations (12.3) and (12.4). The summations in Equations (12.6) and (12.7) are over all the terminal events associated with the jth component. These equations are applicable only if the terminal events are statistically independent.

Example 12.5

A system consists of three components $H1$, $H2$, and $H3$. Terminal event (failure mode) $E1$ is associated with component $H1$, terminal events $E2$, $E3$, and $E5$ are associated with $H2$, and terminal event $E4$ is associated with $H3$. The Vesely–Fussell measure of terminal-event importance with respect to system availability at time $t = 1500$ hours for $E1$, $E2$, $E3$, $E4$, and $E5$ are 0.46, 0.07, 0.53, 0.74, and 0.11, respectively. Rank the components according to their Vesely–Fussell measure of importance with respect to system availability at time $t = 1500$ hours. All the terminal events are statistically independent.

Solution

Vesely–Fussell measures of component importance with respect to system availability at time $t = 1500$ hours are as follows:

$$i^*_{A,1}(t) = i_{A,1}(t) = 0.46$$
$$i^*_{A,2}(t) = i_{A,2}(t) + i_{A,3}(t) + i_{A,5}(t)$$
$$= 0.07 + 0.53 + 0.11 = 0.71$$
$$i^*_{A,3}(t) = i_{A,4}(t) = 0.74$$

We rank the components on the basis of the preceding importance values:

$$H3, H2, H1$$

12.7 COMPUTER PROGRAMS

The computer program IMPORTANCE has the capability to compute the six importance measures for terminal events and the two importance measures for minimal cut sets listed in Section 12.3 (see Appendix I for details).

12.8 DOCUMENTATION

Documentation for the importance analysis should include the following, as appropriate:

1. Minimal cut sets.
2. Terminal event, minimal cut set, and top-event unavailabilities and/or unreliabilities.
3. A list of components and the associated terminal events.
4. Minimal cut set, terminal-event, and/or component importances.
5. Ranking of minimal cut sets, terminal events, and/or components.
6. Assumptions made in the importance analysis.
7. Identification of computer programs and the source of the programs (from whom obtained); include version number and/or version date.
8. Listing of computer input and output.

Some general guidelines on documentation are provided in Section 14.10.

EXERCISE PROBLEMS

12.1. Consider the fault tree shown on the right side of Figure 8.35a. All the terminal events have constant failure rates and constant repair rates. The failure rates and mean time to repair are given in Table 12.1. The terminal events are statistically independent. (i) Determine the minimal cut sets. (ii)

Table 12.1.   Data for Problem 12.1

TERMINAL EVENT	FAILURE RATE (FAILURES PER HOUR)	MTTR (HOURS)
$E1$	10^{-5}	100
$E2$	10^{-6}	10
$E3$	10^{-5}	10
$E4$	5×10^{-6}	40
$E5$	10^{-5}	60

Qualitatively rank the minimal cut sets and the terminal events. (iii) Rank the minimal cut sets and the terminal events according to Vesely–Fussell measure of importance with respect to top-event availability at time $t = 5000$ hours. (iv) Rank the minimal cut sets and the terminal events according to Vesely–Fussell measure of importance with respect to top-event reliability at time $t = 5000$ hours. (v) Rank the minimal cut sets and the terminal events according to Vesely–Fussell measure of importance with respect to top-event availability at time $t = 1000$ hours.

12.2. Consider the fault tree described in Problem 12.1. One at a time, increase the failure rate of each terminal event by a factor of 4 (multiply the failure rates given in Table 12.1 by 4) and determine the corresponding percentage increases in the following: (i) the top-event unavailability at time $t = 5000$ hours, (ii) the top-event unreliability at time $t = 5000$ hours, and (iii) the top-event unavailability at time $t = 1000$ hours. (*Note:* This problem is similar to Example 12.3.)

12.3. Consider the fault tree shown in Figure 9.1. Terminal events $E1$ and $E3$ are failure modes of component $H1$. Terminal events $E2$ and $E4$ are failure modes of component $H2$. The terminal events are nonrepairable, are statistically independent, and have constant failure rates. The failure rates of $E1$, $E2$, $E3$, and $E4$ are λ_1, λ_2, λ_3, and λ_4, respectively. $\lambda_1 = 10^{-5}$ failures per hour, $\lambda_2 = 3 \times 10^{-5}$ failures per hour, $\lambda_3 = 8 \times 10^{-5}$ failures per hour and $\lambda_4 = 10^{-5}$ failures per hour. Rank the minimal cut sets, terminal events, and components according to Vesely–Fussell measure of importance with respect to top-event availability at time $t = 2000$ hours.

REFERENCES

Lambert, H. E. (1975). *Fault Trees for Decision Making in Systems Analysis*. Report No. UCRL-51829. Lawrence Livermore National Laboratory, Livermore, CA.

Lambert, H. E. and B. J. Davis (1980). *The Fault Tree Computer Codes* IMPORTANCE *and* GATE. Lawrence Livermore National Laboratory, Livermore, CA.

Chapter 13
Applications

13.1 INTRODUCTION

Reliability engineering techniques described in the earlier chapters may be applied at any stage during the development, design, manufacture, construction, or operation of engineered systems. This chapter discusses a number of applications. The applications discussed here show how reliability engineering techniques can be used in diverse situations to achieve intended goals in a rational and cost-effective manner. There are yet many more ways in which reliability engineering may be used to improve safety, reliability, availability, performance, and productivity of engineering systems in a cost-effective manner. It is not possible to enumerate and discuss every possible situation where reliability engineering may prove to be helpful. Engineers and engineering managers should keep an open eye to identify opportunities where reliability engineering may be effectively used to improve system design, maintenance, and operation.

In addition to the specific, goal-oriented applications, the very process of conducting a reliability analysis forces the engineers to search for and identify the weaknesses and strengths of the system, and that in itself is useful in improving system performance.

13.2 RELIABILITY ENGINEERING RESULTS

Before discussing specific areas of applications, let us first summarize what types of information and results we get from a system reliability analysis.

1. Qualitative information

   - Potential safety hazards (explosions, toxic chemical release, etc.).
   - Potential component, software, and human failures.
   - Minimal cut sets and minimal path sets.
   - External hazards (lightning, earthquakes, floods, etc.) that can cause component or system failures.
   - Ranking of component, software, and human failures according to their contributions to system-level failures (based on qualitative analyses).

2. Quantitative information

- Component, software, and human failure probabilities.
- External hazard probabilities.
- System availability and reliability.
- Probabilities of accidents and other undesired events.
- Ranking of component, software, and human failures according to their contributions to system-level failures.

Accuracy of the reliability data and results have been discussed in their proper context in Chapters 5, 9, 10, and 11. Engineers who are used to numerical results within 10, 20, or 30% of the exact solution in most other fields of engineering should remember that reliability analysis results are usually accurate at best within an order of magnitude only. As will be indicated in the following sections, even such results could be very useful in many applications.

13.3 RELIABILITY ASSESSMENT

Computation of system reliability, system availability, and accident probabilities is the basic goal of most quantitative reliability analysis projects. An estimate of system reliability (or system availability or accident probabilities) may be required by the owner of the system or by regulatory agencies. Sometimes the buyer may require the seller to meet specific reliability goals; so the seller would have to carry out reliability computations to demonstrate that those reliability goals are met. If frequent system failures occur in an operating plant, the plant owner may want to conduct a reliability investigation to assess system reliability.

Whatever the reason for computing system reliabilities, users of reliability analysis results should keep in mind that the results are accurate at best in an order-of-magnitude sense only. When specific reliability goals have to be met, it should be clearly stated whether the goals should be met at the best-estimate level (as computed using the best available data) or a factor (say, 0.1) should be applied to account for the lack of accuracy. In some very critical projects such as nuclear power plants, it may be advisable to include an uncertainty analysis to determine confidence intervals on the best estimates of system reliability, availability, and accident probabilities (see Section 10.8).

13.4 RELIABILITY IMPROVEMENT

We discuss in this section means of reducing the unreliability (or unavailability or accident probabilities) of systems already designed, built, and/or operating. For one reason or another, the owner or a regulatory agency may require improved system reliability. Reliability improvements in a system already designed but yet to be built are relatively easier to accomplish than reliability improvements in a system already built. Either way, the general concepts and philosophies are the same although the options available may be limited in the latter case.

System reliability may be improved by accomplishing one or more of the following:

1. Improvements in hardware.
2. Improvements in computer control and display software.
3. Improvements in operating procedures.
4. Improvements in maintenance.

Hardware improvements may include the addition of redundancies and/or the strengthening (increasing the reliability) of some components. Hardware improvements are discussed in Section 13.5. Improvements in the reliability of computer software used in control and display (see Section 5.3) may be achieved through additional debugging or by using proven software with a "track record." Software reliability is a rather specialized area, and possible software reliability improvements and costs have to be discussed with software vendors, developers, or consultants.

Because human errors in operation could be a significant contributor to system failures, improvements in operating procedures and/or better training of operators may reduce human errors and thus improve system reliability. Some operating functions may also be automated if the probability of malfunction of the automation is less than the probability of human error. Human errors in design and construction, which affect hardware reliability, may be reduced by implementing stricter quality control in design and construction. Improved maintenance procedures, better training of maintenance personnel, and double-checking of maintenance could reduce maintenance errors and improve system reliability.

The effect of preventive maintenance (periodic maintenance) on system reliability has been discussed in Section 10.5. The cost of preventive maintenance, including the cost of system shutdown for preventive maintenance, and the benefits of improved reliability should be considered in arriving at a cost-effective preventive maintenance procedure and schedule.

13.5 HARDWARE IMPROVEMENTS

System design has already been completed. The system is yet to be built or has already been built. For some reason, the system has to be improved so that its reliability is at least equal to a specified value that is higher than the present system reliability. (*Note:* The methods presented in this section (Section 13.5) are based on the assumption that each component has only one failure mode (terminal event) associated with it. In many problems, components have only one failure mode each, so the methods presented in this section are directly applicable. In some problems, even if a number of failure modes are associated with a component, they may be combined into a single failure mode; methods presented in this section are then applicable. For example, if a pump has two failure modes, namely, 'pump fails to start on demand' and 'pump fails during operation,' they may be combined into a single failure mode, 'pump failure,' in some problems. If some components in the system have multiple failure modes and they may not be combined into a single failure mode, methods presented in this section have to be slightly modified; the general principles are still applicable but the equations used for computing the unreliability of the modified system [for example, Equation (13.2)] need to be changed.)

The first task is to compute the unreliability of the original system, unless this information is already available. Let the original unreliability of the system (the unreliability of the as-designed or as-built system) be U_S, and let the present unreliabilities of its components be u_i, where $i = 1, \ldots, n$, in which n is the number of components. Let the required system reliability (after improvements) be R'_S. So the unreliability of the improved system should not exceed $U'_S = (1 - R'_S)$, where $U'_S < U_S$. System reliability may be improved (i) by adding some component redundancies or (ii) by improving some component reliabilities. These alternatives are discussed in the following subsections.

13.5.1 Addition of Component Redundancies

System reliability may be increased by adding one or more components in parallel to some of the already existing components. Consider the system configuration (block diagram) shown in Figure 13.1a. Failures of components $H1$, $H2$, $H3$, $H4$, and $H5$ are statistically independent. System reliability can be increased by adding one or more components as shown in Figure 13.1b. In this case we have added redundancies to components $H1$ and $H4$. If necessary, we may add more than one redundant component to a single component; for example, we could have added $H1'$ and $H1''$ in parallel to $H1$. The added redundancy may be active redundancy or standby redundancy. The added redundant components may or may not be identical to the existing component; for example, $H1'$ could have identical reliability as $H1$ or a different reliability.

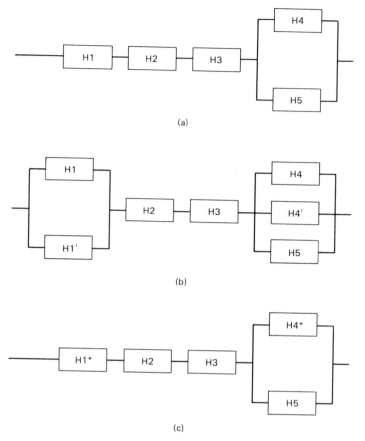

Figure 13.1. Original and modified system configurations: (a) original system; (b) modified system; (c) modified system (alternate form of representation).

The unreliability of the original system is given by

$$U_S = f(u_1, u_2, u_3, u_4, u_5) \tag{13.1}$$

where $f(\cdot)$ is a function that transforms component reliabilities to system reliability. This function may not be known explicitly in closed form; it represents the process of computing the system reliability through a fault tree analysis or other methods of system reliability analysis; Equation (13.3) is an example of $f(\cdot)$.

The modified system configuration may be represented by Figure 13.1c also. In this case, the combined effect of $H1$ and $H1'$ is represented by a single "element" $H1^*$, and the combined effect of $H4$ and $H4'$ is represented by $H4^*$. Comparing Figures 13.1a and 13.1c, it is evident that the form of the block diagram is the same except that $H1$ is replaced by $H1^*$ and $H4$ is replaced by $H4^*$. Because the system block diagram is the same, the geometry (structure) of the corresponding fault trees will also be the same. So the functional relationship $f(\cdot)$ in Equation (13.1) will not change, and the unreliability of the modified system is given by

$$U_S^* = f(u_1^*, u_2, u_3, u_4^*, u_5) \qquad (13.2)$$

where u_1^* represents the unreliability of the parallel system containing $H1$ and $H1'$, and u_4^* represents the unreliability of the parallel system containing $H4$ and $H4'$.

There are numerous ways by which the required system unreliability U_S' may be achieved. The questions before us are (i) To which components shall we add redundancies? and (ii) What should be the reliability of the added redundancy? (In most cases, identical components are added as redundancy, so the unreliability of the redundant component is the same as the unreliability of the original component.)

One method of answering the first question is through an importance ranking of the components. Components are ranked according to the Vesely–Fussell measure of importance (see Section 12.6). The highest-ranked component shall receive priority in adding redundancy. For example, in the system shown in Figure 13.1a, let the importance ranking of the components be as follows:

$$H2, H1, H5, H4, H3$$

First we examine if it is infeasible to add redundancy to any of the components. It may not always be physically possible (feasible) to add redundant components because of space limitations in the system layout; this is particularly true if the system has already been built. Those components for which addition of redundancies is infeasible should be removed from the ranking. Let us say that it is not possible to add redundancies to components $H5$. So the revised ranking is

$$H2, H1, H4, H3$$

If the cost of adding a redundancy is approximately the same for all of these components, the obvious choice is to add redundancy to $H2$, which is ranked first. If the system reliability requirement is still not met after adding redundancy to $H2$, redundancies are added to $H1$, $H4$, and $H3$, in that order, until the required system reliability is achieved. In the preceding discussion, the term "cost" includes not only the hardware cost of the added redundant component but also the additional expenses in operation and maintenance over the life of the system. The proper method is to transform the expected additional operational and maintenance expenses over the life of the system to its *present worth* and add it to the hardware cost. Methods of calculating the present worth can be found in books on engineering economics, for example, Grant, Ireson, and Leavenworth (1982).

If the cost of adding redundancies is substantially different for the different components, cost should be considered in prioritizing the components.

Suppose we are interested in improving the reliability of a system consisting of n components $H1, H2, \ldots, Hn$. The cost of adding a redundant component is different for the different components. We may consider the different alternatives (add redundancy to $H1$, add redundancy to $H2$, etc.), the effect of each alternative on system reliability, and the cost of each alternative. The alternative that provides the required system reliability and has the lowest cost is the best choice.

Although this procedure is straightforward, the amount of computation involved could be considerable if the number of components (n) is large. Component importance ranking may be used as an aid in selecting the most attractive alternatives for further study. Let $n = 10$, and let the Vesely–Fussell measure of importance for the jth components be i_j^*; the Vesely–Fussell measure is with respect to system reliability (or availability as appropriate) at time T, where T is the mission duration, the interval between periodic maintenances, or some other suitable period of time.

$$i_1^* = 0.07 \qquad i_2^* = 0.24 \qquad i_3^* = 0.01 \qquad i_4^* = 0.04 \qquad i_5^* = 0.89$$

$$i_6^* = 0.16 \qquad i_7^* = 0.08 \qquad i_8^* = 0.01 \qquad i_9^* = 0.03 \qquad i_{10}^* = 0.32$$

The importance measures of $H1$, $H3$, $H4$, $H7$, $H8$, and $H9$ are very small compared to the others, and we may drop these components as possible candidates for adding redundancy. We are left with $H2$, $H5$, $H6$, and $H10$. We may compute the effect of adding redundancy to each of these components and the cost of each such alternative. We will then select the least-cost alternative that meets the system reliability requirements. In this example, we are able to reduce the alternatives from 10 to 4, and thus are able to reduce the necessary computations by over 50%. We may expect even higher reductions if the number of components is larger.

Example 13.1

A system is divided into a series of three components, as shown in Figure 13.2a. Each component has only the failure mode (terminal event) associated with it. The unavailabilities of components $H1$, $H2$, and $H3$ at time $t = 8000$ hours are $q_1 = 0.001$, $q_2 = 0.04$, and $q_3 = 0.03$. The component failure are statistically independent. The cost of adding an identical, active parallel component redundancy to $H1$, $H2$, and $H3$ are approximately the same. The required system availability at time $t = 8000$ hours is 0.995. Develop a new system configuration by adding suitable active parallel redundancies. (*Note:* It is possible to add redundancy to all three components.)

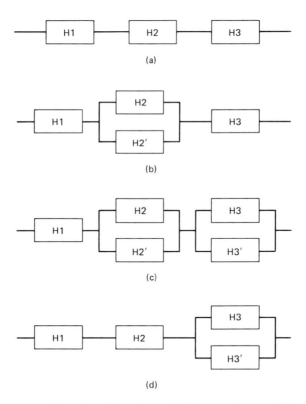

Figure 13.2. System configurations (Example 13.1): (a) original system; (b) modified system with redundancy added to $H2$; (c) modified system with redundancies added to $H2$ and $H3$; (d) modified system with redundancy added to $H3$.

Solution

The required system availability = 0.995. So, acceptable system unavailability = 0.005. Any system with an unavailability less than or equal to 0.005 is acceptable.

First, we calculate the unavailability of the present system (Figure 13.2a). The fault tree for this system is shown in Figure 13.3. The minimal cut sets are

$$M1 = E1 \qquad M2 = E2 \qquad M3 = E3$$

The system unavailability is given by

$$Q_S = f(q_1, q_2, q_3) = q_1 + q_2 + q_3 = 0.001 + 0.04 + 0.03$$
$$= 0.071 \tag{13.3}$$

This unavailability is greater than 0.005, so the present system is unacceptable.

The Vesely–Fussell measures of component importance with respect to system availability at $t = 8000$ hours are

$$i_{A,1}^* = \frac{Q_1}{Q_S} = \frac{q_1}{Q_S} = 0.014$$

$$i_{A,2}^* = \frac{Q_2}{Q_S} = \frac{q_2}{Q_S} = 0.56$$

$$i_{A,3}^* = \frac{Q_3}{Q_S} = \frac{q_3}{Q_S} = 0.42$$

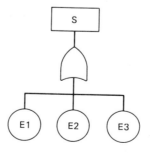

Figure 13.3. Fault tree for system shown in Figure 13.2a.

So the Vesely–Fussell importance ranking is

$$H2, H3, H1$$

Because the cost of adding redundancy is approximately the same for all three components, consider the effect of adding redundancy in the same order as the Vesely–Fussell importance ranking (that is, in the order of $H2$, $H3$, $H1$).

If we add an identical, active parallel component to $H2$ (see Figure 13.2b), we have

$$q_2^* = q_2 \cdot q_2 = 0.04 \times 0.04 = 0.0016$$

where q_2^* is the unavailability of the parallel redundant unit consisting of $H2$ and $H2'$. The unavailability of the modified system (Figure 13.2b) is

$$Q_S^* = q_1 + q_2^* + q_3 = 0.001 + 0.0016 + 0.03$$
$$= 0.0326$$

The unavailability of this system is greater than 0.005, so this modification is unacceptable.

Let us add a redundant component in parallel to $H3$ also. The resulting system configuration is shown in Figure 13.2c.

$$q_3^* = q_3 \cdot q_3 = 0.03 \times 0.03 = 0.0009$$
$$Q_S^* = q_1 + q_2^* + q_3^* = 0.001 + 0.0016 + 0.0009 = 0.0035$$

This unavailability is acceptable. So the modified system configuration shown in Figure 13.2c is recommended.

Example 13.2

Consider the same system described in Example 13.1 (Figure 13.2a). The costs of adding active parallel redundancy to $H1$, $H2$, and $H3$ are $C1$, $C2$, and $C3$, respectively; $C1 = \$65,000$, $C2 = \$80,000$, and $C3 = \$50,000$. The required system availability is 0.95. Develop a new system configuration by adding suitable active parallel redundancies. (*Note:* It is possible to add redundancy to all three components.)

Solution

The present system (original system) unavailability and component ranking are the same as those computed in Example 13.1. That is, the unavailability of the present system = 0.071. The component ranking is $H2$, $H3$, $H1$.

The required system availability = 0.95. So, acceptable system unavailability = 0.05.

First, let us consider the effect and cost of adding an identical component $H2'$ in parallel to $H2$ (see Figure 13.2b). The unavailability of the modified system (Q_S^*) has already been computed in Example 13.1:

$$Q_S^* = 0.0326$$

This is less than the acceptable unavailability of 0.05, so this modification is acceptable. The cost of this modification is \$80,000.

Let us consider another alternative. We add an identical component $H3'$ in parallel to $H3$ (see Figure 13.2d. The unavailability of the unit consisting of $H3$ and $H3'$ in parallel is

$$q_3^* = q_3 \cdot q_3 = 0.03 \times 0.03 = 0.0009$$

The unavailability of the modified system is

$$Q_S^* = q_1 + q_2 + q_3^* = 0.001 + 0.04 + 0.0009 = 0.0419$$

This is less than the acceptable unavailability of 0.05. So this modification is also acceptable. The cost of this modification is \$50,000.

We need not consider the third alternative of adding an identical component parallel to $H1$ because the cost of such an addition is higher than the cost of the modification discussed in the preceding paragraph.

So the preferred modification is as shown in Figure 13.2d.

13.5.2 Component Reliability Improvements

Component reliability can be increased by either strengthening the component against its failure mechanisms or by replacing the original component by a higher-reliability component.

Let u_i and u_i^* be the unreliabilities of the ith component before and after component reliability improvements. We have

$$u_i^* \leq u_i \qquad i = 1, 2, \ldots, n \qquad (13.4)$$

where n is the number of components in the system. The inequality in Equation (13.4) should hold for at least one component; otherwise, there will be no change in system unreliability.

The unreliability of the original system is

$$U_S = f(u_1, u_2, \ldots, u_n) \qquad (13.5)$$

where $f(\cdot)$ is as defined under Equation (13.1). System unreliability after component improvements is

$$U_S^* = f(u_1^*, u_2^*, \ldots, u_n^*) \tag{13.6}$$

As long as Equation (13.4) is satisfied by all pairs of u_i and u_i^* and the inequality in Equation (13.4) is satisfies by at least one of the pairs, we will have $U_S^* < U_S$. Thus the reliability of the modified system is greater than that of the original system. The u_i^* should be such that the reliability of the modified system $R_S^* = (1 - U_S^*)$ is greater than or equal to the required system reliability.

There are numerous combinations of u_i^* values that will yield the necessary system reliability. We have to choose one such combination. How do we do that? Ideally, we should consider the cost of each improvement (the cost of changing each component unavailability from u_i to u_i^*) and choose that combination of u_i^* that provides the required reliability at the lowest cost of component improvements. As discussed in Section 13.5.1, the cost of improvements should include not only hardware costs but also added expenses of maintenance and operation, if any. The cost of improving the ith component will usually be a monotonically increasing function of $(u_i^* - u_i)$. That is, if the cost of improving the ith component is given by

$$C = g_i(u_i^* - u_i) \tag{13.7}$$

then $g_i(x)$ is a monotonically increasing function of x. This function could be linear or nonlinear. We may not always have a smooth cost function; it could be a step function in some cases because the reliability of some components could be increased only in steps (see Example 13.3).

In most plant design applications, formal optimization techniques are not used to determine the lowest-cost combination of u_i^* that provides the required reliability, because of the complexity of solving the optimization problem and the lack of precise data on component reliabilities and component costs as a function of reliability. Usually a trial-and-error procedure is used (see Example 13.3). We may not get the optimal modification but, when properly carried out, we should get a system modification that is near-optimum. (*Note:* A brief discussion of reliability optimization is provided in Section 13.7.)

Example 13.3

The top event (T) of the fault tree shown in Figure 13.4 represents a catastrophic accident. The component failure modes (terminal events) $E1$, $E2$, $E3$, and $E4$ are statistically independent and nonrepairable. The

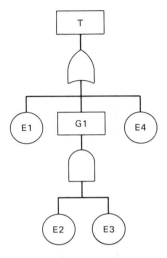

Figure 13.4. Fault tree for Example 13.3.

probabilities (= unreliabilities = unavailabilities) of $E1$, $E2$, $E3$, and $E4$ at time $t = 12,000$ hours are $P[E1] = 0.001$, $P[E2] = 0.09$, $P[E3] = 0.05$, and $P[E4] = 0.0002$. The cost of reducing these probabilities is shown in Figure 13.5. Maximum possible reduction in the probabilities of $E1$, $E2$, $E3$, and $E4$ are 60, 50, 60, and 30%, respectively. Determine a strategy to decrease the top-event probability (= top-event unreliability = top-event unavailability) to at least 0.004 by decreasing the probabilities of one or more failure modes. Space limitations rule out the addition of redundant components.

Solution

First, we calculate the top-event probability. The minimal cut sets are

$$M1 = (E1)$$
$$M2 = (E2, E3)$$
$$M3 = (E4)$$

The minimal cut set probabilities are

$$P[M1] = P[E1] = 0.001$$
$$P[M2] = P[E2]P[E3] = 0.09 \times 0.05 = 0.0045$$
$$P[M3] = P[E4] = 0.0002$$

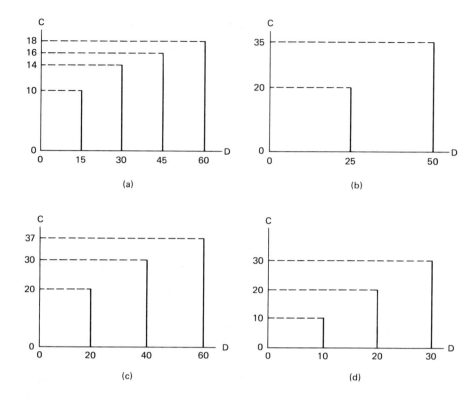

C — Cost (in thousands of dollars).
D — Percentage decrease in failure probability.

Figure 13.5. Cost of reliability improvement (Example 13.3): (a) failure mode $E1$; (b) failure mode $E2$; (c) failure mode $E3$; (d) failure mode $E4$.

We calculate the top-event probability using the small probability approximation.

$$P[T] = P[M1] + P[M2] + P[M3] = 0.001 + 0.0045 + 0.0002$$

$$= 0.0057$$

The Vesely–Fussell measures of importance of the terminal events with respect to top-event probability (top-event unavailability) at $t = 12,000$

hours are

$$i_{A,1} = \frac{0.001}{0.0057} = 0.18$$

$$i_{A,2} = \frac{0.0045}{0.0057} = 0.80$$

$$i_{A,3} = \frac{0.0045}{0.0057} = 0.80$$

$$i_{A,4} = \frac{0.0002}{0.0057} = 0.04$$

Importance ranking of the terminal events is

$$(E2, E3, E1, E4) \quad \text{or} \quad (E3, E2, E1, E4)$$

The importance of $E4$ is very small compared to $E1$, $E2$, and $E3$, and the cost of decreasing the failure probability of $E4$ is not much smaller than the corresponding costs for the other terminal events. So we drop $E4$ as a possible candidate for improvement.

Let us consider reducing the failure probability of $E2$ by 50%. The revised terminal-event probability is

$$P[E2^*] = 0.045$$

The revised probability of $M2$ is

$$P[M2^*] = P[E2^*]P[E3] = 0.045 \times 0.05 = 0.00225$$

There is no change in the probabilities of $M1$ and $M3$. So,

$$P[T^*] = P[M1] + P[M2^*] + P[M3]$$
$$= 0.001 + 0.00225 + 0.0002$$
$$= 0.00345$$

This is less than the acceptable probability of 0.004. So this modification is acceptable. The cost is \$35,000 (see Figure 13.5).

Let us consider reducing the failure probability of $E3$ by 60%. Revised values of $P[E3*]$, $P[M2*]$, and $P[T*]$ are

$$P[E3*] = 0.05\left[1 - \tfrac{60}{100}\right] = 0.02$$
$$P[M2*] = 0.09 \times 0.02 = 0.0018$$
$$P[T*] = 0.001 + 0.0018 + 0.0002$$
$$= 0.003$$

This is less than the acceptable top-event probability, so this modification is acceptable. The cost is $37,000 (see Figure 13.5).

Let us see if we may decrease the failure probability of $E3$ by 40% and still achieve the required top-event probability. If we decrease $P[E3]$ by 40%, we get

$$P[E3*] = 0.03$$
$$P[M2*] = 0.09 \times 0.03 = 0.0027$$
$$P[T*] = 0.001 + 0.0027 + 0.0002$$
$$= 0.0039$$

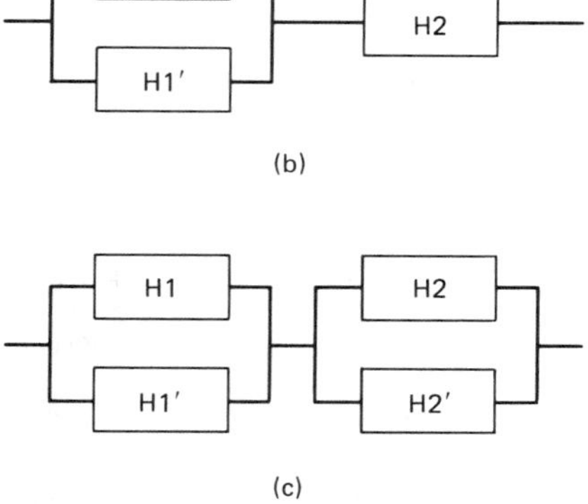

(a)

(b)

(c)

Figure 13.6. System configurations (Example 13.4): (a) original system; (b) modified system with redundancy added to $H1$; (c) modified system with redundancies added to $H1$ and $H2$.

This is less than the acceptable top-event probability of 0.004, so a 40% reduction in the failure probability of $E3$ is acceptable. The cost is $30,000 (see Figure 13.5).

Even the maximum possible reduction of 60% in the failure probability of $E1$ will not achieve the required top-event probability. So we do not consider this alternative.

On the basis of the preceding calculations, we recommend that the failure probability of $E3$ be reduced by 40% at a cost of $30,000.

13.5.3 Combination Approach

When both component reliability improvements and additions of redundant components are possible, we may consider (i) component reliability improvements, (ii) addition of redundant components, and (iii) combinations of both, and recommend the most economical alternative.

Example 13.4

Consider the system configuration shown in Figure 13.6a. The availabilities of components $H1$ and $H2$ are q_1 and q_2, respectively; $q_1 = 0.02$ and $q_2 = 0.03$. Component failures are statistically independent. Improve the system availability to 0.98. Component availability improvements and addition of redundant components are possible for both $H1$ and $H2$. The cost of adding an identical active redundant component to $H1$ is $240,000, and the cost of adding an identical active redundant component to $H2$ is

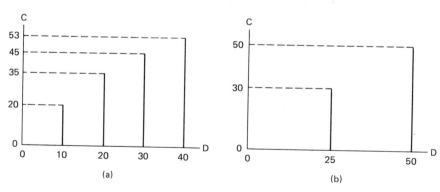

C = Cost (in thousands of dollars).
D = Percentage decrease in unavailability.

Figure 13.7. Cost of availability improvement (Example 13.4): (a) component $H1$; (b) component $H2$.

$390,000. The cost of component availability improvement is shown in Figure 13.7. Maximum possible reductions in the unavailabilities of $H1$ and $H2$ are 40 and 50%, respectively.

Solution

The required system availability = 0.98. So, the acceptable level of system unavailability = $1 - 0.98 = 0.02$.

The unavailability of the present system (before any modifications) is

$$Q_S = q_1 + q_2 = 0.02 + 0.03 = 0.05 \qquad (13.8)$$

Through Vesely–Fussell importance measures, we can show that component ranking is $H1$, $H2$.

So, let the first consider the addition of an identical active redundant component in parallel to $H1$ (Figure 13.5b). Considering $H1$ and $H1'$ (in parallel) as a single unit, the unavailability of this unit is

$$q_1^* = q_1 \cdot q_1 = 0.02 \times 0.02 = 0.0004$$

The unavailability of the modified system is obtained by replacing q_1 by q_1^* in Equation (13.8).

$$Q_S^* = q_1^* + q_2 = 0.0004 + 0.03 = 0.0304$$

This value is higher than the acceptable level of 0.02.

Because the Vesely–Fussell importance measure of $H2$ is less than that of $H1$, addition of an identical redundant component to $H2$ will not yield a system unavailability lower than 0.0304. So, let us consider the case of adding redundancy to both $H1$ and $H2$ (see Figure 13.6c).

$$q_2^* = q_2 \cdot q_2 = 0.03 \times 0.03 = 0.0009$$
$$Q_S^* = q_1^* + q_2^* = 0.0004 + 0.0009$$
$$= 0.0013$$

This is lower than the acceptable system unavailability of 0.02, so it is acceptable.

Cost of the modification = cost of adding $H1'$ + cost of adding $H2'$
$$= \$240,000 + \$390,000$$
$$= \$630,000$$

Let us now consider another alternative. Add a redundant component $H1'$ in parallel to $H1$, and improve the availability of $H2$. Let us reduce the availability of $H2$ by 50%. We have

$$q_1^* = 0.0004 \text{ (as calculated before)}$$
$$q_2^* = 0.03\left[1 - \tfrac{50}{100}\right] = 0.015$$

The unavailability of the modified system is

$$Q_S^* = q_1^* + q_2^* = 0.0004 + 0.015 = 0.0154$$

This unavailability is less than the acceptable level, so the modification is acceptable.

Cost of the modification = cost of adding $H1'$

+ cost of reducing $H2$ unavailability by 50%

= $240,000 + $50,000

= $290,000

There is yet another alternative: add a redundant component in parallel to $H2$ and reduce the unavailability of $H1$. The cost of such a modification will be greater than $290,000 because the cost of adding a redundant component to $H2$ alone is $390,000. So we do not explore this alternative further.

Another alternative is to reduce the unavailabilities of $H1$ and $H2$ (component improvements). The maximum reduction possible in $H1$ unavailability is 40% and the maximum reduction possible in $H2$ is 50%. System unavailability calculations will show that these component unavailability reductions will not reduce the system unavailability to 0.02 or less. So this alternative is unacceptable.

Based on the preceding calculations, we recommend that an identical, active redundant component be added in parallel to $H1$ and the unavailability of $H2$ be reduced by 50%. The cost of such a modification is $290,000. The unavailability of the modified system is 0.0154.

13.6 DESIGN TO RELIABILITY

Reliability assessment becomes an integral part of the design process if specific reliability goals are to be met during system design. Meeting specific reliability goals may require an iterative process in system design. The cycle of iteration would be preliminary design, to reliability assessment, to design changes, to reliability assessment, to design changes, to reliability assessment, ... until the reliability goals are met. We shall discuss one way of integrating specific reliability goals into the design. (The *Note* at the end of the fist paragraph of Section 13.5 regarding components with single and multiple failure modes is valid for this section also.)

First, a conceptual design of the system is made which indicates the components and their arrangements (system configuration). Because system reliability is a function of component reliabilities, our aim is to allocate a specific reliability to each component so that the overall system reliability is equal to or greater than the required system reliability.

Let there be n components and let the component unreliabilities be u_i, where $i = 1, \ldots, n$. Let the required system reliability be R'_S. The corresponding system unreliability is $U'_S = (1 - R'_S)$. Component unreliabilities u_i, which are yet to be determined, must satisfy the following condition:

$$U'_S \geq f(u_1, u_2, \ldots, u_n) \qquad (13.9)$$

where $f(\cdot)$ is as defined under Equation (13.1).

There are numerous combinations of u_i that can satisfy Equation (13.9). We shall determine a practically feasible and cost-effective combination.

The process of determining u_i is called *reliability allocation*. As we noted earlier, there are numerous combinations of u_i that can satisfy Equation (13.9). Each such acceptable combination has a cost associated with it because the cost of each component is a function of its reliability (the higher the reliability, the higher the cost). There are formal *reliability optimization techniques* which will find the most economical combination of u_i values that will satisfy Equation (13.9) at minimum cost. Reliability optimization is discussed in Section 13.7. Formal reliability optimization techniques are rarely used in plant system design because of the complexity of the procedure and the lack of precise data on component reliabilities and component costs as a function of reliability. Rather, a trial-and-error procedure is used in reliability allocation. The aim is to allocate reliabilities so that system reliability requirements are met and the cost is as low as possible (it may not be the absolute minimum as in formal optimization procedures). The cost of a system includes not only the cost of building a system (hardware costs) but also operating and maintenance expenses during the life of the system.

The trial-and-error procedure of reliability allocation consists of four steps.

Step 1

On the basis of what level of reliability may reasonably be achieved for each component, assign a reliability to each component. This may be based on previous experience with such components.

Step 2

Compute the system reliability (use fault tree analysis, if necessary).

Step 3

If the system reliability, as computed in Step 2, is equal to or greater than the required reliability, we have successfully completed the reliability allocation. If not, go to Step 4.

Step 4

Because the required system reliability is not achieved, improvements need to be made. The reliability improvement techniques discussed in Section 13.5 may be used here. As discussed in that section, add redundant components and/or improve component reliabilities so that system reliability goals are met.

13.7 RELIABILITY OPTIMIZATION

Formal reliability optimization techniques can be used in reliability improvement (Section 13.5) and reliability allocation (Section 13.6).

Let the system be composed of n components, and let the unreliability of the ith component be u_i, where $i = 1, 2, \ldots, n$. Let the required system reliability be R'_S. The corresponding system unreliability is $U'_S = (1 - R'_S)$. System unreliability should be equal to or less than this value.

The cost of each component is a function of its reliability. Let $C_i(u_i)$ be the cost of the ith component at an unreliability of u_i. The notation u_i within brackets indicates that C_i is a function of u_i.

The mathematical problem of reliability optimization is to minimize the lifetime cost of the system while achieving the required system reliability. As discussed in Section 13.5, lifetime cost includes not only initial hardware and installation costs but also lifetime maintenance and operating expenses. (We may choose to consider only the initial hardware and installation costs if we wish to optimize only the initial cost.)

The problem is posed mathematically as follows: Determine the values of u_i so that

$$\sum_{i=1}^{n} C_i(u_i) = \text{minimum} \tag{13.10}$$

subject to the constraint that

$$U'_S \geq U_s = f(u_1, u_2, \ldots, u_n) \tag{13.11}$$

where $f(\cdot)$ is as defined under Equation (13.1).

A corollary to the preceding optimization problem may also be posed. Here, we have a fixed amount of money and we wish to design a system that has the highest possibility reliability at that budget. The problem may be posed as follows: Determine the values of u_i so that

$$U_S = f(u_1, u_2, \ldots, u_n) = \text{maximum} \qquad (13.12)$$

subject to the constraint that

$$\sum_{i=1}^{n} C_i(u_i) = B \qquad (13.13)$$

where B is the available money (budget).

A number of optimization techniques are available to solve these problems [Equations (13.10) and (13.11) or Equations (13.12) and (13.13)]. Some of the available methods are:

1. Dynamic programming
2. Geometric programming
3. Linear programming
4. Discrete maximum principle
5. Sequential unconstrained minimization technique
6. Generalized reduced gradient method
7. Lagrangian multipliers method
8. Generalized Lagrangian functions method

We will not discuss these methods here. Interested readers are referred to Tillman, Hwang, and Kuo (1980).

These optimization techniques are seldom used in plant systems design because of the mathematical complexity of the procedures and, more important, the lack of precise data on component reliabilities and component costs as a function of their reliabilities. The trial-and-error methods are discussed in Sections 13.5 and 13.6 are often used. Although a trial-and-error procedure may not provide an "exactly" optimized system, a trial-and-error procedure should provide cost-effective systems that are close to the optimum.

13.8 OPTIMAL SENSOR LOCATIONS

Consider a fault tree whose top event (T) is either 'system failure' or an 'accident.' Let there be six terminal events and let the minimal cut sets be

$$M1 = (E6)$$
$$M2 = (E2, E3)$$
$$M3 = (E2, E1)$$
$$M4 = (E3, E4, E5)$$

The top event can occur if and only if any one of these minimal cut sets occurs (if any one of the minimal cut sets fails). A minimal cut set is said to have occurred if all the terminal events contained in the minimal cut sets have occurred. Because $E6$ is part of a first-order minimal cut set, the occurrence of $E6$ results in the top event. Because $E1$, $E2$, $E3$, $E4$, and $E5$ are parts of second- and higher-order minimal cut sets, the occurrence of just one of them does not result in the top event. For example, the occurrence of $E2$ does not result in the top event unless either $E3$ (which is contained along with $E2$ in $M2$) or $E7$ (which is contained along with $E2$ in $M3$) has already occurred. If $E2$ occurs and we know that it has occurred, we could possibly repair (or replace) it before either $E3$ or $E7$ occurs and thus prevent a possible top-event occurrence. In other words, a top-event occurrence could possibly be prevented if the repair time of $E2$ is less than the time between the occurrence of $E2$ and the occurrence of either $E3$ or $E7$, and $E2$ can be repaired while the system is operating. (Although the mean time to repair $E2$ is known, the time between an $E2$ occurrence and the occurrence of either $E3$ or $E7$ is a random variable and is not known.) Still, an attempt to repair $E2$ as soon as it occurs is usually a wise decision. The key is to show that $E2$ has occurred soon after the occurrence of $E2$. Because $E2$ alone does not result in the top event, how do we know that $E2$ has occurred (unless it is visible to the operator)? One method is to place a sensor or alarm that will alert the operator as soon as $E2$ has occurred. (Such a provision may not always be possible but it should be possible with at least some terminal events in large systems.) Provision of alarms or sensors to each and every possible terminal event (component) may be very expensive if there are a large number of terminal events (components). Can we prioritize the terminal events so that sensors are placed for the top-ranked terminal events only? What are the criteria for prioritizing? Terminal events that have the greatest tendency of occurring *prior to* the top-event occurrence should be ranked the highest. Terminal events such as $E6$ (member of a first-order minimal cut set) that occur almost simultaneously with the top event should get the lowest ranking because they do not occur prior to the top event but occur almost simultaneously with the terminal event. There is no need to have a sensor for such terminal events because we cannot prevent top-event occurrence once such terminal events have occurred.

Consider the minimal cut set $M2$. This minimal cut set can occur in two ways:

1. $E2$ occurs first and then $E3$.
2. $E3$ occurs first and then $E2$.

$E2$ occurs prior to the top event in the first case, and $E2$ occurs almost simultaneously with the top event in the second case. (We are neglecting the situation where both $E2$ and $E3$ occur simultaneously; the probability of such situations is very small if $E2$ and $E3$ are statistically independent.) Consider the minimal cut set $M4$. The possible ways in which $M4$ can occur include:

1. $E3$ occurs prior to $E4$ or $E5$ or $E4$ and $E5$.
2. $E3$ occurs after $E4$ and $E5$.

$E3$ occurs prior to the top event in the first case, and $E3$ occurs almost simultaneously with the top event in the second case.

We are interested in prioritizing the terminal events according to their tendency to occur prior to the top event. The terminal event with the greatest tendency is ranked first, the terminal event with the next highest tendency is ranked second, and so on. According to Lambert (1975), the *sequential contributory importance* (see Section 12.3) provides the basis for such a ranking. The sequential contributory importance can be computed using the computer program IMPORTANCE (see Appendix I). (*Note:* The sequential contributory importance of the ith terminal event at time t is defined as the expected number of top-event occurrences during the time interval from 0 to t, caused by the occurrences of the minimal cut sets containing the ith terminal event, with the ith terminal event occurring prior to the top event [Lambert (1975)].)

Example 13.5

There are six terminal events $(E1, E2, \ldots, E6)$ in a fault tree. Their sequential contributory measures of importance at time $t = 7200$ hours have been computed using the computer program IMPORTANCE [Lambert and Davis (1980)]. Results are $I1 = 0.04$, $I2 = 0.07$, $I3 = 0.11$, $I4 = 0.006$, $I5 = 0.14$, and $I6 = 0.09$, where I1, $I2$, $I3$, $I4$, $I5$, and $I6$ are the sequential contributory measures of importance of terminal events $E1$, $E2$, $E3$, $E4$, $E5$, and $E6$, respectively. It is possible to place alarms for all of the terminal events except for $E3$. The cost of placing the alarms is approximately the same for those terminal events. We have funds for only three alarms. Determine the terminal events for which alarms should be provided.

Solution

First we rank the terminal events according to their sequential contributory measure of importance:

$$E5, E3, E6, E2, E1, E4$$

It is not possible to place an alarm for $E3$, so we remove it from the ranking. The revised ranking is

$$E5, E6, E2, E1, E4$$

We have funds for only three alarms, so we recommend that alarms be provided for $E5$, $E6$, and $E2$.

13.9 POSTFAILURE INSPECTION STRATEGIES

The development of postfailure inspection strategies does require some system reliability computations. Much or all of the computations may be carried out a priori as a contingency measure and kept on file or stored in a computer disk. The rest of the computations may be carried out on-site (after a failure) if the system fault tree and reliability data can be accessed from an on-site computer terminal.

We shall first discuss *standby-system* failures and then *operating-system* failures. A system is called a standby system if the system is in a standby mode (idle mode) and is required to operate only on demand. For example, a fire-extinguisher system in an industrial facility is normally in a standby mode and is required to operate only when there is a fire. Standby redundant systems are another example. If the operation of a system is very critical to safety, two such systems may be built and one is kept in standby mode as a redundant system (backup system); it will be required to operate only if the other system (the operating system) fails.

13.9.1 Standby Systems

Components in standby systems can fail even when the system is in the standby mode. Standby systems are tested at regular intervals in order to assure that they are in operational readiness. If a system is found to have failed, we need to inspect the various components and repair those that have failed. The repair crew may repair all of the field components immediately or may choose to repair first only a sufficient number of components to make the system operational; the remaining failed components may be repaired later. Consider, for example, the system shown in Figure 13.8. If we find that components $H1$, $H2$, $H3$, and $H5$ have failed, it is not necessary to repair all four of these components to restore the system to operational readiness; it is sufficient to repair either $H1$ or $H2$. The other three components may be repaired later.

The post failure inspection scenario we pose is as follows: we find that the standby system is inoperational either during the periodic testing or when it was switched on-line because it was needed. There are n compo-

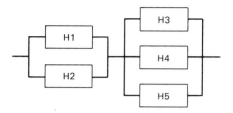

Figure 13.8. System configuration.

nents in the system and we do not know which components have failed. What is the order in which the components should be inspected in order to identify the failed components as quickly as possible?

A rational postfailure inspection procedure is described here. (The procedure is illustrated in Example 13.6.) It consists of seven steps.

Step 1

Rank the terminal events (component failure modes) according to the Vesely–Fussell measure of importance.

Step 2

Inspect the terminal events (components associated with the terminal events) in the order of ranking until a failed terminal event is found. Let the terminal event be the ith-ranked terminal event. (*Note:* Terminal events ranked higher have been inspected and found to have not failed.)

Step 3

If the ith-ranked terminal event is part of a first-order minimal cut set, that first-order minimal cut set is the cause for system failure. Repair of the ith terminal event will restore the system to operational readiness (if there are no other failed minimal cut sets). The other terminal events may be inspected after the ith-ranked terminal event is repaired or while the ith terminal event is being repaired.

Step 4

If the ith-ranked terminal event does not belong to a first-order minimal cut set, develop a *conditional sublist* for further inspection. (If the ith-ranked terminal event has failed, we shall not go to the $(i + 1)$th-ranked terminal event next; rather, we will inspect according to the conditional sublist.)

Step 5

Procedure for generating the sublist on the condition that the ith-ranked terminal event has failed is as follows: Because the ith-ranked terminal event has occurred, set its probability of occurrence as equal to 1, and rank all the minimal cut sets that contain the ith-ranked terminal event, according to the Vesely–Fussell measure of importance. Terminal events in the minimal cut set thus ranked first shall be inspected first according to their probabilities (higher-probability terminal events in that minimal cut set are inspected first).

Step 6

If all the terminal events in that minimal cut set have not failed, we shall inspect the terminal events in the second-ranked minimal cut set, then the third, etc., until all the minimal cut sets containing the ith-ranked terminal event are inspected. The minimal cut set in which all the terminal events have failed is the cause for the system failure (if there is no other failed minimal cut sets). Repair at least one terminal event in the minimal cut set and restore the system to operational readiness. There could be other failed terminal events also. So the remaining terminal events could be inspected and repaired as the system is being restored to operational readiness or later.

Step 7

If the system does not become operational even after the failed minimal cut set is repaired, what shall we do? It is possible that there are additional failed minimal cut sets. Such minimal cut sets should be identified using the same procedure described in Steps 1–6. The system will become operational once all the failed minimal cut sets are repaired.

Example 13.6

A standby system was routinely inspected at $t = 8000$ hours and found to be inoperational. The minimal cut sets of the fault tree representing 'system failure' are

$$M1 = (E4)$$

$$M2 = (E1, E3)$$

$$M3 = (E2, E3)$$

$$M4 = (E3, E5, E6)$$

The component unavailabilities at $t = 8000$ hours are

$$q_1 = 0.01 \qquad q_2 = 0.02 \qquad q_3 = 0.01$$

$$q_4 = 0.01 \qquad q_5 = 0.09 \qquad q_6 = 0.1$$

All the terminal events are statistically independent.

Develop the order in which the terminal events should be inspected. Consider a scenario in which $E1$, $E3$, $E5$, and $E6$ have failed; of course, we do not know of these failures until we inspect them.

Solution

The unavailabilities of the minimal cut sets are computed first.

$$P[M1] = P[E4] = 0.01$$

$$P[M2] = P[E1]P[E3] = 0.01 \times 0.01 = 0.0001$$

$$P[M3] = P[E2]P[E3] = 0.02 \times 0.01 = 0.0002$$

$$P[M4] = P[E3]P[E5]P[E6] = 0.01 \times 0.09 \times 0.1 = 0.00009$$

where $P[X]$ represents the unavailability of X.

The system unavailability $P[S]$ is computed using the small probability approximation.

$$P[S] = P[M1] + P[M2] + P[M3] + P[M4]$$

$$= 0.01 + 0.0001 + 0.0002 + 0.00009$$

$$= 0.01039$$

The Vesely–Fussell measures of importance of the terminal events with respect to system unavailability at $t = 8000$ hours are computed as follows:

$$i_{A,1} = \frac{0.0001}{0.01039} = 0.01$$

$$i_{A,2} = \frac{0.0002}{0.01039} = 0.02$$

$$i_{A,3} = \frac{0.0001 + 0.0002 + 0.00009}{0.01039} = 0.04$$

$$i_{A,4} = \frac{0.01}{0.01039} = 0.96$$

$$i_{A,5} = \frac{0.00009}{0.01039} = 0.009$$

$$i_{A,6} = \frac{0.00009}{0.01039} = 0.009$$

Although $E5$ and $E6$ have the same value of importance, we may rank $E6$ before $E5$ because it has a higher probability. So the ranking is

$$E4, E3, E2, E1, E6, E5$$

and we inspect the terminal events in the same order.

We inspect $E4$ first and it has not failed. Next we inspect $E3$ and we find that it has failed.

We may repair $E3$ and continue with our inspection. What component shall we inspect after $E3$? We shall not inspect $E2$ even though it is ranked next to $E3$. Instead, we shall generate a conditional sublist.

$E3$ is a member of minimal cut sets $M2$, $M3$, and $M4$. So we rank these minimal cut sets after setting $P[E3] = 1$.

$$P[M2] = P[E1]P[E3] = 0.01 \times 1 = 0.01$$

$$P[M3] = P[E2]P[E3] = 0.02 \times 1 = 0.02$$

$$P[M4] = P[E3]P[E5]P[E6] = 1 \times 0.09 \times 0.1 = 0.009$$

So the ranking of $M2$, $M3$, and $M4$ according to the Vesely–Fussell measure of importance with respect to system availability at $t = 8000$ hours is

$$M3, M2, M4$$

There is just one more terminal event (other than $E3$) in $M3$. This is $E2$. So $E2$ is inspected and found to have not failed.

Next we go to $M2$. Here we have $E1$. So $E1$ is inspected and we find that it is failed. Now we have identified a minimal cut set, namely, $M2$, in which all the terminal events have failed. We start repair on $E1$ and continue with the inspection of the other terminal events in the order of the original ranking.

Going back to the original ranking of the terminal events, the third-ranked terminal event is $E2$. We have already inspected it as part of the sublist for $E3$, and we have found that it has not failed. So we go to the fourth-ranked terminal event $E1$. We have already inspected it and found that it has failed. $E1$ appears in only one minimal cut set ($M2$) along with $E3$. We have already inspected $E3$ and found that it has failed.

We go to the fifth-ranked terminal event $E6$. We inspect $E6$ and find that it has failed. $E6$ appears in the minimal cut set $M4$ along with $E3$ and $E5$. $E3$ has already been inspected and found to have failed. So we inspect $E5$ and find that it has failed. So $M4$ is also a minimal cut set which contains terminal events that have all failed.

Summarizing, our systematic inspection has identified that terminal events $E1$, $E3$, $E5$, and $E6$ have failed. Correspondingly, the minimal cut sets $M2$ and $M4$ have failed. It is not very rare in standby systems to find that more than one minimal cut set have failed. Although the system becomes inoperational as soon as one minimal cut set has failed, we notice it only during the periodic inspection or when the operation of the standby system is demanded. During the interval between the one minimal cut set failure and the periodic inspection or demand, additional minimal cut sets may fail. This is the reason why there could be more than one failed minimal cut set present in a standby system. The situation in operating systems is different. As soon as one minimal cut set is failed, the system fails and stops, and we notice it. Two or more minimal cut sets may fail simultaneously on very rare occasions; the probability of such a happening is very small compared to the probability of the failure of just one minimal cut set.

13.9.2 Operating Systems

The procedure for developing the best strategy for the postfailure inspection of operating systems is the same as the procedure used for standby systems, with just one difference. According to Lambert (1975), it is better to use the sequential contributory measure of importance instead of the Vesely–Fussell measure of importance in Step 1 of Section 13.9.1. We still use the Vesely–Fussell measure of importance in the sublist generation (Step 5).

13.9.3 Inspection under Partial Knowledge

In Sections 13.9.1 and 13.9.2, we discussed inspection strategies when we have no knowledge (information) about which terminal events have failed. Here, we discuss inspection strategies when we have some knowledge (information) about terminal-event failures or intermediate-event failures (component failures or subsystem failures).

13.9.3.1 Partial knowledge of component failures. Some terminal-event failures (component failures) are noticed by the operator immediately. For example, it may be easily noticed if a motor or pump fails to operate. If we know that one or more terminal events have occurred and the system has also failed, how do we proceed with the inspection of the other terminal events?

1. If at least one of the observed terminal-event failures is a member of a first-order minimal cut set, then that terminal event is responsible for the system failure (unless another minimal cut set failure also exists[1]).
2. If we notice n terminal-event failures and they all belong to an nth-order minimal cut set, then that minimal cut set is responsible for the system failure (unless another minimal cut set failure also exists[1]). Example 13.7 illustrates this scenario.
3. If neither of the preceding two situations is true, we may develop a conditional sublist as discussed in Step 5 of Section 13.9.1 and carry out the inspection according to the sublist. Examples 13.8 and 13.9 illustrate this procedure.]

Example 13.7

An operating system has stopped operating (failed) and the operator notices that terminal events $E2$, $E5$, and $E7$ have failed. What is the best inspection strategy? Minimal cut sets of the system are

$$M1 = (E1)$$
$$M2 = (E3, E4)$$
$$M3 = (E2, E5, E6)$$
$$M4 = (E2, E5, E7)$$
$$M5 = (E2, E4, E7, E8)$$

---

[1]Simultaneous existence of two or more minimal cut set failures is a very rare occurrence in operating systems.

Solution

Examining the minimal cut sets, we notice that the three failed terminal events ($E2$, $E5$, and $E7$) belong to a third-order minimal cut set, namely, $M4$. So this minimal cut set is the cause for system failure. Repairing at least one of these three terminal events ($E2$, $E5$, and $E7$) should put the system back in operation unless some other minimal cut set has occurred simultaneously. Such simultaneous occurrences are very rare in operating systems.

It does not mean that there are no other terminal-event failures. For example, it is possible that $E3$ has also failed, but it will not cause system failure by itself. Once $E2$, $E5$, or $E7$ is repaired, the system will be operable even if $E3$ is in a failed state.

Example 13.8

An operating system has failed and the operator notices that terminal events $E2$, $E4$, and $E5$ have failed. What is the best inspection strategy?

The minimal cut sets for the system failure are

$$M1 = (E1, E8)$$

$$M2 = (E2, E3)$$

$$M3 = (E6, E7)$$

$$M4 = (E4, E5, E6)$$

$$M5 = (E2, E7, E8)$$

The unavailabilities of the terminal events at the time of system failure are

$$q_1 = 0.01 \quad q_2 = 0.01 \quad q_3 = 0.01$$

$$q_4 = 0.01 \quad q_5 = 0.01 \quad q_6 = 0.02$$

$$q_7 = 0.01 \quad q_8 = 0.015$$

The terminal events are statistically independent.

Consider the scenario that, as the inspection progresses, we find that $E3$ has failed in addition to $E2$, $E4$, and $E5$. (Of course, we do not know that $E3$ has failed until we inspect it.)

Solution

First, we need to generate a sublist under the condition that terminal events $E2$, $E4$, and $E5$ have failed. Toward this end, we set the unavailabilities of $E2$, $E4$, and $E5$ as 1.0 and then rank the minimal cut sets in which these terminal events are present.

$$Q_2 = q_2 \cdot q_3 = 1 \times 0.01 = 0.01$$
$$Q_4 = q_4 \cdot q_5 \cdot q_6 = 1 \times 1 \times 0.02 = 0.02$$
$$Q_5 = q_2 \cdot q_7 \cdot q_8 = 1 \times 0.01 \times 0.015 = 0.00015$$

So the ranking of these minimal cut sets is

$$M4, M2, M5$$

Examining $M4$, there is just one more terminal event ($E6$) other than $E4$ and $E5$. So we inspect $E6$ and find that it has not failed. Next, we consider $M2$. Here also we have one more terminal event ($E3$) other than $E2$. We inspect $E3$ and find that is has failed. This means that the terminal cut set $M2$ is the cause for the system failure. Repair of $E2$ or $E3$ (which together constitute the minimal cut set $M2$) should put the system back in operation.

Example 13.9

This problem is the same as Example 13.8 with just one difference: Instead of the scenario considered in Example 13.8, consider the scenario that, as the inspection progresses, we will find that $E1$ and $E8$ have failed in addition to $E2$, $E4$, and $E5$.

Solution

The ranking of the minimal cut sets is the same as in Example 13.8.

Examining the minimal cut set $M4$, which is ranked first, there is one more terminal event ($E6$) other than $E4$ and $E5$. So we inspect $E6$ and find that it has not failed. Next, we consider $M2$. Here also we have one more terminal event ($E3$) other than $E2$. We inspect $E3$ and find that it has not failed. We then consider the minimal cut set $M5$. We have two terminal events ($E7$ and $E8$) other than $E2$. We inspect $E8$ first because it has a higher unavailability than $E7$. We find that it has failed. We inspect $E7$ and find that it has not failed. Because not all of the terminal events in $M5$ have failed, $M5$ is not the cause of system failure.

The preceding inspections show that $M2$, $M4$, or $M5$ have not failed in spite of the fact that we know that at least one terminal event in these minimal cut sets has failed. So we move to the remaining two minimal cut sets, namely $M1$ and $M3$. We already know from our inspections so far that $E7$ has not failed. That means $M3$, which contains $E7$, has not failed.

The only remaining minimal cut set is $M1$, which consists of $E1$ and $E8$. We know from our previous inspections that $E8$ has failed. So we inspect $E1$ and find that it has failed. Therefore the minimal cut set $M1$ is the cause for system failure. Repairing $E1$ or $E8$ should put the system back in operation.

13.9.3.2 Partial knowledge of subsystem (intermediate event) failures. Consider the fault tree shown in Figure 13.9. The top event 'pressure system shut-off' occurs because either 'excessive temperature' or 'excessive pressure.' If either one of these two intermediate events occurs, the pressure system is automatically shut off to avoid a catastrophic accident (pressure system rupture). If we know that the system was shut off because of excessive pressure, we may develop our inspection strategy on the basis of that knowledge; we need to consider only the right branch of the fault tree under the intermediate event 'excessive pressure' (Figure 13.9). This right branch of the tree is now treated as the full tree with 'excessive pressure' as the top event. Minimal cut sets for this fault tree are generated and the sequential contributory measures of importance for the terminal events are computed and ranked, and the procedure described in Section 13.9.2 is used to develop the inspection strategy.

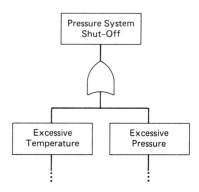

Figure 13.9. A sample fault tree.

Had we not known that the pressure system shutoff was due to 'excessive pressure,' we would have had to consider the complete fault tree (Figure 13.9) in the development of an inspection strategy. Our partial knowledge that 'excessive pressure' (intermediate event) is the cause for system shutoff has reduced our effort in developing the inspection strategy by about half.

13.9.4 Time Constraints

In the previous sections (Sections 13.9.1–13.9.3), we have not taken into consideration the effect of inspection time on the development of a rational inspection strategy. Inspection time is different for different terminal events. It depends on the accessibility and complexity of the component associated with the terminal event, and it may range from only a fraction of an hour to many hours. The procedures described in the previous sections provide the best inspection strategy if the inspection times are approximately the same for all of the terminal events. If the inspection times are considerably different, we may take into account the effect of inspection times through a rigorous mathematical analysis [Lambert (1975)]. This analysis is rather involved. We may avoid such an analysis in many applications and, instead, use the procedures discussed in the previous sections, along with some judgement on the part of the inspection crew. This approach is illustrated in Example 13.10.

Example 13.10

Consider the problem solved in Example 13.6. The time required to inspect the terminal events $E1$, $E2$, $E3$, $E4$, $E5$, and $E6$ is 2.0, 1.0, 24.0, 1.0, 1.5, and 2.0 hours, respectively. Develop a rational inspection strategy.

Solution

The terminal-event ranking according to the Vesely–Fussell measure of importance is (see Example 13.6)

$$E4, E3, E2, E1, E6, E5$$

Because the inspection time for $E3$ is considerably higher than that for the other terminal events, we may choose to inspect it last even though it is ranked second. This is a judgement call. Although the inspection times are different among the other terminal events also, the difference is not much, so we do not change the order of inspection of the other terminal events.

Therefore the suggested order of inspection is

$$E4, E2, E1, E6, E5, E3$$

This order of inspection is different from that indicated in Example 13.6, where we did not consider the effect of inspection times.

Example 13.11

Consider the problem solved in Example 13.6. The time required to inspect the terminal events $E1$, $E2$, $E3$, $E4$, $E5$, and $E6$ is 12.0, 1.0, 15.0, 3.0, 1.5, and 2.0 hours, respectively. Develop a rational inspection strategy.

Solution

The terminal-event ranking according to the Vesely–Fussell measure of importance is (see Example 13.6)

$$E4, E3, E2, E1, E6, E5$$

Inspection times for $E2$, $E4$, $E5$, and $E6$ are approximately the same, whereas the inspection times for $E1$ and $E3$ are much higher. So we decide that we will inspect $E1$ and $E3$ after inspecting the other terminal events.

What shall be the order of inspection of $E2$, $E4$, $E5$, and $E6$? We use the same order in which they appear in the Vesely–Fussell ranking. So we inspect first $E4$, $E2$, $E6$, and $E5$. What shall be the order of inspection of $E1$ and $E3$? Because their inspection times are approximately the same, we decide to inspect them in the order in which they appear in the Vesely–Fussell ranking. Although the inspection time for $E3$ is slightly more than that for $E1$, we choose to inspect $E3$ first because it appears before $E1$ in the ranking; this is a judgement call.

So the suggested order of inspection is

$$E4, E2, E6, E5, E3, E1$$

13.10 MANAGEMENT DECISIONS

Management decisions regarding repairs, replacements, spare parts inventory, etc., may be made on the basis of reliability analysis results. Example 13.12 illustrates one such application. The principles used in that example are applicable in other decision-making situations also.

Example 13.12

A system S is part of an industrial plant. This system contains sophisticated components that require special repair tools. Preventive maintenance of these components is carried out once a year and the components are considered as good as new after the maintenance. Preventive maintenance is conducted by a contractor who brings the necessary tools (the expensive repair tools) and takes them back after the maintenance. If such tools are not on-site, components cannot be repaired immediately after a component failure but have to wait until the next preventive maintenance or until the system itself fails (note that one component failure does not necessarily mean system failure). So the components are, from a practical point of view, treated as nonrepairable components for the duration of preventive maintenance (one year). The fault tree of the system is shown in Figure 13.10, where $E1$, $E2$, $E3$, and $E4$ are component failures, $G1$ and $G2$ are intermediate events, and S is system failure. If the component failures lead to system failure, the system failure will not only damage the system but will also damage some other interfacing systems in the plant and will thus require plant shutdown. There will be no personnel injury or death. The cost of repairing all damages and the cost of the resulting plant shutdown is $750,000. The plant manager has to decide whether to keep

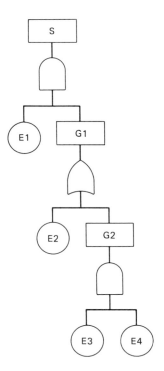

Figure 13.10. Fault tree for Example 13.12.

Table 13.1.   Component Failure and Repair Data
(Example 13.12)

TERMINAL EVENT	FAILURE RATE (FAILURES PER HOUR)	MTTR (HOURS)
E1	1×10^{-5}	20
E2	2×10^{-5}	50
E3	2×10^{-5}	10
E4	3×10^{-5}	10

the necessary repair tools on-site at a cost of $10,000 per year so that components can be repaired as soon as they fail. (Plant operators are capable of repairing the components if the necessary repair tools are available.) What is the best decision from an economical point of view. (*Note:* In this illustrative example, we consider only two options: to repair the terminal-event failures as soon as they occur or to wait until the scheduled annual, preventive maintenance. Other possible options are not considered here.) Component failure data are given in Table 13.1.

Solution

Our procedure for making a rational decision is as follows. We will first calculate the expected number of system failures per year under two options: (i) Necessary repair tools are not available on-site (so treat the components as nonrepairable); (ii) necessary tools are available on-site (so treat the components as repairable). The probable annual cost of system failure (= expected number of system failures per year × cost of damage due to system failure) is then computed for the two options. Because the expected number of failures under the first option should be higher than the expected number of failures under the second option, the probable annual cost of system failure is higher for the first option. Keeping the tools on-site is the more rational decision if the difference between the probable costs under the first and second options is greater than the cost of keeping the necessary repair tools on-site ($10,000).

The minimal cut sets for the top event (system failure) of the fault tree in Figure 13.10 are

$$M1 = (E1, E2)$$

$$M2 = (E1, E3, E4)$$

(i) *Option 1 (treat all of the components as nonrepairable):* Let q_i be the unavailability of the ith terminal event at time $= 1$ year, and let Q_i be the unavailability of the ith minimal cut set at time $= 1$ year. (We consider the unavailabilities at the end of one year because the duration between successive preventive maintenance is one year.)

$$q_i = \lambda_i t$$

where λ_i is the failure rate of the ith component and $t = 1$ year $= 365 \times 24 = 8760$ hours.

$$q_1 = 8.76 \times 10^{-2} \qquad q_2 = 17.52 \times 10^{-2}$$
$$q_3 = 17.52 \times 10^{-2} \qquad q_4 = 26.28 \times 10^{-2}$$

The minimal cut set unavailabilities at the end of one year are

$$Q_1 = q_1 q_2 = 1.53 \times 10^{-2}$$
$$Q_2 = q_1 q_3 q_4 = 4.0 \times 10^{-3}$$

System unavailability at the end of one year is

$$Q_S = Q_1 + Q_2 = 1.93 \times 10^{-2}$$

Because all the components are nonrepairable, system unavailability is equal to system unreliability. Also, system unreliability is equal to the expected number of failures. Because we have considered $t = 1$ year, what we have is the expected number of failures per year:

$$\text{expected number of failures per year} = 1.93 \times 10^{-2}$$
$$\text{probable annual cost of system failure}$$
$$= (1.93 \times 10^{-2}) \times (750,000) = \$14,475$$

(ii) *Option 2 (treat all of the components as repairable):* The component unavailabilities at the end of one year are given by

$$q_i = \lambda_i \tau_i$$

where λ_i and τ_i are the failure rate and MTTR of the ith component.

$$q_1 = 2 \times 10^{-4} \qquad q_2 = 10^{-3}$$
$$q_3 = 2 \times 10^{-4} \qquad q_4 = 3 \times 10^{-4}$$

The minimal cut set unavailabilities at the end of one year are given by

$$Q_1 = q_1 q_2 = 2 \times 10^{-7}$$

$$Q_2 = q_1 q_3 q_4 = 1.2 \times 10^{-11}$$

The probability density function of the time to first failure of the jth minimal cut set is f_j:

$$f_1 = Q_1 \left[\frac{\lambda_1}{q_1} + \frac{\lambda_2}{q_2} \right] = 1.4 \times 10^{-8} \text{ per hour}$$

$$f_2 = Q_2 \left[\frac{\lambda_1}{q_1} + \frac{\lambda_3}{q_3} + \frac{\lambda_4}{q_4} \right] = 3 \times 10^{-12} \text{ per hour}$$

The unreliability of the jth minimal cut set at $t = 1$ year ($= 8760$ hours) is U_j:

$$U_1 = \int_0^t f_1 \, dt = f_1 t = 1.2 \times 10^{-4}$$

$$U_2 = \int_0^t f_2 \, dt = f_2 t = 2.6 \times 10^{-8}$$

System unreliability U_S is given by

$$U_S = U_1 + U_2 = 1.2 \times 10^{-4}$$

For a repairable system, system unreliability is a lower bound to the expected number of system failures N_S:

$$N_S \geq U_S$$

U_S is a very good approximation to N_S if the system is very reliable [Lambert and Davis (1980)]. So, approximately

$$N_S = U_S = 1.2 \times 10^{-4} \text{ per year}$$

probable annual cost of system failure $= (1.2 \times 10^{-4}) \times (750,000) = \$90$

Let the annual reduction in the probable cost of system failure by keeping the necessary repair tools on-site be D,

$$D = 14475 - 90 = \$14,385$$

This reduction in the probable cost of system failure is higher than the annual cost of keeping the necessary repair tools on-site ($10,000). So it is advisable that the necessary repair tools be kept on-site so that components can be repaired as soon as they fail.

EXERCISE PROBLEMS

13.1. Consider a series system consisting of two electronic components $H1$ and $H2$. Both components are nonrepairable during system operation; both components have constant failure rates and are statistically independent. The failure rates of $H1$ and $H2$ are 4×10^{-5} failures per hour and 3×10^{-5} failures per hour, respectively. The system is part of a helicopter. The helicopter is maintained at the end of every mission, at which time components $H1$ and $H2$ are brought back to "as good as new" status. Reliability of the system should be at least 0.96 at the end of each mission. Computations show that this goal is not met. It is possible to add active parallel redundancies to either or both $H1$ and $H2$. The active redundant component to $H1$ could be identical component with a failure rate of 4×10^{-5} failures per hour (cost = $10,000) or a functionally similar but lower-reliability component with a failure rate of 8×10^{-5} failures per hour (cost = $6000). The active redundant component of $H2$ could be an identical component with a failure rate of 3×10^{-5} failures per hour (cost = $5000) or a functionally similar but lower-reliability component with a failure rate of 6×10^{-5} failures per hour (cost = $3500). Develop a new system that meets the reliability requirement at the lowest possible cost.

13.2. Consider the system configuration shown in Figure 8.29. A, B, and C are three components, which are statistically independent. According to the original design, failure-on-demand probabilities of components A, B, and C are $P[A] = 0.1$, $P[B] = 0.09$, and $P[C] = 0.02$. Costs of components A, B, and C are $500, $400, and $1000, respectively. A client orders 10,000 of these systems but requires that the system failure-on-demand probability should not exceed 0.025. Because the original design does not meet this requirement, the reliability engineering department was asked to design a system that would meet the requirement while at the same time not changing the system configuration (Figure 8.29). The reliability engineer may replace one or more of the original components by higher-reliability components. The higher-reliability components have the following failure-on-demand probabilities. $P[A] = 0.05$, $P[B] = 0.05$, and $P[C] = 0.01$. Costs of the higher-reliability models of components A, B, and C are $1000, $1000, and $1600, respectively. Determine the combination of components that would meet the reliability requirement at the lowest possible cost.

13.3. Consider a series system consisting of three components $H1$, $H2$, and $H3$. The component failures are statistically independent. Steady-state unavailabilities of $H1$, $H2$, and $H3$ are 0.02, 0.03, and 0.05, respectively. Improve the system availability such that the steady-state availability of the system is

Table 13.2.   Data for Problem 13.4

TERMINAL EVENT	UNAVAILABILITY AT $t = 5000$ HOURS
$E1$	3×10^{-1}
$E2$	3×10^{-1}
$E3$	4×10^{-1}
$E4$	1×10^{-2}
$E5$	2×10^{-2}
$E6$	4×10^{-2}
$E7$	8×10^{-2}
$E8$	2×10^{-1}
$E9$	1×10^{-2}

at least 0.95 and the cost of improvements is a minimum. It is possible to decrease the unavailability of $H1$ by 25, 50, or 75%, at a cost of $3000, $5000, or $6500, respectively. It is not possible to add any parallel redundancy to $H1$. It is possible to decrease the unavailability of $H2$ by 30 or 60%, at a cost of $5000 or $9000, respectively. It is also possible to add an active parallel redundancy of an identical component (steady-state unavailability = 0.03), at a cost of $50,000. It is not possible to decrease the unavailability of $H3$ but it is possible to add an active parallel redundancy of an identical component (steady-state unavailability = 0.05), at a cost of $45,000.

13.4. The fault tree of a standby system is as shown in Figure 11.2. The top event T represents 'system failure.' The terminal events are statistically independent and have constant failure rates and constant repair rates. The system was inspected at time $t = 5000$ hours and was found to be in a failed state. Determine the order in which the terminal events should be inspected (the best inspection strategy). Consider the scenario in which terminal events $E3$, $E4$, $E5$, and $E8$ have failed. (Of course, the inspection crew does not know of these failures until they inspect them.) Terminal-event unavailabilities time $t = 5000$ hours are given in Table 13.2.

13.5. Consider the standby system described in Problem 13.4. Determine the order in which the terminal events should be inspected (the best inspection strategy), under the scenario that $E1$, $E2$, and $E8$ have failed. (Of course, the inspection crew does not know of these failures until they inspect them.)

13.6. The fault tree of a system is as shown in Figure 8.20. For the purposes of this problem, assume that E, F, G, J, K, M, N, P, Q, and R are terminal events (instead of intermediate events as indicated in Figure 8.20). The top event T represents 'system failure.' The terminal events are statistically independent. The system fails during operation at time $t = 4400$ hours, and the operator notices that terminal events G and K have failed. What is the best inspection strategy? Consider the scenario in which terminal events F

and J have failed, in addition to G and K. (Of course, the inspection crew does not know of the failures of F and J until they inspect them.) The unavailabilities of the terminal events at time $t = 4400$ hours are as follows: $q_E = q_F = q_G = 0.05$, $q_J = q_K = 0.02$, $q_M = 0.08$, and $q_N = q_P = q_Q = q_R = 0.05$.

13.7. Consider the system described in Problem 13.6. The system fails during operation at time $t = 4400$ hours, and the operator notices that terminal events G, K, M, and N have failed. What is the best inspection strategy? Consider the scenario in which terminal events F and J have failed, in addition to G, K, M, and N. (Of course, the inspection crew does not know of the failures of F and J until they inspect them.)

REFERENCES

Grant, E. L., W. G. Ireson, and R. S. Leavenworth (1982). *Principles of Engineering Economy*. John Wiley and Sons, Inc., New York.

Lambert, H. E. (1975). *Fault Trees for Decision Making in Systems Analysis*. Report No. UCRL-51829. Lawrence Livermore National Laboratory, Livermore, CA.

Lambert, H. E. and B. J. Davis (1980). *The Fault Tree Computer Codes IMPORTANCE and GATE*. Lawrence Livermore National Laboratory, Livermore, CA.

Tillman, F. A., C. L. Hwang, and W. Kuo (1980). *Optimization of Systems Reliability*. Marcel Dekker, Inc., New York.

Chapter 14
Implementation and Management

14.1 INTRODUCTION

Reliability engineering has been an integral part of many defense system projects for many years. Reliability engineering has also been integrated into the design of many aerospace systems. Recent years have also seen the application of reliability engineering principles in the probabilistic risk assessment of nuclear power plants. More recently, reliability engineering has found applications in the design, construction, operation, and maintenance of power, petrochemical, manufacturing, and other industrial facilities. Reliability engineering programs (reliability engineering projects) are rather new to the corporate, project, and engineering managers responsible for the design, construction, operation, and maintenance of these facilities.

The scope of this chapter is to provide some guidelines on the implementation and management of a reliability program within a corporate or project structure. Although the guidelines provided here are developed with the power, petrochemical, manufacturing, and related industries in mind, the general principles are applicable to defense, aerospace, and other industries also.

Developing, implementing, and managing a reliability program is not an exact science, so the discussions in this chapter should be treated as general guidelines rather than as hard-and-fast rules.

What is a reliability engineering program? A *reliability engineering program* is the systematic reduction, elimination, and/or control of potential hardware, software, and human failures through the life of a plant (or product) so that the reliability, availability, and/or safety of the plant (or product) is an optimum within the cost and functional constraints of the plant (or product).

Ideally, a reliability engineering program should start at the planning stage and continue through the conceptional design, detailed design, manufacturing, construction, testing, and operating phases of the plant (or product). If, however, a project has already been underway (say, it is past the conceptual design, detailed design, or construction phase), it is still possible to implement a reliability engineering program to continue the project in such a way as to achieve the best possible reliability, availability, and/or safety within the limitations of the existing situation. It is best to start as early in the project as possible but it is "better late than never."

Because reliability engineering programs are rather new in the power, petrochemical, and manufacturing industries, there is occasional resistance to it from engineers and engineering managers. Cooperation between reliability engineering personnel and the other engineering personnel is vital to the success of any reliability engineering program. This is best achieved by a visible and strong commitment on the part of top corporate and project managers and by a general indoctrination of project personnel on the need for reliability engineering.

14.2 CHARTER OF THE RELIABILITY PROGRAM

A well-defined reliability program charter is essential to the success of the program. Each project should have an individualistic reliability program charter. Ideally, the program charter should be approved by a senior corporate official or the project manager.

The program charter should include the program objectives, management structure, controls, reviews, budget, and schedules.

1. What are the overall objectives of the program? The objectives could be to meet specific reliability requirements, to compute reliability, availability, and/or accident probabilities, to identify safety hazards and make recommendations to reduce or eliminate such hazards, to identify system failure modes and make recommendations on changes in design, operation, and/or maintenance to minimize system failures, etc. (one or more of these objectives may apply). (*Note:* Specific numerical goals on reliability, availability, and/or accident probabilities may be set by the customer, regulatory agencies, or the project management team. Not every project will have such specific numerical requirements.)

2. What is the management structure of the program? Specific details about interfaces between the reliability program manager (program leader) and other managers should be specified. Major interfaces with customers and component suppliers should also be noted. The organizational structure under the reliability program manager is seldom specified in the program charter.

3. What are the program control requirements? Some reliability programs require that there be external reviews of the program results and recommendations by one or more experts. Whether the reviewers should be from inside the company (but outside the program team) or from outside the company should be specified. Also, the level of review (level of detail) should be indicated.

4. Specific means by which the quality of the reliability program is controlled and assured should be stated. *Quality assurance* could be achieved through internal checks (within the program team) or by external audits (usually by audit teams from outside the program but from inside the company).

5. Budget and schedule, including specific milestones, should be specified. Some reliability programs run through the whole life of the plant—from inception to dismantlement or mothballing. In such long-term projects, budget, schedule, and milestones should be specified for each phase of the program (design phase, construction phase, operating phase). There may be changes and updates as the project progresses from one phase to another.

6. Who is responsible for the reliability program? In long-term programs that run through from inception to dismantlement, responsibility for the reliability program may shift from the plant designer or constructor to the plant operator. Interfacing at the juncture of transfer of responsibility may be discussed in the program charter. The general identification of documents to be transferred, and the liabilities and responsibilities of each party should be spelled out.

Program charters seldom specify the analysis methods, computer software, and data to be used. This is left to the program manager, who will establish these details depending on program objectives, budget, and schedule.

14.3 RELIABILITY PROGRAM PLAN

The reliability program plan is developed by the program manager. Unlike the program charter, it need not be approved by the project manager or a corporate official. The program plan outlines the details of implementing the program charter.

The program plan should include the following information, as appropriate.

1. Scope of work
2. Organizational structure and external interfaces
3. Task breakdown
4. Budget allocation, manpower allocation, and schedule
5. Data sources
6. Analysis techniques and computer software

7. Quality assurance
8. Documentation
9. Engineering and management reviews
10. Component acquisition
11. Failure reporting

Some guidance on each of these items is provided in the following sections. These guidelines will be helpful in developing a program plan and in implementing a reliability program.

14.4 SCOPE OF THE PROGRAM

The scope of the program should be tied to program objectives specified in the program charter. The program plan will be more elaborate and detailed than the charter. The program plan should provide specific details on what is expected from the reliability program. One or more of the following may be expected from the program. Identify each clearly in the program plan.

1. Identification of potential safety hazards and recommendations to reduce or to eliminate them.
2. Identification of potential hardware, software, and human failure modes.
3. Identification of system failure modes (combinations of hardware, software, and human failures that will produce system-level failures).
4. Computation of the probability of potential accidents that could cause personnel injuries and/or substantial property damage.
5. Computation of system availability and reliability.
6. Set reliability, availability, and/or safety criteria for the plant.
7. Recommendations for the reduction of system unavailability, system unreliability, and/or accident probabilities. Recommendations may include changes in design, operating procedures, and/or maintenance procedures.

14.5 ORGANIZATIONAL STRUCTURE

The program charter specifies the place of the reliability program manager within the organizational structure of the project. The program plan should contain the organizational structure of the reliability program team under the reliability program manager.

The reliability program team may be divided into a number of task groups, each group headed by a group leader.

Educational qualifications, experience, and expertise requirements for the reliability program members should be specified in the program plan. During staffing, these requirements must be met as closely as possible. Engineers from a variety of disciplines (including reliability engineering, systems design, component testing, human factors, cost–benefit analysis, plant operation, and plant maintenance) may be included in the team, depending on the size of the project and manpower allocation to the reliability program. If the manpower is limited, the program should be staffed with reliability engineers, each with a good understanding of one or more of the previously noted disciplines. In addition, engineers from the other disciplines may participate in the reliability program on a part-time basis. Close interface and communication with the staff of the other engineering disciplines in the project (design team, operating personnel, maintenance personnel) are essential for a successful reliability program.

Personnel and, to some extent, the organizational structure may change from one phase of the project to another. All personnel involved in the reliability program during the plant-design phase may not remain the program during the plant-construction or plant-testing phase; some new personnel may be brought in and some existing personnel may be transferred out.

The responsibility, accountability, and functions of each person (particularly the group leaders) should be clearly identified in the program plan.

14.6 DATA SOURCES

If a quantitative reliability analysis is necessary to meet the scope of the program, hardware failures probabilities, software error probabilities, and human error probabilities are required as data. The accuracy and validity of the quantitative system reliability analysis are dependent on these data. The program plan should identify data sources that may be used in the project. Both plant-specific and generic data sources may be used, as appropriate.

As the project progresses, engineers working on quantitative reliability analysis may identify other sources (that is, other than those listed in the program plan) that may be useful. Such sources should be used *only if and after* they are approved for use by the program manager or a group leader designated by the program manager. If additional data sources are approved during the project, the appropriate sections of the program plan should be updated to include those data sources in the approved list.

14.7 ANALYSIS PROCEDURES

Analysis procedures to be used in the project should be specified in the program plan. The specification should be as specific as possible. Reference may be made to books, reports, and/or papers that describe the procedures to be used.

As the project progresses, engineers working in the project may identify other methods (that is, other than those listed in the program plan) that may be useful. Such methods should be used *only if and after* they are approved for use by the program manager or a group leader designated by the program manager. The appropriate sections of the program plan should be updated to reflect this.

14.8 COMPUTER SOFTWARE

Computer software to be used in the project should be specified in the program plan. The computer programs may be in-house programs, public-domain programs, and/or commercially available programs. Whenever possible, the version number or version date of the computer programs should be specified. Unless the program manager is satisfied that the computer programs have already been verified sufficiently to meet the project needs, additional verifications should be undertaken as part of the reliability program or additional verifications should be requested from the developer, vendor, or other appropriate organizations. The program plan should specify whether the programs authorized for use have already been adequately verified or need further verification.

As the project progresses, engineers working in the project may identify other computer programs (that is, other than those listed in the program plan) that may be useful. Such software should be used *only if and after* they are approved for use by the program manager or a group leader designated by the program manager. The appropriate sections of the program plan should be updated to include those newly approved computer programs.

14.9 QUALITY ASSURANCE

The quality of all the reliability program work should be assured through independent verifications of the calculations, assumptions, approximations, and data. Such independent verifications are carried out by engineers who have not participated in that particular analysis; these checkers may be part of the program team (for example, an engineer working on the analysis of system A may check the analysis of system B).

The independent verification may be a detailed step-by-step checking of the data, methods, assumptions, approximations, and calculations or a broad-brush checking of the validity of the data, methods, assumptions, and approximations, and the reasonableness of the final results. The former is preferred if budget and manpower permit.

If step-by-step verifications of calculations are carried out, such a verification may be either *complete checking* (each and every sheet of the analysis is checked) or *spot-checking* (portions of the analysis—selected by the checker—are checked).

If disagreements between the originator of the analysis and the checker arise, it should be resolved by the group leader or someone designated by the program manager.

The program plan should specify the following:

1. Type of verification (step-by-step checking, broad-brush checking, or other).
2. Complete checking or spot-checking (if spot-checking, specify the approximate percentage of analysis that needs to be spot-checked).
3. Who has the authority to resolve any disagreements between originator and checker.

14.10 DOCUMENTATION

Usually, two types of document are generated: summary reports and detailed calculations. *Summary reports* provide a concise summary of the reliability program activities, results, and recommendations. We recommend that interim summary reports be prepared at the end of each major task (say, after the preliminary hazard analysis, after the failure modes and effects analysis, after the qualitative fault tree analysis, etc.). Final reports may be prepared at the end of major milestones, such as at the end of systems design, at the end of preoperation testing, etc., as applicable. Interim reports may be updated as necessary and combined to provide the final reports. Each interim and final report should be dated and assigned a version number.

The program plan should specify the following:

1. When the interim reports are to be prepared.
2. When the final reports are to be prepared.
3. Who has the authority to originate and approve interim reports.
4. Who has the authority to originate and approve final reports (for example, group leaders may be authorized to originate the report and the program manager may be authorized to approve the report).

Detailed calculations are generated as part of the analysis. We recommend that the following procedures be adhered to in the preparation of calculation sheets.

1. All calculations should be clearly presented so that they could be followed by other engineers without difficulty.
2. Identify references, if any, that describe the analysis techniques used.
3. State all assumptions and their justifications (cite references, if applicable).
4. Each calculation sheet should contain the name of the originator and checker, if any.
5. Each calculation sheet should contain page number and date.
6. Revised calculation sheets should be identified by the revision number (revision 1, revision 2, etc.) and revision date.
7. Computer programs should be identified with version number or version date and from where obtained.
8. Computer input and output listings should be included as appendices to the calculations.

We recommend that calculation sheets by preserved until the plant is decommissioned. If changes are made in hardware or operation–maintenance procedures, their impact on system reliability, availability, or safety may be studied quickly and easily if the original calculations are available. Also, if any mistakes are uncovered at a later date, corrections can be made quickly and the impact of the mistakes on computed system reliability, availability, or safety can be determined quickly. Some organizations store a microfilm copy of the calculations in addition to or in lieu of the paper copies.

An ideal calculation will follow the preceding recommendations. However, company philosophies on documentation may differ. The program plan should specify the procedure to be followed in generating calculation sheets (all or some of the recommendations discussed in the preceding paragraphs may be followed).

Good calculations are characterized by completeness, reproducibility, and verifiability. Calculation documents should be *complete*. They should include not only all of the steps of the calculation but also all of the assumptions, approximations, and references. Calculations should be *reproducible*. Any engineer of comparable experience and knowledge as the calculation originator should be able to reproduce the calculations using the same data and references. Calculation documents should be *verifiable*. An engineer of comparable experience and knowledge as the calculation originator should be able to check the calculations without necessarily having to discuss it with the originator.

14.11 ENGINEERING AND MANAGEMENT REVIEWS

Engineering and management reviews of the results and recommendations of reliability programs are optional. Because reliability engineering applications are relatively new in the power, petrochemical, and manufacturing industries, such reviews are advisable, at least in major projects. These reviews are conducted in addition to the quality assurance discussed in Section 14.9.

Engineering reviews: Engineering reviews are known as *peer reviews.* These are conducted by personnel from outside the project (sometimes, outside the company) who have the necessary experience and knowledge. The purpose of engineering reviews is not to check the calculations (this might have been done as part of the quality assurance) but to provide an overall review of how the reliability program is executed in relation to the charter and plan of the reliability program. Engineering reviews concentrate on the appropriateness of the data sources, methods, and computer software used in the project and also on the validity of the recommendations. Engineering reviews also examine the results to see if they seem reasonable.

Management reviews: Management reviews are conducted by either the project manager or by experts from within or outside the company appointed by the project manager. Reviewers examine the program charter, program plan, final results, recommendations, and engineering review reports (if any) and submit their findings to the project manager. Management reviews may be conducted at the end of system design and/or at other appropriate milestones.

The program plan should specify if engineering and/or management reviews are required and when the reviews should be conducted.

14.12 COMPONENTS ACQUISITION

Quantitative system reliability analyses require component failure probabilities as data. Failure probabilities of commonly used off-the-shelf components may be based on information from plant-specific or generic data sources. Component suppliers may be required, as part of the contractual agreement, to provide any data base they may have on the reliability of their components. Some component suppliers may not have such data bases or may hesitate to provide access to such data bases. If the latter is the case, the component supplier may be required to confirm if the failure probabilities used in systems reliability analyses agree with their confidential data.

Some component suppliers, particularly electrical and electronics component manufacturers, have extensive reliability programs of their own, including component reliability analysis, life testing, and reliability screening. In such cases, they may be contractually required to provide reliability data on component supplied by them.

Some projects may require custom-made, major equipment. Because no operating experience or very little operating experience may be there for such equipment, detailed reliability analyses may have to be conducted to estimate their reliabilities. Such reliability analyses may be conducted by the manufacturer or the buyer.

The program plan should specify what reliability information needs to be secured from the different component suppliers. The program plan should also specify who (program manager, group leader, an engineer) should interface with component suppliers and approve the information received. This will ensure smooth interfaces with component suppliers.

Any information supplied by component suppliers should be included in the reliability program documentation as appendices, unless the information is confidential and is not publicized.

14.13 FAILURE REPORTING

If there are component, software, human, or system failures during testing or operation of the plant, relevant information about the failures such as the cause, effect, remedial actions, and repair time should be recorded in a systematic manner (see Section 5.2.4). Such records will form the basis for plant-specific reliability data bases.

Some industries have industrywide data maintained by industry groups that require group members to report failures in their plants to the industry data bases.

The program plan should specify the following:

1. Who is responsible for maintaining failure records? (plant office or corporate office, which group, etc.).
2. Format in which failure information is to be reported by operating personnel.
3. Format in which failure information received from operating personnel is to be recorded and maintained.
4. Should failure information be communicated to any outside agency or industry data base? Format in which failure information needs to be communicated? Which department is responsible for this communication?

14.14 CONCLUDING REMARKS

We have provided some general guidelines on implementing and managing reliability programs. The discussions in the preceding sections relate to a reliability program for the design, construction, and operation of a power, petrochemical, or manufacturing plant. The general principles are equally applicable for implementing a reliability program for product development, design and manufacture of industrial equipment, defense systems, or consumer products.

There are nine characteristics to a successful reliability program:

1. Commitment of upper-level corporate and project management.
2. A well-defined program charter.
3. A clear and complete program plan.
4. Qualified program staff.
5. Well-structured program organization with established reporting hierarchy, internal and external interfaces, accountability, and responsibilities.
6. Adequate budget.
7. Good quality assurance, engineering reviews, and management reviews.
8. Good communication between the reliability program team and other engineering teams.
9. Documentation that is complete, reproducible, and verifiable.

Appendix I
Computer Programs for Reliability Analysis

A number of computer programs used in system reliability analysis have been mentioned in the various chapters of this book. An alphabetical listing and brief descriptions of those programs are provided in this appendix.

Many of the programs listed here are updated periodically by the developers or program vendors. Also, computational efficiency, maximum size of the problem that can be solved, method of analysis, and analytical assumptions may vary from program to program. It is prudent to get the latest information on one or more programs before purchasing or using a program. Assure that the program has the needed capabilities and is computationally efficient.

Most programs are verified by the developers and documentation on verification may be available. If the user is not satisfied that the program has been adequately verified, the user should carry out his or her own verification of the program.

The following list is not exhaustive; there are other computer programs that may be equally good and useful. New computer programs may also become available from time to time.

1. *Program Name:*   BACFIRE
   Capabilities: Common cause analysis
   Reference: Cate and Fussell (1977)
   Developers: C. L. Cate and J. B. Fussell
      University of Tennessee
      Knoxville, Tennessee.

2. *Program Name:*   BOUNDS
   Capabilities: Uncertainty analysis
   Reference: Lee and Apostolakis (1976)
   Developers: Y. T. Lee and G. E. Apostolakis
      University of California
      Los Angeles, California.

3. *Program Name:*   CAT
   Capabilities: Fault tree construction and plotting
   Reference: Salem, Apostolakis, and Okrent (1977)
   Developers: S. L. Salem, G. E. Apostolakis, and D. Okrent
                        University of California
                        Los Angeles, California.

4. *Program Name:*   COMCAN-II
   Capabilities: Common cause analysis
   Reference: Rasmuson, Marshall, Wilson, and Burdick
                        (1979)
   Developers: D. M. Rasmuson, N. H. Marshall, J. R.
                        Wilson, and G. R. Burdick
                        EG & G Idaho, Inc.
                        Idaho Falls, Idaho.

5. *Program Name:*   FTAP
   Capabilities: (i) Minimal cut set determination
                        (ii) Minimal path set determination
   Reference: Willia (1978)
   Developer: R. R. Willia
                        University of California
                        Berkeley, California.

6. *Program Name:*   ICARUS
   Capabilities: Computation of system unavailability, and op-
                        timum test interval
   Reference: Vaurio and Sciandone (1979)
   Developers: J. K. Vaurio and D. Sciandone
                        Argonne National Laboratory
                        Argonne, Illinois.

7. *Program Name:*   IMPORTANCE
   Capabilities: Importance ranking of minimal cut sets and
                        terminal events (all the importance measures
                        discussed in Chapter 12)
   Reference: Lambert and Gilman (1977)
   Developers: H. E. Lambert and F. M. Gilman
                        Lawrence Livermore National Laboratory
                        Livermore, California.

8. *Program Name:*   MICSUP
   Capabilities: (i) Determination of minimal cut sets
                        (ii) Determination of minimal path sets

Reference:	Pande, Spector, and Chatterjee (1975)
Developers:	P. K. Pande, M. E. Spector, and P. Chatterjee
	University of California
	Berkeley, California.

9. *Program Name:* MOCARS
   Capabilities: Uncertainty analysis
   Reference: Matthews (1977)
   Developer: S. D. Matthews
   EG & G Idaho, Inc.
   Idaho Falls, Idaho.

10. *Program Name:* PL-MOD
    Capabilities:
    (i) Minimal cut set determination
    (ii) Computation of system unavailability
    (iii) Importance ranking of terminal events and minimal cut sets (Vesely–Fussell measure)
    Reference: Olmos and Wolf (1977)
    Developers: J. Olmos and J. Wolf
    Massachusetts Institute of Technology
    Cambridge, Massachusetts.

11. *Program Name:* RAS
    Capabilities:
    (i) Determination of minimal cut sets
    (ii) Computation of system unreliability
    (iii) Importance ranking of minimal cut sets and terminal events (Vesely–Fussell measure only)
    Note: This program can analyze phased missions (a system with different phases of operation; each phase may have different minimal cut sets)
    Reference: Rasmuson, Marshall, and Burdick (1977)
    Developers: D. M. Rasmuson, N. H. Marshall, and G. R. Burdick
    EG & G Idaho, Inc.
    Idaho Falls, Idaho.

12. *Program Name:* SAMPLE
    Capabilities: Uncertainty analysis
    Reference: Nuclear Regulatory Commission (1975)
    Developer: Nuclear Regulatory Commission
    Washington, DC.

13. *Program Name:* SETS
    Capabilities:
        (i) Fault tree plotting
        (ii) Determination of minimal cut sets
        (iii) Determination of minimal path sets
        (iv) Computation of top-event probability
        (v) Common cause analysis
    Reference: Worrell and Stack (1978)
    Developers: R. B. Worrell and D. W. Stack
    Sandia National Laboratory
    Albuquerque, New Mexico.

14. *Program Name:* SPASM
    Capabilities: Uncertainty analysis
    Reference: Leverenz (1981)
    Developer: Electric Power Research Institute
    Palo Alto, California.

15. *Program Name:* WAMBAM
    Capabilities: Computation of top-event probability
    Reference: Leverenz and Kirch (1976)
    Developer: Electric Power Research Institute
    Palo Alto, California.

16. *Program Name:* WAMCOM
    Capabilities: Common cause analysis
    Reference: Putney (1981)
    Developer: Electric Power Research Institute
    Palo Alto, California.

17. *Program Name:* WAMCUT
    Capabilities:
        (i) Fault tree plotting
        (ii) Determination of minimal cut sets
        (iii) Computation of minimal-cut-set, intermediate-event, and top-event probabilities
    Reference: Leverenz and Kirch (1978)
    Developer: Electric Power Research Institute
    Palo Alto, California.

18. *Program Name:* WAMCUT-II
    Capabilities:
        (i) Fault tree plotting
        (ii) Determination of minimal cut sets
        (iii) Computation of intermediate-event and top-event probabilities

Reference: Putney and Kirch (1981)
Developers: B. F. Putney and H. R. Kirch
                  Science Applications Inc.
                  Palo Alto, California.

REFERENCES

Cate, C. L. and J. B. Fussell (1977). *BACFIRE—A Computer Program for Common Cause Failure Analysis.* Report No. NERS-77-02. University of Tennessee, Knoxville.

Lambert, H. E. and F. M. Gilman (1977). *The IMPORTANCE Computer Code.* Report No. UCRL-76269. Lawrence Livermore National Laboratory, Livermore.

Lee, Y. T. and G. E. Apostolakis (1976). *Probability Intervals for the Top Event Unavailability of Fault Trees.* Report No. UCLA-ENG-7663. University of California, Los Angeles.

Leverenz, F. L. (1981). *SPASM: A Computer Code for Monte Carlo System Evaluation.* Report No. EPRI-NP-1685. Electric Power Research Institute, Palo Alto, CA.

Leverenz, F. L. and H. R. Kirch (1976). *User's Guide for the WAMBAM Computer Code.* Report No. EPRI-217-2-5. Electric Power Research Institute, Palo Alto, CA.

Leverenz, F. L. and H. R. Kirch (1978). *WAMCUT—A Computer Code for Fault Tree Evaluation.* Report No. EPRI-NP-803. Electric Power Research Institute, Palo Alto, CA.

Matthews, S. D. (1977). *MOCARS: A Monte Carlo Simulation Code for Determining Distribution and Simulation Limits.* EG & G Idaho, Inc., Idaho Falls.

Nuclear Regulatory Commission (1975). *Reactor Safety Study: An Assessment of Accident Risks in U.S. Commercial Nuclear Power Plants* (WASH-1400). Nuclear Regulatory Commission, Washington, DC.

Olmos, J. and J. Wolf (1977). *A Modular Approach to Fault Tree and Reliability Analysis.* Report No. MITNE-209. Massachusetts Institute of Technology, MA.

Pande, P. K., M. E. Spector, and P. Chatterjee (1975). *Computerized Fault Tree Analysis: TREEL and MICSUP.* Report No. ORC-75-3. University of California, Berkeley.

Putney, B. F. (1981). *WAMCOM: Common Cause Methodologies using Large Fault Trees.* Report No. EPRI-NP-1851. Electric Power Research Institute, Palo Alto, CA.

Putney, B. F. and H. R. Kirch (1981). *WAMCUT-II—A Fault Tree Evaluation Program.* Report No. SAI-SR-234-81-PA. Science Applications, Inc., Palo Alto, CA.

Rasmuson, D. M., N. H. Marshall, and G. R. Burdick (1977). *User's Guide for the Reliability Analysis System (RAS).* EG & G Idaho, Inc., Idaho Falls.

Rasmuson, D. M., N. H. Marshall, J. R. Wilson, and G. R. Burdick (1979). *COMCAN-II—A Computer Program for Automated Common Cause Failure Analysis.* EG & G Idaho, Inc., Idaho Falls.

Salem, S. L., G. E. Apostolakis, and D. Okrent (1977). A new methodology for the computer-aided construction of fault trees. *Annals of Nuclear Energy* **4** 417–433.

Vaurio, J. K. and D. Sciandone (1979). *Unavailability Modeling and Analysis of Redundant Safety Systems.* Report No. ANL-79-87. Argonne National Laboratory, Argonne, IL.

Willia, R. R. (1978). *Computer-Aided Fault Tree Analysis.* Report No. ORC-78-14. University of California, Berkeley.

Worrell, R. B. and D. W. Stack (1978). *A SETS User's Manual for the Fault Tree Analyst.* Report No. SAND 77-2051. Sandia National Laboratory, Albuquerque, NM.

Appendix II
Component Reliability Data Sources

An alphabetical list and brief descriptions of some generic sources for component reliability data are provided in this appendix. The list is not exhaustive; there are other data sources that may be equally good and useful. Also, new data sources may become available from time to time.

The data sources listed here include published books, reports, papers, and computerized data bases.

(*Note:* No generic data sources are available for software error probabilities. A number of reports and papers containing human error probabilities are identified in Section 5.4.)

1. *Name:*   FARADA (FAilure RAte DAta)
   Description: Reports are issued periodically. These reports contain component failure probabilities derived from the National Aeronautics and Space Administration and the Department of Defense experience.
   Source: Fleet Missile Systems Analysis and Evaluation Group
   Department of Defense
   Corona, California.

2. *Name:*   GIDEP Data
   Description: The Government Industry Data Exchange Program (GIDEP) was established by the Joint Military Logistic Commanders of the US Army, Navy, and Air Force Logistic Commands, the Air Force Systems Command, the National Aeronautics and Space Administration and the Canadian Military Electronics Standards Agency. This is a computerized data base. Periodic reports are also issued.
   Source: GIDEP Operations Center
   Fleet Missile Systems Analysis and Evaluation Group
   Department of Defense
   Corona, California.

3. *Name:*   IEEE Data
   Description: Failure probabilities of a variety of electrical and electronic components are provided. Failure probabilities are based on the opinions of over 200 experts. The experts have made their estimates on the basis of operating data from a number of industries.
   Reference: Institute of Electrical and Electronics Engineers (1977).

4. *Name:*   LER Data
   Description: Licensee Event Reports (LER) are issued periodically. These reports contain summaries of component failures in US nuclear power plants.
   Source: Nuclear Regulatory Commission
                  Washington, DC.

5. *Name:*   MIL-HB 217C Data
   Description: This military handbook (MIL-HB) provides failure probabilities for mostly electronic components used in military applications. The handbook is updated periodically.
   Source: Rome Air Development Center
                  Department of Defense
                  Griffis Air Force Base, New York.

6. *Name:*   NERC Data
   Description: Failure data from US power plants are published annually by the National Electric Reliability Council (NERC). (This information was previously published by the Edison Electric Institute (EEI), and the responsibility changed hands in 1979.)
   Source: National Electric Reliability Council
                  New York, New York.

7. *Name:*   NPRD (Nonelectric Parts Reliability Data)
   Description: Reports are issued periodically. These reports contain failure probabilities of nonelectronic parts used in military applications.
   Source: Rome Air Development Center
                  Department of Defense
                  Griffis Air Force Base, New York.

8. *Name:*   NPRDS Data
   Description: The Nuclear Plant Reliability Data System
                    (NPRDS) is developed and maintained by the
                    Southwest Research Institute on behalf of the
                    Nuclear Regulatory Commission and the nuclear
                    power industry. It contains failure data from oper-
                    ating US nuclear power plants. This is a computer-
                    ized data base. Periodic reports are also issued.
   Source: Southwest Research Institute
                  San Antonio, Texas.

9. *Name:*   PVP Data
   Description: This is a technical paper that contains pressure
                    vessel and piping (PVP) failure data from the
                    United States, Europe, and Japan. Similar papers
                    are published by Dr. Bush once every few years.
   Reference: Bush (1985).

10. *Name:*   *Reactor Safety Study*
    Description: The *Reactor Safety Study* contains failure proba-
                    bilities of a variety of mechanical, electrical and
                    electronic components. Although the data are
                    compiled for use in nuclear power plant applica-
                    tions, data from nuclear and nonnuclear power
                    plants, chemical plants, and military applications
                    are included. The *Reactor Safety Study* also con-
                    tains a list of component reliability data sources.
    Reference: Nuclear Regulatory Commission (1975).

REFERENCES

Bush, S. H. (1985). Statistics of pressure vessel and piping failures. In *Pressure Vessel and Piping Technology—A Decade of Progress*, C. Sundararajan, ed., American Society of Mechanical Engineers, New York, pp. 875–894.

Institute of Electrical and Electronics Engineers (IEEE) (1977). *IEEE Guide to the Collection and Presentation of Electrical, Electronic and Sensing Component Reliability Data for Nuclear Power Generating Stations.* John Wiley and Sons, Inc., New York.

Nuclear Regulatory Commission (1975). *Reactor Safety Study: An Assessment of Accident Risks in U.S. Commercial Nuclear Power Plants* (WASH-1400). Nuclear Regulatory Commission, Washington, DC.

Author Index

Allen, D. J., 204
Apostolakis, G. E., 204, 399–400

Barlow, 332
Bastani, F. B., 128
Benjamin, J. R., 54, 63
Birnbaum, 332
Burdick, G. R., 400–401
Bush, S. H., 106, 406

Cate, C. L., 399
Chatterjee, P., 401
Chu, B. B., 315
Cornell, C. A., 54, 63
Crosetti, P. A., 229

Davis, B. J., 333
Davis, D. J., 62
Dhillon, B. S., 69
Dougherty, E. M., 129
Drenick, R. F., 62

Erdmann, R. C., 229, 278
Esary, J. D., 248, 254

Fleming, K. B., 315
Fragola, J. R., 129
Freund, J. E., 13, 43
Fussell, J. B., 204, 229, 263, 269, 332–334,
    341, 374, 399

Gangadharan, A. C., 213, 227, 229
Gatelby, W., 278
Gauss, 43
Gaver, D. P., 315
Gilman, F. M., 400
Grant, E. L., 351
Gumbel, 71
Guttmann, H. E., 128–129
Gyftopolos, E. P., 279

Haugen, E. B., 95
Henley, E. J., 204, 279, 316
Henry, E. B., 229

Herd, G. R., 62
Hollingsworth, G. E., 116
Husseiny, A. A., 128
Hwang, C. L., 366

Iannino, A., 127–128
Ireson, W. G., 351

Joos, D. W., 128

Kirch, H., 229, 402, 403
Kumamoto, H., 204, 316
Kuo, W., 366

Lambert, H. E., 129–130, 143, 152, 207, 333,
    368, 379, 400
Leavenworth, R. S., 351
Lee, Y. T., 399
Levenbach, G. J., 116
Leverenz, F. L., 229, 278, 402
Linstone, H. A., 117

Mann, N. R., 95
Marshall, N. H., 229, 315, 400–401
Matthews, S. D., 401
Meister, D., 128
Musa, J. D., 127–128

Narum, R. E., 278

Okrent, D., 204, 400
Okumoto, K., 127–128
Olkin, 315
Olmos, J., 401

Pande, P. K., 400
Papazaglou, I. A., 279
Park, K. S., 129
Pershing, A. V., 116
Peters, G. A., 128
Poisson, 42–43
Polk, R., 279
Pontecorro, A., 128
Powers, G. J., 204

Proschan, F., 248, 254, 332
Putney, B. F., 294, 402–403

Raabe, P. H., 315
Ramamoorthy, C. V., 128
Rao, M. S. M., 204, 213, 227, 229
Rasmuson, D. M., 400–401
Recht, J. L., 128
Rumble, R. T., 278

Sabri, Z. A., 128
Salem, S. L., 204, 400
Schafer, R. E., 95
Sciandone, D., 400
Semanderes, S. N., 227
Shooman, M. L., 68
Singhpurwalla, N. D., 95
Spector, M. E., 401
Stack, D. W., 402

Stoddard, D., 278
Sundararajan, C., 95, 213, 227, 229
Swain, A. D., 128–129

Tillman, F. A., 366
Tompkins, F. C., 204
Turoff, M., 117

Vaurio, J. K., 400
Vesely, W. E., 229, 278, 332–334, 341, 374

Widawsky, W. H., 279
Willia, R. R., 400
Williams, R. L., 278
Wilson, J. R., 400
Wolf, J., 401
Wong, P. Y., 227
Worrell, R. B., 402

Subject Index

Acquisition, component, 396–397
Allocation, reliability, 364
Availability. *See also* Unavailability
  asymptotic, 71, 81
  of components, 51, 71, 83–84
  difference with reliability, 77
  interval, 71, 80–82
  limiting interval, 81
  point, 71, 80
  steady-state, 71, 81
  of systems, 89, 233, 252–254, 262–265, 277
Average rate, 42
Average value, 30

Barrier, 321–323
Bathtub curve, 67, 69–70
Binary bit, 227
Binomial distribution, 38–39
Boolean algebra, 8, 221–223
Boolean arithmetic model, 278
Boolean indicated cut set, 229
Boolean operations. *See* Boolean algebra
Bound
  lower, 194, 240, 247, 253–254, 309
  upper, 194, 240, 247, 248, 254, 309–310
Break-in period, 67
Break-in phase, 67

Charter, of reliability program, 389–390
Coefficient
  of skewness, 32
  of variation, 31
Coherency, 234–236, 262, 269–274
Combination testing, 223–224
Common cause, 3, 182, 187–190, 286–288
  analysis, 4, 286–326
    preliminary, 290–291
    qualitative, 291–296, 315–317
    quantitative, 296–317
  candidate, 287, 289, 293
  event, 286
  failure, 97, 117–120, 187–188, 250, 286–287
  significant, 287, 293

Common link, 289, 296
Complementary tree, 204–205
Component
  definition of, 48
  duplicate, 91
  nonrepairable, 48, 50
  primary, 91
  repairable, 48, 50, 71
Computer program, 4, 204, 230, 281, 325, 343, 393, 399–403
Confidence
  band, 197
  interval, 107
  range, 107
Correction factor, 116
Correlation coefficient, 35
Covariance, 34
Criticality, 142, 146, 148, 150
Cumulative distribution function
  of binomial distribution, 39
  definition of, 24, 27
  of exponential distribution, 43
  of failure, 60–61
  joint, 33–34
Cumulative failure probability, 51, 60–61, 63
Cut set, 212–214

Data
  collection of, 136–137, 148
  hardware failure, 94–95, 129
  historical, 95–96
  human error, 94, 128–130
  human failure. *See* human error
  repair, 101
  software error, 94, 127–128
  software failure. *See* Software error
  source. *See* Data base)
Data base, 392, 404–406
  combination of, 109–111
  computerized, 98
  expert opinion, 106, 108–109
  generic, 106, 114–116, 404–406
  human error, 128, 404
  plant-specific, 98–106

primary consolidated, 106–109
raw, 106–107
secondary consolidated, 106, 108, 109
software error, 404
software failure. *See* Software error
Data source. *See* Data base
Deductive analysis, 2–3, 151
Delphi method, 117
Dependence, 15, 35, 98, 157, 190–195, 203, 238, 244, 249–250
Descriptors, of a random variable, 30–32
Design to reliability, 4, 363–365
Detection time, 71–72, 95, 98
Discrete distributions. *See* Probability distribution for discrete variables
Distributions. *See* Probability distribution; Cumulative distribution function
Documentation, 130–131, 143, 151–152, 206–207, 230, 281–282, 326, 343, 394–395
Dual tree, 204–205
Duration, test-maintenance, 95

Element of a sample space, 6
Environmental conditions, 65, 96, 108, 157, 188
Error
  gross, 188
  human, 2, 94, 97, 128–129. *See also* Failure, human
  software, 2, 94, 127–128. *See also* Failure, software; Reliability, software
Event(s)
  basic, 172
  common mode, 287
  complement of, 18
  compound, 6
  definition of, 6
  house, 175–179, 182
  intermediate, 156, 170, 233, 248, 378
  intersection of, 19
  mutually exclusive, 18
  neutral, 287, 316
  simple, 6
  success, 251
  terminal, 154, 170–183
  top, 154, 169–170, 233, 333
  triggering, 137. *See also* Trigger mechanism
  undesired, 153–154
  undeveloped, 171–172
  union of, 19

Expectation, 30, 41, 44
Expected life, 53
Expected number of failures
  of components, 53, 79
  of systems, 89, 233, 261, 269
Experiment, 5
Expert opinion, 106, 108–109, 117, 128
Exponential distribution, 43–44, 54, 62
Exponential failure law, 62
Exponential law, 61–62
Exponential rule, 116
Extreme value distribution, 43

Failure
  combined primary-secondary, 270–273
  command, 173–174, 235, 269
  common cause, 97, 117–120, 187–188, 250, 286–287
  definition of, 49
  dynamic, 129
  human, 1, 94, 97, 129–130. *See also* Error, human
  modes of, 3, 49, 98, 146, 157, 181
  on-demand, 82, 110
  primary, 96, 173, 235, 250, 269
  probability density function of. *See* Probability density function, failure
  probability of. *See* Probability, of failure
  propagating, 97, 188–190, 250
  quasistatic, 129
  reporting, 397
  secondary, 96–97, 173–174, 235, 250, 269
  signal, 173
  software, 1, 94. *See also* Error, software; Reliability, software
  switching, 93
Failure modes and effects analysis, 2–3, 136, 146–153
Failure modes and effects table, 147, 149–151
Failure rate
  of components, 56–57, 62, 95, 100, 110, 118–120
  constant, 62
  during an interval, 56
  instantaneous, 57
  mean, 110
  of systems, 89, 233, 260, 263–264
Fault, 48–49
Fault tree
  analysis, 2–3, 136, 143, 147, 151
    qualitative, 3, 153–154, 199, 212–233, 331

quantitative, 3, 153–154, 199, 212,
    233–282, 331
construction, 3, 153–209
    computerized, 204
    data for, 130
    reduction, 199–204
    simplification, 199–204

Gamma distribution, 71
Gates, 158–169
    AND, 155, 158, 237–238
    AND–NOT, 168–169
    exclusive OR, 166–167
    inhibit, 168, 243
    m-out-of-n, 164
    OR, 154, 158, 238–240
    priority AND, 167–168
    tabular AND, 164
    tabular OR, 163
Gaussian distribution, 43, 54. See also
    Normal distribution
Geometric-mean formula, 310
Gumbel distribution, 71

Hazard
    analysis. See Preliminary hazard analysis
    function, 57–58, 60–61, 68–70
    rate, 57, 61
Hazardous element, 136–139
Hazardous source. See Hazardous element
Hazards and effects table, 136, 138–142
Historical data, 95
Human error. See Error, human
Human failure. See Failure, human
Hypergeometric distribution, 38

Identity, 10
Idle mode, 66
Implementation, 388–398
Importance
    analysis, 4, 331–343. See also Ranking
    measures, 332
        Barlow–Proschan, 332
        Birnbaum, 332
        for components, 341–342
        criticality, 332
        for minimal cut sets, 332–334
        sequential contributory, 332, 368, 374
        for terminal events, 332, 334–335
        upgrading function, 332
        Vesely–Fussell, 332–335, 341–342, 350,
            370, 374
    ranking. See Ranking

Inclusion–exclusion principle, 240, 247, 254
Independence. See Dependence
Inductive analysis, 2
Infant mortality, 67
Inhibit condition, 168
Inspection, 4, 369, 375
Intersection, 9, 221

Joint probability density function. See
    Probability density function, joint
Joint probability distribution. See Probability
    distribution, joint
Joint probability mass function. See
    Probability mass function, joint

k-out-of-n system, 90–91

Life cycle, 71
Lognormal distribution, 43

Maintenance
    improvements in, 347
    interval, 72
    outage, 196–198
    periodic, 104, 277–278, 347
    preventive. See periodic
    unavailability. See Test-maintenance,
        unavailability
Management, 4, 380, 388–398
Markov process, 278, 315
Mean, 30–31
    arithmetic, 110
    of exponential distribution, 44
    failure rate, 110
    geometric, 110, 310
    repair time. 72, See also Mean time to
        repair
Mean square value, 31
Mean time between failures (MTBF)
    of components, 80
    of systems, 89
Mean time between repairs (MTBR), of
    components, 80
Mean time to failure (MTTF)
    of components, 53, 61, 80
    of systems, 89, 233, 265
Mean time to repair (MTTR)
    of components, 71–72, 77, 100, 110
    of systems, 89, 233, 263, 265
Median, 31, 110
Minimal cut set, 199, 212, 214–216, 219–230,
    233–234, 245–248, 250–251, 254, 287,
    289, 294–296, 315–317

Minimal path set, 212, 216–219, 233, 251–254
Mission, 48
Mode
  failure. *See* Failure mode
  idle, 66
  off-line, 92
  on-line, 92
  operating, 66, 92
  of a random variable, 31
  standby, 66, 92
Moments of random variables, 30–32, 34–35
Monte Carlo, 229, 279
m-out-of-*n* gate, 164
Multinomial distribution, 38
Mutually exclusive, 18, 182, 192, 198,
    248–249

Normal distribution, 71. *See also* Gaussian
    distribution
Normal useful life, 67
Null set, 227
Null string, 227
Null tree, 319

Occurrence rate, 42, 118
Operating conditions, 65, 95–96, 108, 157,
    188
Operating mode, 66, 92
Operating procedure, 347
Operating system, 369, 374
Optimization, 1, 356, 364–366
Organizational structure, 391–392

Path set, 204, 212, 216–218
Parallel system, 90–93
Pedigree, 96
Plan, reliability program, 390–391
Point estimate, 109, 280–281
Point value. *See* Point estimate
Poisson distribution, 42–43
Poisson process, 42–43
Power rule, 116
Preliminary hazard analysis, 2, 136–145, 147,
    153
Prime number, 224–227
Prime of life, 62, 67
Probability
  axioms of, 17–18
  conditional, 15, 20, 118
  definition. *See* interpretation
  derived, 14

  of an event, 14, 18
  of failure, 51, 94, 129, 150, 153, 236
  of failure-on-demand, 82, 100, 110
  human error, 94, 128–129, 404
  human failure, 94, 129–130
  interpretation
    frequentist, 13
    objective, 12
    subjective, 13
  range, 107
  of a sample space, 18
  software error, 94, 127–128, 404
  software failure, 94
  theorems, 18–21
  of an undesired event, 153
  of a variable, 14
Probability density function
  basic properties of, 29
  definition of, 15, 27
  exponential, 44
  failure, 52–53, 57–58, 60–61
  joint, 33–34
Probability distribution
  basic properties of, 29
  binomial, 38–39
  for continuous variables, 27–29, 43
  for discrete variables, 23–26, 38
  exponential, 43–44
  joint, 33–34
  multivariate, 33
  Poisson, 41–43
  univariate, 33
Probability mass function
  basic properties of, 29
  binomial, 38–39
  definition of, 15, 23
  joint, 33
  Poisson, 42
Productivity, 1, 345
Quality
  assurance, 4, 393–394
  control, 67

Random function, 15
Random process, 15
Random variable, 14
Ranking
  of components, 4, 350
  importance, 331–332
  of minimal cut sets, 219–220, 331–332
  of terminal events, 212, 220, 331–332,
    334–335, 370–371
Rare event approximation, 239

Redundancy, 90–92, 348–351, 361
Reliability. *See also* Unreliability
    allocation, 364
    assessment, 4, 346
    of components, 51–52, 60–61, 63–64, 66,
        68–71, 77–78, 83–85
    design to, 4, 363–365
    function, 60–61
    improvement, 4, 346–351, 355–356, 361
    optimization, 356, 364–366
    of software, 94, 127–128, 247. *See also*
        Probability, software error;
        Probability, software failure of
        systems, 89, 233, 263–265, 277–278
    testing, 95
Repair
    rate, 72–73, 77, 89, 233
    time, 71–72, 100. *See also* Mean time to
        repair;
    Mean time between repairs; Mean repair
        time
Review, 396

Safety, 1, 245
Sample point, 6
Sample space, 6
    conditional, 7
    continuous, 7
    discrete, 7
    multidimensional, 6
    one-dimensional, 6
Sensors, 4, 366–368
Series system, 89–90
Set
    basic laws of, 11–12
    collectively exhaustive, 9, 20
    complement of a, 10
    empty, 6, 18
    equality of, 10
    mutually exclusive, 9, 18, 20
    null, 227
    operations, 8–9
    sub, 6, 214, 227
    super, 227
    theory, 5–12
Skewness, coefficient of, 32
Small probability approximation, 239, 246,
    248
Software
    error. *See* Error, software
    failure. *See* Failure, software
    reliability. *See* Reliability, software
Spare parts, 1, 98, 279

Square root
    formula, 310
    method, 310
Standard deviation, 31, 44
Standby, 66, 196, 369–371
State
    failed, 49
    normal, 49
    success, 233
Stochastic process, 15
Success tree, 204
Switch, 92–93
Symbols, 158
System
    configurations, 89–93
    definition, 89
    description, 156–157
    diagram, 154
    k-out-of-n, 90–91
    nonrepairable, 89, 236
    parallel, 90–93
    reliability. *See* Reliability, system
    repairable, 89, 261
    series, 89–90

Testing
    component, 95
    and maintenance. *See* Test-maintenance
    outage, 196
    periodic, 104
Test-maintenance
    duration, 95
    unavailability, 82, 106, 110, 196
Time
    of first occurrence, 44
    interval, 50
    to repair. *See* Repair time
Transfer symbols, 183, 185–187
Trigger mechanism, 139–140

Unavailability. *See also* Availability
    of components, 51, 71, 83–84, 262
    interval, 81
    of minimal cut sets, 263–264
    point, 71
    steady-state, 71, 84
    of systems, 89, 236, 247–248, 253, 262, 264
    testing and maintenance, 82, 106, 110, 196
    test-maintenance. *See* testing and
        maintenance
Uncertainty, 108, 280–281
Union, 8, 221–222

Unreliability. *See also* Reliability
  of components, 51, 71
  of systems, 89, 236, 247–248, 263–265, 267

Variance, 31, 41
Venn diagram, 8–10

Wear-in period, 67
Wear-out
  period, 67
  phase, 67
Weibull distribution, 70
Weighting factor, 110, 117